Current Topics in Behavioral Neurosciences

Volume 23

About this Series

Current Topics in Behavioral Neurosciences provides critical and comprehensive discussions of the most significant areas of behavioral neuroscience research, written by leading international authorities. Each volume offers an informative and contemporary account of its subject, making it an unrivalled reference source. Titles in this series are available in both print and electronic formats.

With the development of new methodologies for brain imaging, genetic and genomic analyses, molecular engineering of mutant animals, novel routes for drug delivery, and sophisticated cross-species behavioral assessments, it is now possible to study behavior relevant to psychiatric and neurological diseases and disorders on the physiological level. The *Behavioral Neurosciences* series focuses on "translational medicine" and cutting-edge technologies. Preclinical and clinical trials for the development of new diagnostics and therapeutics as well as prevention efforts are covered whenever possible.

More information about this series at http://www.springer.com/series/7854

David J.K. Balfour · Marcus R. Munafò
Editors

The Neurobiology and Genetics of Nicotine and Tobacco

Springer

Editors
David J.K. Balfour
Ninewells Hospital
University of Dundee Medical School
Dundee
UK

Marcus R. Munafò
School of Experimental Psychology
University of Bristol
Bristol
UK

ISSN 1866-3370 ISSN 1866-3389 (electronic)
Current Topics in Behavioral Neurosciences
ISBN 978-3-319-37919-7 ISBN 978-3-319-13665-3 (eBook)
DOI 10.1007/978-3-319-13665-3

Springer Cham Heidelberg New York Dordrecht London

Softcover reprint of the hardcover 1st edition 2015

Printed on acid-free paper

Springer International Publishing AG Switzerland is part of Springer Science+Business Media (www.springer.com)

Preface

When one of us (DJKB) first started studying the psychopharmacology of nicotine some 40 years ago the numbers of researchers interested in the topic was small and could probably be accommodated around a large dinner table. Our understanding of the potential hazards of smoking was at a fairly early stage as was our understanding of the neural mechanisms that mediated the behavioral responses to nicotine. At that time smoking was considered to be a habit, not an addiction, and was still widely accepted. Readers who are not old enough to remember those times may be familiar with the television series, *Mad Men*. That series gives you an impression of how acceptable smoking was. Even into the 1980s, the fact that neurones within the brain expressed nicotinic receptors was still debated among some researchers. We have come a long way since that time, and now it is not unusual to have 1,000 delegates or more at conferences on nicotine and tobacco, and sessions dedicated to nicotine are not uncommon at many neuroscience conferences. Moreover, public health policy is now driven by a sound evidence base relating both to the toxicity of primary and second-hand (also known as environmental) tobacco smoke and the plethora of neuroscience studies that have established nicotine as one of the most widely studied recreational drugs. The primary purpose of the chapters in this book and its companion volume is to explore the extent to which the wide range of approaches adopted to investigate the behavioral responses to nicotine and the molecular and neural mechanisms that mediate these effects have opened our eyes to the properties of this unique and fascinating drug.

The chapters in the book are divided into two sections. The first considers the molecular and genetic factors which influence the responses to nicotine and tobacco. This section of the volume addresses the nature of the receptors in the brain that mediate the responses to nicotine and how an understanding of the genetic variation within these receptors, the neurotransmitter systems that respond to them and the metabolic clearance of the drug impacts on nicotine and tobacco dependence and its successful treatment. The second section focuses on the evidence that nicotine may exert potentially beneficial effects within the brain, particularly with regard to memory and attention. These cognitive effects of the drug may explain, in part at least, why some people with underlying psychopathologies of cognition, such as schizophrenia or attention deficit hyperactivity disorder (ADHD), are particularly vulnerable to tobacco dependence and resistant to treatment. The chapters in this section of the book also seek to relate our understanding of the structure of neuronal nicotinic receptors to their effects on attention and learning and memory. The last chapter of this section considers the way in which learning theory can be recruited to explain some of the important elements of nicotine dependence.

We hope that the volumes *The Neurobiology and Genetics of Nicotine and Tobacco* and *The Neuropharmacology of Nicotine Dependence* will provide readers with a contemporary overview of current research on nicotine psychopharmacology and its role in tobacco dependence from leaders in this field of research and that they will prove valuable to those who are developing their own research programs in this important topic.

Dundee, Scotland David J.K. Balfour
Bristol Marcus R. Munafò

Contents

Structure of Neuronal Nicotinic Receptors 1
Francesca Fasoli and Cecilia Gotti

Genetics of Smoking Behaviour 19
Jennifer J. Ware and Marcus R. Munafò

Pharmacogenetics of Nicotine and Associated Smoking Behaviors 37
Julie-Anne Tanner, Meghan J. Chenoweth and Rachel F. Tyndale

Heterogeneity Across Brain Regions and Neurotransmitter Interactions with Nicotinic Effects on Memory Function 87
Edward D. Levin, Brandon J. Hall and Amir H. Rezvani

Nicotinic Receptors and Attention 103
Britta Hahn

Nicotinic Receptors, Memory, and Hippocampus 137
Munir Gunes Kutlu and Thomas J. Gould

A Hierarchical Instrumental Decision Theory of Nicotine Dependence 165
Lee Hogarth and Joseph R. Troisi II

Smoking Abstinence and Neurocognition: Implications for Cessation and Relapse 193
F. Joseph McClernon, Merideth A. Addicott and Maggie M. Sweitzer

Index 229

Structure of Neuronal Nicotinic Receptors

Francesca Fasoli and Cecilia Gotti

Abstract Nicotinic acetylcholine receptors (nAChRs) are a critical component of the brain's cholinergic neurotransmission system that modulates important physiological processes and whose dysfunction has been observed in patients with neurodegenerative diseases and mental illness. nAChRs are a heterogeneous family of receptor subtypes consisting of pentameric combinations of α and β subunits, and are widely expressed throughout the central and peripheral nervous system. nAChR subtypes share a common basic structure but their biophysical and pharmacological properties depend on their subunit composition, which is therefore central to understanding receptor function in the nervous system and discovering new subtype-selective drugs. We briefly review some recent findings concerning the structure and function of nAChRs, particularly the native subtypes.

Keywords Neuronal nicotinic receptor subtypes · Subunit composition · Stoichiometry · Nicotine · αBungarotoxin · Acetylcholine binding sites

Contents

1 Introduction 2
2 Overall Structure of Nicotinic Subtypes 2
2.1 Orthosteric ACh Binding Sites 5
2.2 Accessory Subunits 8
2.3 Subunit Stoichiometry 9
3 Native Subtypes 9
3.1 Techniques for Studying Native nAChRs 9
3.2 Autoradiographic Studies 10
3.3 Biochemical and Pharmacological Studies of Native Subtypes 11
4 Conclusions 14
References 14

F. Fasoli · C. Gotti (✉)
Department of Medical Biotechnologies and Translational Medicine, Consiglio Nazionale Delle Ricerche, Institute of Neuroscience, University of Milan, Via Vanvitelli 32, 20129 Milan, Italy
e-mail: c.gotti@in.cnr

D.J.K. Balfour and M.R. Munafò (eds.), *The Neurobiology and Genetics of Nicotine and Tobacco*, Current Topics in Behavioral Neurosciences 23,
DOI 10.1007/978-3-319-13665-3_1

1 Introduction

The cholinergic system is one of the most important and phylogenetically oldest nervous pathways. Acetylcholine (ACh) is the neurotransmitter that is synthesized, stored, and released by cholinergic neurons, and the key molecules that transduce the ACh message are the muscarinic metabotropic receptors and the ionotropic neuronal nicotinic acetylcholine receptors (nAChRs) (Picciotto et al. 2012). nAChRs are a heterogeneous family of ion channels ubiquitously expressed in the central nervous system (CNS) that by responding to the endogenous neurotransmitter acetylcholine (ACh) and nicotine, the most widespread drug of abuse, are implicated in a variety of physiological processes (reviewed in Hurst et al. 2013).

nAChRs activation excites target cells and mediates fast synaptic transmission in autonomous ganglionic neurons and in some brain areas, but anatomical and functional evidence suggests that the nAChRs in brain are preferentially located at preterminal and presynaptic sites where they regulate the release of both excitatory and inhibitory neurotransmitters (reviewed in Albuquerque et al. 2009; Jensen et al. 2005).

nAChRs and nicotinic mechanisms contribute to cognitive function and their decline or dysfunction has been observed in neurodegenerative diseases and mental illness. In addition genetic studies have linked nAChRs to epilepsy and schizophrenia, and studies of mutant (knockout or knockin) mice have shown that they are involved in pain mechanisms, anxiety, and depression (reviewed in Changeux 2010; Drenan and Lester 2012; Gotti and Clementi 2004; Hurst et al. 2013; Picciotto et al. 2001). nAChRs are particularly important in two critical periods of brain life: early pre- and post-natal circuit formation, and age-related cell degeneration. They are involved in neuronal survival, as it has been shown that nicotinic agonists are neuroprotective in both in vivo and in vitro models. Furthermore, it is becoming evident that the perturbation of cholinergic nicotinic neurotransmission can lead to various diseases during development, adulthood, and aging (Picciotto and Zoli 2008). As a number of comprehensive reviews (Albuquerque et al. 2009; Changeux 2009; Gotti et al. 2009) have described the structure and function of nAChRs, the aim of this paper is to provide a brief overview of the most recent findings on the structure of the brain nAChR subtypes.

2 Overall Structure of Nicotinic Subtypes

nAChRs were initially identified by means of ligand binding assays using radioactive ligands, which showed that ^{125}I-αBungarotoxin (αBgtx) and ^{3}H-nicotine bind to receptors with different anatomical and pharmacological distributions (Clarke et al. 1985).

The pharmacological heterogeneity of nAChRs revealed by these ligand studies was later confirmed and extended by means of the molecular cloning of a family of genes encoding 12 subunits and by studies of their expression in heterologous systems (reviewed in Albuquerque et al. 2009; Dani and Bertrand 2007; Gotti et al. 2006).

nAChRs belong to the large superfamily of homologous receptors, which also include muscle-type AChRs and GABAA, glycine, and serotonin (5-HT) ionotropic receptors (reviewed Gotti et al. 2009; Miller and Smart 2010). Like all of the other members of the ligand-gated ion channel superfamily, the nAChR subunits have a relatively hydrophilic extracellular amino terminal portion that carries the ACh binding site and faces the synaptic cleft, followed by three hydrophobic transmembrane domains (M1–M3), a large intracellular loop, and then a fourth hydrophobic transmembrane domain (M4) (Fig. 1a) (reviewed in Albuquerque et al. 2009). The subunits are arranged around a central pore and site-directed mutagenesis experiments have shown that the transmembrane M2 domain lines the ion channel and identified the residues important for the ion selectivity, permeability, and channel gating of the receptors.

In vertebrates the genes that have been cloned so far (CHRNA2–CHRNA10, and CHRNB2–CHRNB4) code for subunits that are divided into two subfamilies of nine α subunits (α2–α10) and three β (β2–β4) expressed in the nervous system, cochlea, and a number of non-neuronal tissues (Gotti and Clementi 2004). All the nine α subunits have adjacent cysteines (analogous to cysteines 192–193 of the α subunit of the muscle-type AChR), whereas the β subunits do not have cysteines. The different combinations of nAChR subunits have led to the formation of a heterogeneous family of pentameric subtypes with different structural, functional, and pharmacological properties. Two main classes of nAChR subtypes have been identified: the αBgtx-sensitive receptors, which are made up of the α7, α8, α9 and/or α10 subunits can form homomeric or heteromeric receptors, and the αBgtx-insensitive receptors, which are heteromeric receptors, formed by α and β subunits that bind agonists with high affinity but not αBgtx (Fig. 1c) (reviewed in Gotti et al. 2009).

Studies of heterologous systems have shown that α7 subunits can also form functional channels with the subunits present in non-αBgtx binding receptors such as the α5 (Girod et al. 1999), β2 (Khiroug et al. 2002), β3 (Palma et al. 1999) and β4 subunits (Criado et al. 2012). It is presumed that both homomeric and heteromeric nAChRs have a pentameric structure with the subunits organized around a central channel: the homo-oligomeric receptors have five identical (orthosteric) ACh binding sites per receptor molecule (Fig. 1b) (Palma et al. 1996) located at the interface between two adjacent subunits, whereas hetero-oligomeric receptors have two or three α subunits and three or two β subunits, and therefore two orthosteric binding sites per receptor molecule located at the interface between the α and β subunits (Taly et al. 2009) (see Fig. 1b). Each orthosteric ACh binding site has a principal (or "plus") and a complementary or "minus" component. In heteromeric nAChRs, the principal component is carried by the α2, α3, α4, and α6 subunits with the complementary site carried by the β2 or β4 subunits, whereas each subunit in the homomeric receptors contributes to both the principal and complementary

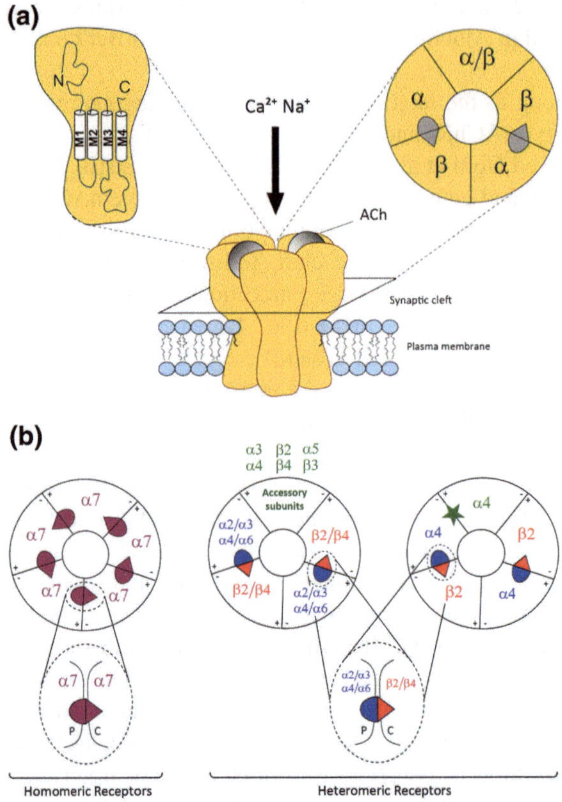

Fig. 1 **a** *Left* Diagram showing the putative transmembrane topology of nAChR subunits. The extracellular amino terminal portion is followed by three hydrophobic transmembrane domains (M1–M3), a large intracellular loop, and then a fourth hydrophobic transmembrane domain (M4). *Middle* Pentameric arrangement of nAChR subunits in an assembled receptor. *Right* Localization and schematic organization of the ACh binding sites in a heteromeric receptor. **b** Structure of heteropentameric and homopentameric neuronal subtypes. The pentameric arrangement of nAChR subunits in an α7 homopentameric subtype (*right*), heteromeric receptor subtypes (*middle*), and the $(\alpha4)_3(\beta2)_2$ subtype (*right*). The localisation of the subunit interfaces of the orthosteric binding sites is indicated, with the primary component (*P*) carried by the α subunits, and the complementary component (*C*) carried by α or non-α subunits. In addition to the two orthosteric sites, the $(\alpha4)_3(\beta2)_2$ subtype has a binding site at the α4/α4 interface (*star*)

components, which are present on opposite sides of the same subunit (reviewed in Corringer et al. 2000; Taly et al. 2009) (see Fig. 1b).

Heteropentamer neuronal nAChRs have the fifth subunit that does not contribute to the orthosteric site (and this subunit is called accessory subunit). In heterologous systems, α5 and β3 subunits only form functional channels when they are co-expressed with a primary and complementary subunit (Groot-Kormelink et al. 1998; Ramirez-Latorre et al. 1996) thus indicating that they can only function as accessory subunits, whereas the α3 or α4 and β2 or β4 subunits can form orthosteric ligand binding sites or assemble in the accessory position to produce receptors with different stoichiometries (see Fig. 1b).

However, recent studies have revealed further complexity in the definition of binding sites and the possible subunit involved. The use of concatenated receptors whose DNA was linked covalently led to the expression of receptors whose subunits were in a specified order, and whose pharmacological properties were similar to those of the non-linked receptors (Nelson et al. 2003; Zhou et al. 2003). Studies of the concatameric $(\alpha4)_3$ $(\beta2)_2$ subtype have shown that in addition to the two orthosteric binding sites at the α4/β2 interface they also have an additional binding site at the α4/α4 interface (Moroni et al. 2008) as shown in Fig. 1b left.

Jin et al. (2014) have very recently, using the concatameric approach, expressed either the dimeric constructs of α4 and β2 subunits expressed with a free α5 subunit and concatameric pentameric, receptors incorporating a single copy of α5, in different positions, and found that the α5 subunit can occupy the position of a non-binding subunit, or replace a β2 subunit participating in an orthosteric binding site. However, functional receptors apparently cannot be formed with α5 subunits in both canonical binding sites.

2.1 Orthosteric ACh Binding Sites

Much of our knowledge of agonist binding sites comes from studies of muscle AChRs in which the use of affinity labelled reagents and subunit chimeras and/or site-directed mutagenesis have shown that it is the large extracellular amino terminal domain that contributes to the ACh binding domain pocket (Bartos et al. 2009; Taly et al. 2009). Moreover, recently a significant contribution to the identification of the ligand binding site in nAChRs has also been made by analyzing the crystal structure of the ACh binding protein from the freshwater snail (Celie et al. 2005; Rucktooa et al. 2009). This homopentameric soluble protein (AChBP), which is 210 residues long, binds ACh and is secreted by snail glial cells into cholinergic synapses and has an affinity spectrum resembling that of homomeric α7 or α9 receptors.

Many amino acid residues contribute to the orthosteric ACh binding site. They are grouped into short sequences that form loops A, B, and C (the principal component) and D, E, and F (the complementary component) (Fig. 2a). Inside the compact structure of an AChBP, at the center of the interface of the ligand binding domain (LBD) of two adjacent subunits, loops A, B, D, and F form a hydrophobic pocket to which agonists bind that is closed by loop C (Fig. 3d). Studies of the co-crystals of AChBPs using nicotinic ligands have shown that the conserved residues in the binding site are tyrosine (Y) 93 (loop B), tryptophan 149 (W) and Y151 (loop A), Y190 and Y198 (loop C), W55 and E57 (loop D), and the disulfide bridge between cysteines (C) 192 and 193 (the numbering refers to the amino acids present in Torpedo AChRs) (Changeux and Taly 2008). The aromatic residues and disulfide bridge of the ACh binding site are electronegative and neutralize the

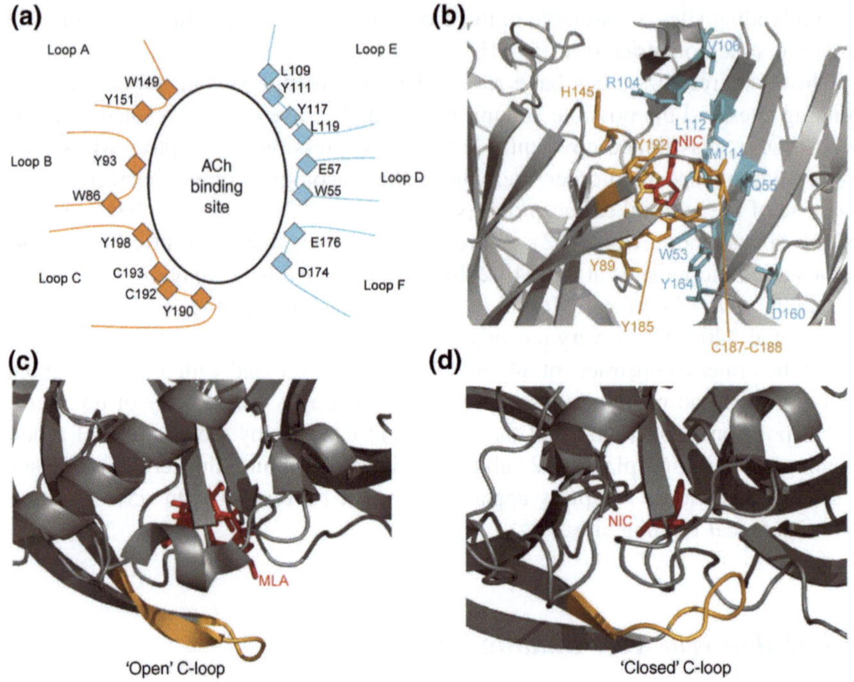

Fig. 2 Structure of the ACh binding site. **a** Diagram of an ACh binding site showing the amino acids in the loops that participate in its formation. Loops A, B, C are provided by the α subunit, and loops D, E, F by the adjacent subunit. **b–d** The X-ray structure of the binding site of the AChBP co-crystallographed in the presence of nicotine (**b** and **d**) and the antagonist methyllycaconitine (MLA) (**c**). Agonists bind in a pocket containing the binding site that promotes extensive contact between the ligand and protein. As in the case of other antagonists, the binding of methyllycaconitine leads to an open conformation of loop C (*yellow*) that interferes with receptor activation and channel opening (**c**). Figure is reproduced with permission from (Changeux and Taly 2008)

positive charge carried by the majority of nicotinic ligands. The binding of ACh therefore gives rise to a non-covalent interaction between the cation and the electron-rich π system of the W residue in loop B. In general, the tertiary or quaternary ammonium charges of nicotinic ligands bind to the center of the pocket formed by the hydrophobic aromatic residues of loops A, B, C, and D. This interaction is maintained in the family of Cys-loop receptors, although the loops involved depend on the receptor type (Miller and Smart 2010).

X-ray studies of AChBPs have shown that, in the absence of the agonist or presence of the antagonist, loop C does not cover the hydrophobic pocket, whereas in the presence of the agonist, the binding site has a closed conformation with loop C covering it (Figs. 3c, d). The role of the movement of loop C in activating the receptor was determined by the studies of a chimera obtained from the fusion of most AChBPs to the transmembrane regions of the serotonin (5-HT3) receptor.

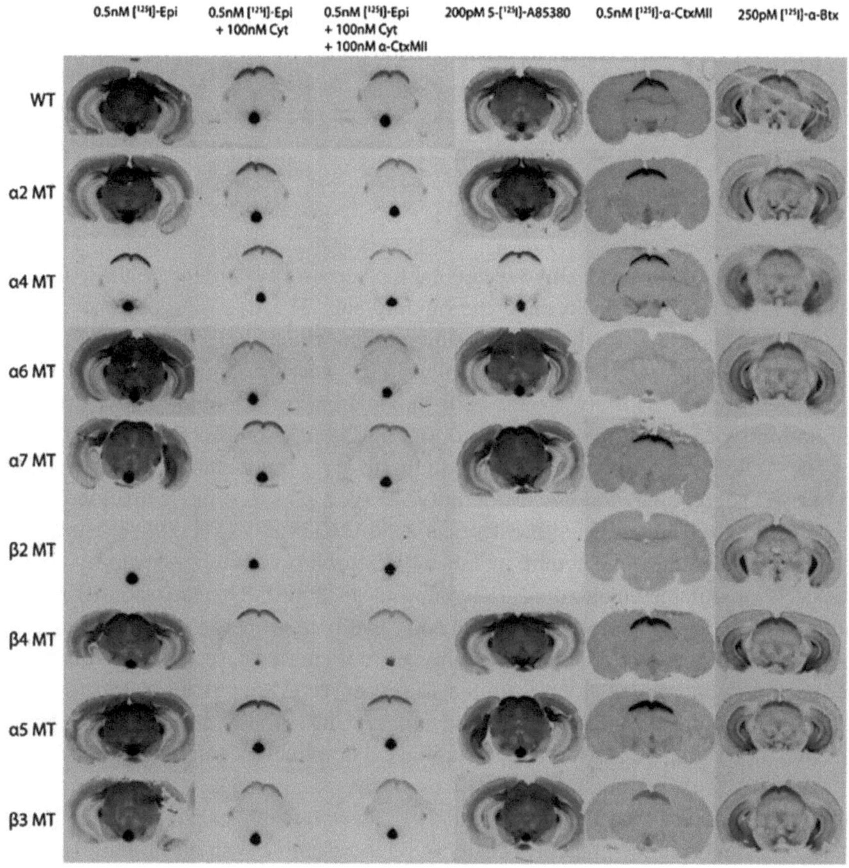

Fig. 3 Autoradiographic images of coronal mouse brain sections at approximately −3.5 μm Bregma. Autoradiograms of ^{125}I-epibatidine (total binding, binding in the presence of 100 nM cytisine and binding in the presence of 100 nM cytisine and 100 nM αconotoxin MII), ^{125}I-A85380, ^{125}I-α-conotoxin MII and ^{125}Iα-Bungarotoxin to WT mice and each of the nAChR KO mice for sections at a level of approximately 3.5 mm Bregma. Figure is reproduced with permission from (Baddick and Marks 2011)

This chimera functions as an ACh receptor in which ACh binding determines the closure of loop C and the activation of the receptor (Mukhtasimova et al. 2009).

An electrophysiology study has also identified feedback communication between the movement of loop C and the opening of the channel. This study also demonstrated that the binding of one molecule of ACh opens the channel for a short time, whereas the binding of two molecules prolongs the opening (Mukhtasimova et al. 2009). It seems, therefore, that the closed conformation of loop C gives rise to a state of channel pre-activation and that the duration of channel opening depends on the number of closed conformations of loop C in the receptor. Single-channel analyses of ACh receptors has shown that full agonist binding easily allows the transition from pre-activation to complete activation of the receptor and the consequent closing of loop C.

2.2 Accessory Subunits

In heteromeric αBgtx-insensitive receptors, the accessory subunits are those that do not directly participate in forming the binding site. The role of accessory subunit has been investigated in the α4β2*[1] subtypes in which the presence of different accessory subunits (α5, β3, α4, β2) changes their pharmacological and biophysical properties, their sensitivity to allosteric modulators, and their sensitivity to up-regulation (Kuryatov et al. 2008; Moroni et al. 2006, 2008; Tapia et al. 2007).

The $(\alpha4\beta2)_2\alpha5$ subtype has the highest Ca^{2+} permeability, whereas the $(\alpha4\beta2)_2\beta2$ subtype has the greatest affinity for ACh and nicotine activation, and is also the most sensitive to nicotine desensitization (Kuryatov et al. 2008). Moreover, the presence of the α5 subunit in the α4β2* subtype confers sensitivity to the allosteric modulator galantamine (Kuryatov et al. 2008). The inclusion of the α5 subunit also affects the pharmacological and functional properties of other subtypes: for example in the α3β2* and α3β4* subtypes, it increases desensitization and Ca^{2+} permeability, and alters agonist-stimulated responses (Tapia et al. 2007).

The β3 subunit co-assembles with several nAChR subunit combinations, but in all cases other than α3β3β4, it appears to have a dominant negative effect that leads to the absence of the functional expression of the assembled β3* receptor complex (Broadbent et al. 2006; Palma et al. 1999). However, our ex vivo studies (Gotti et al. 2006) indicate the great propensity of β3 to assemble with α6 subunit, and α6* receptor expression in β3 knockout mice is decreased in the cell bodies and nerve terminals of dopaminergic neurons. This decrease suggests that the β3 subunit is important for the formation of the majority of α6β2* or α4α6β2* receptors, and that its loss causes defects in nAChR assembly, degradation, and/or trafficking. The exclusive role of the β3 subunit as an accessory subunit has been confirmed using fluorescently labelled α6 and β3 subunits and the FRET technique, which has shown that only a single β3 subunit is incorporated in pentameric α6β2* receptors (Drenan et al. 2008). So the accessory subunit in the fifth position may influence many characteristics of nAChRs including their agonist sensitivity, channel kinetics, Ca^{2+} permeability, assembly, interactions with chaperone proteins, trafficking, and cell localization (reviewed in Colombo et al. 2013).

[1] The native subtypes are identified by their known subunits; if these are followed by an asterisk, it means that other unidentified subunits may also be present.

2.3 Subunit Stoichiometry

As neuronal nAChRs are pentameric, they can have considerable molecular differences in subunit composition or different subunit stoichiometries even if they have the same subunit composition. Heterologously expressed α4β2 and α3β4 subtypes can exist in two stoichiometries, made of either two or three copies of the α subunit in the channel pentamer. The two stoichiometries of the α4β2 subtype can be distinguished because the subtype with two α4 subunits, $(\alpha 4)_2(\beta 2)_3$ subtype, is activated at much lower ACh concentrations and is more sensitive to other agonists than the subtype with three α4 subunits, $(\alpha 4)_3(\beta 2)_2$ subtype (Moroni et al. 2006). Nicotinic ligands may also have different efficacy toward the two stoichiometries: for example, the selective α4β2 ligand sazetidine binds with high affinity to both stoichiometries but it is a full agonist of $(\alpha 4)_2(\beta 2)_3$ nAChRs, while had an efficacy of only 6 % on$(\alpha 4)_3(\beta 2)_2$ nAChRs (Zwart et al. 2008). The agonist sensitivity of the two α3β4 stoichiometries is similar, but only the subtype with two α3 subunits is susceptible to enhancement by low zinc concentrations, and the two stoichiometries have markedly different single-channel conductance and kinetics (Krashia et al. 2010).

The two stoichiometries $(\alpha 4)_2(\beta 2)_3$ and $(\alpha 4)_3(\beta 2)_2$ exist functionally in rodent cortical and thalamic brain preparations (Gotti et al. 2008), whereas the different α4β2 and α3β4 stoichiometries expressed in heterologous systems depend on the expression system (Krashia et al. 2010; Moroni et al. 2006).

Recent studies have shown that the stoichiometries of the α4β2 and α3β4 subtypes are both modulated by exposure to nicotine, mainly by acting intracellularly during α4β2 receptor assembly, nicotine favors the formation of the $(\alpha 4)_2$ $(\beta 2)_3$ stoichiometry, which is up-regulated both intracellularly and at the plasma membrane of heterologous cells and neurons (Colombo et al. 2013).

Nicotine also changes the stoichiometry of the α3β4 subtype. Recent data from our laboratory have shown that, during α3β4 assembly, nicotine binding favors an $(\alpha 3)_2(\beta 4)_3$ stoichiometry that is more stable and preferably released from the endoplasmic reticulum for transport to the plasma membrane (Mazzo et al. 2013).

3 Native Subtypes

3.1 Techniques for Studying Native nAChRs

The approaches currently used to localize and identify nAChR subtypes include techniques for localizing subunit mRNA (in situ hybridization (ISH) and single-cell PCR) or protein (immunoprecipitation and immunocytochemistry), receptor autoradiography at regional or cellular level, techniques for assessing subtype composition and pharmacology (binding in tissue homogenates, immunoprecipitation, immunopurification, and Western blotting), and functional assays (neurotransmitter release from slices or synaptosomes and electrophysiological techniques).

Information concerning the subunit associations of most of the nAChR subtypes identified in neurons comes from immunoprecipitation and immunopurification studies using subunit-specific antibodies (Ab), whose specificity has been tested in the tissues of wild-type (WT) and knockout (KO) mice, and brain areas. This was possible because the native subtypes extracted from the membrane were radiolabelled with nicotinic ligands, and the selection of the antigens was made by receptor binding. The immunoprecipitation of native radiolabelled receptors also ensured that only pentameric receptors were immunoprecipitated because analysis of the sucrose gradients of the different native subtypes solubilized using a non-denaturing detergent (such as Triton x-100) shows that they preserve their pentameric assembly, unlike transfected subtypes in which high affinity binding sites can also be detected with a sedimentation coefficient that is not exclusively compatible with pentameric subunit assembly (Kuryatov et al. 2000).

The immunochemical localization of neurotransmitter receptors and microscopic subunit protein co-localization are important criteria when localizing native subtypes and defining their subunit composition. However, these techniques cannot be used in the case of nAChR subunits because most of the available anti-subunit antibodies (Abs) are nonspecific, which means that immunocytochemistry labelling leads to similar staining patterns in tissues obtained from the WT and KO mice (Moser et al. 2007).

The major drawback of immunoprecipitation and immunopurification is that the spatial resolution obtained is only at regional level.

In summary, although very few nicotinic ligands are subtype specific, the combined use of ligand binding assays and autoradiographic studies of brain tissues from WT and subunit KO mice has provided critical data for identifying and defining the subunit composition, localization, and pharmacology of native subtypes.

3.2 Autoradiographic Studies

The earliest autoradiographic ligand binding studies demonstrated that the binding of ^{125}I-αBgtx was distinct from that obtained using the radioactive agonist ligands ^{3}H-nicotine, ^{3}H-ACh, and ^{3}H-cytisine (Clarke et al. 1985; Marks and Collins 1982) and a number of studies showed that ^{3}H-cytisine and ^{3}H-ACh bind with nM affinity to the same sites bound by ^{3}H-nicotine. Badio and Daly (1994) subsequently identified a new ligand epibatine that binds with very high affinity (pM) heteromeric receptors to the same sites as those bound by ^{3}H-nicotine. Since then, further saturation binding studies to rat and chick membranes have shown that epibatidine also binds to low affinity sites and that some of this binding was competed by the presence of αBgtx (Marks et al. 2006).

Over the last 10 years many more ligands specific for the different subtypes have been discovered, including A85380 (which is specific for β2-containing receptors) and αconotoxin MII, which is specific for the α3β2 and α6β2 subtypes. The use of these old (^{125}I-αBgt and ^{125}I-epibatidine) and new ligands (^{125}I-αconotoxin MII, and ^{125}I-A85380) and brain slices obtained from WT and α2, α4, α6, α7, β2, β4, α5, and β3 KO mice has provided a clear picture of the localization of the subtypes (Baddick and Marks 2011) (see Fig. 3).

Deletion of the α7 subunit completely and selectively eliminates ^{125}I-αBgtx binding, but deletion of any of the other subunits had no effect. Deletion of the β2 subunit completely eliminates ^{125}I-A85380 binding throughout mouse brain, whereas deletion of the α4 subunit eliminates most ^{125}I-A85380 binding, but residual ^{125}I-A85380 binding sites were found in the dopaminergic pathways, visual tracts, medial habenula, and interpeduncular nucleus probably because of the presence of the α3β2* subtype. The binding of ^{125}Iα-conotoxin MII is eliminated in most brain regions by deleting α6 or β2 subunit, and reduced by deleting the α4 or β3 subunit (Baddick and Marks 2011).

Inhibition of high-affinity ^{125}I-epibatidine binding sites by cytisine is a useful experimental means of separating subsets of cytisine-sensitive receptors mainly represented by α4β2 subtype and the cytisine-resistant ^{125}I-epibatidine binding sites of the α6β2* and α3β4* subtypes. When 100 nM cytisine is included in the incubation of ^{125}I-epibatidine with slices of WT mice, labelling was almost exclusively observed in the medial habenula, fasiculus retroflexus, interpeduncular nucleus, and inferior colliculus; this binding was lost in β4 KO mice, thus further indicating the presence of an α3β4 subtype in these areas.

3.3 Biochemical and Pharmacological Studies of Native Subtypes

Vertebrate neuronal nAChR subunits may co-assemble in many possible combinations, and many more subtypes have been heterologously expressed than those identified in vivo. It seems that native nAChRs are assembled into functional pentamers with a relatively restricted number of subunit combinations.

3.3.1 Non αBungarotoxin-Sensitive Receptors

Various immunological and affinity ligand immunopurification studies have shown that the large majority of nAChRs in the nervous system are heteromeric and contain one type of α and one type of β subunit (reviewed in Gotti et al. 2006) α4β2 receptors account for 90 % of the high affinity neuronal nAChRs in mammalian brain, whereas α3β4 receptors are the major subtype in the autonomic ganglia, adrenal medulla, and subsets of CNS neurons in the medial habenula, dorsal medulla, pineal gland, and retina. Some α4β2 and α3β4 subtypes may also contain the α5 subunit, whose presence is believed to increase the rate of channel desensitization and calcium permeability (reviewed in Fucile 2004).

In addition to these major central and peripheral nAChR subtypes, many other native nAChR subtypes with more complex subunit compositions have been identified in the rodent mesostriatal pathway (ventral tegmental area, nucleus accumbens, and caudate putamen) and visual pathways (retina, superior colliculus, and lateral geniculate nucleus) including the α6β2β3* receptors that consist of the α6β2β3 and α4α6β2β3 subtypes (Gotti et al. 2006). The two ortostheric ACh binding sites are identical in the α6β2β3 subtype, but different in the α4α6β2β3 subtype, which has both an α6β2 and an α4β2 interface (Champtiaux et al. 2003).

Data from our laboratory have shown that the subunit composition of α6* receptors in the different subsystems of mesostriatal dopamine neurons are partially heterogeneous. The dopamine terminals of the nigrostriatal pathway exclusively express α4α6β2β3 receptors, whereas those of the mesolimbic pathway express a majority of α6β2β3 receptors. In addition, a minor population of α4β2β3 receptors is only expressed in the caudate putamen nuclei (Gotti et al. 2010). Furthermore, binding studies in striatal tissue have shown that α4α6β2β3 is the subtype that is preferentially vulnerable to nigrostriatal damage as it is lost in the striatum of animal models of Parkinson's disease and human patients (Quik and Wonnacott 2011).

As shown above in autoradiographic studies in WT and KO mice, both habenula and interpeduncular nucleus express high levels of almost all known heteromeric nAChR subunits, and the highest level of nAChRs in the CNS. Biochemical and immunoprecipitation studies have confirmed the heterogeneity of the nAChR subtypes expressed in the habenula-interpeduncular nucleus pathway, and found that in this pathway both rat and mouse contain two major and distinct populations of β2* and β4*-containing receptors (Grady et al. 2009). The β2* population in the habenula contains the α4β2* and α3β2* subtypes, some of which also contain the accessory α5 or β3 subunits. In the interpeduncular nucleus, β2* nAChRs exist as three populations of approximately equal size: α2β2*, α3β2*, and α4β2*.

In agreement with binding studies, our immunoprecipitation studies found that the β4* nAChR population in both regions is mainly associated with the α3 subunit, and a significant fraction of the α3β4* nAChRs contain accessory, mostly β3 subunits. The β3 subunits are associated with α6β2* nAChR in the mesostriatal dopaminergic and visual pathways but the presence of β3 subunits not associated with α6 subunits in the habenula-interpeduncular nucleus pathway is in line with the findings of in situ hybridization studies showing high levels of β3 but not α6 subunit mRNA in the habenula (Champtiaux et al. 2003; Cui et al. 2003). This is a novel subtype because ganglionic α3β4*nAChRs contain α5 as an accessory subunit (Conroy and Berg 1995).

3.3.2 αBungarotoxin-Sensitive Receptors

The availability of the αBgtx specific ligand has allowed the purification and characterization of the αBgtx receptors present in the brains of various species: they are pentamers of only α7 subunits in rat (Cui et al. 2003; Drisdel and Green 2000), but may be homomeric α7 or α8 or α7-α8 receptors in chick (Gotti et al. 1994; Liu et al. 2009). It has recently been shown that α7 and β2 subunits are co-expressed in rat basal forebrain cholinergic neurons, and form a novel heteromeric α7β2 subtype that has different biophysical and pharmacological properties from those of the homomeric α7 receptors expressed by ventral tegmental area neurons. This α7β2 subtype is highly sensitive to functional inhibition by oligomeric forms of amyloid Aβ1-42, and may be relevant in Alzheimer's disease (Liu et al. 2009).

Using the α7-selective ligand αBgtx to affinity purify α7-containing receptors, we have recently biochemically identified an α7β2 subtype in the basal forebrain of humans and WT mice (Moretti et al. 2014). Western blotting with subunit-specific antibodies showed that the α7β2* nAChRs were expressed in WT but not in β2 KO mice, and that they represent only 3 % of all β2-containing receptors. We found that an α7β2 receptor is also present in human basal forebrain, but not in the cerebellum (Moretti et al. 2014).

The distribution of native subtypes identified by biochemical, pharmacological, and functional studies in rodent brain are shown in Fig. 4 (Gotti et al. 2006).

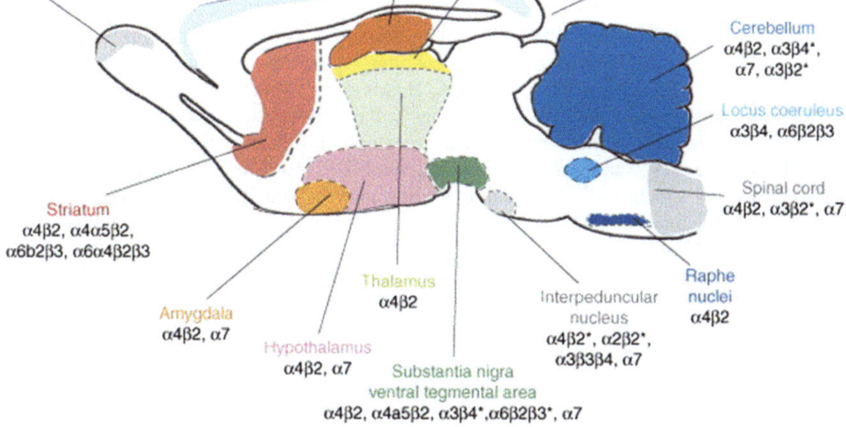

Fig. 4 Regional distribution of nAChR subtypes in rodent brain. *Source* Figure is modified and reproduced, with permission, from (Gotti et al. 2006)

4 Conclusions

There has been remarkable progress in our understanding of the structure of nicotinic receptor subtypes and their involvement in a variety of nervous functions and disorders. Much has been learned concerning the role of the most widely expressed subtypes, although further studies are necessary in order to identify the in vivo presence of different stoichiometries and identify and localise minor receptor subtypes in different brain areas.

Elucidating the molecular structure and subunit composition of native nAChRs is crucial for the development of nAChR subtype-selective compounds, whose benefits are associated with minimal adverse effects. The therapeutic use of nicotinic drugs is further complicated by the recent findings showing the ubiquitous expression of nAChRs in extra-neuronal tissues (Wessler and Kirkpatrick 2008). The subunit composition and the signaling mechanisms of these non-neuronal receptors have not yet been defined, and this will be an area of intensive future investigation.

Acknowledgments Francesca Fasoli is a Ph.D. student supported by the Fondazione Vollaro.

Funding was received from the CNR Research Project on Aging, Regione Lombardia Projects NUTEC ID 30263049 and MbMM-convenzione n°18099/RCC.

References

Albuquerque EX, Pereira EF, Alkondon M, Rogers SW (2009) Mammalian nicotinic acetylcholine receptors: from structure to function. Physiol Rev 89:73–120

Baddick CG, Marks MJ (2011) An autoradiographic survey of mouse brain nicotinic acetylcholine receptors defined by null mutants. Biochem Pharmacol 82:828–841

Badio B, Daly JW (1994) Epibatidine, a potent analgetic and nicotinic agonist. Mol Pharmacol 45:563–569

Bartos M, Corradi J, Bouzat C (2009) Structural basis of activation of Cys-loop receptors: the extracellular-transmembrane interface as a coupling region. Mol Neurobiol 40:236–252

Broadbent S, Groot-Kormelink PJ, Krashia PA, Harkness PC, Millar NS, Beato M, Sivilotti LG (2006) Incorporation of the beta3 subunit has a dominant-negative effect on the function of recombinant central-type neuronal nicotinic receptors. Mol Pharmacol 70:1350–1357

Celie PH, Klaassen RV, van Rossum-Fikkert SE, van Elk R, van Nierop P, Smit AB, Sixma TK (2005) Crystal structure of acetylcholine-binding protein from Bulinus truncatus reveals the conserved structural scaffold and sites of variation in nicotinic acetylcholine receptors. J Biol Chem 280:26457–26466

Champtiaux N, Gotti C, Cordero-Erausquin M, David DJ, Przybylski C, Lena C, Clementi F, Moretti M, Rossi FM, Le Novere N, McIntosh JM, Gardier AM, Changeux JP (2003) Subunit composition of functional nicotinic receptors in dopaminergic neurons investigated with knock-out mice. J Neurosci 23:7820–7829

Changeux JP (2009) Nicotinic receptors and nicotine addiction. C R Biol 332:421–425

Changeux JP (2010) Nicotine addiction and nicotinic receptors: lessons from genetically modified mice. Nat Rev Neurosci 11:389–401

Changeux JP, Taly A (2008) Nicotinic receptors, allosteric proteins and medicine. Trends Mol Med 14:93–102

Clarke PB, Schwartz RD, Paul SM, Pert CB, Pert A (1985) Nicotinic binding in rat brain: autoradiographic comparison of [3H]acetylcholine, [3H]nicotine, and [125I]-alpha-bungarotoxin. J Neurosci 5:1307–1315

Colombo SF, Mazzo F, Pistillo F, Gotti C (2013) Biogenesis, trafficking and up-regulation of nicotinic ACh receptors. Biochem Pharmacol 86:1063–1073

Conroy WG, Berg DK (1995) Neurons can maintain multiple classes of nicotinic acetylcholine receptors distinguished by different subunit compositions. J Biol Chem 270:4424–4431

Corringer PJ, Le Novere N, Changeux JP (2000) Nicotinic receptors at the amino acid level. Annu Rev Pharmacol Toxicol 40:431–458

Criado M, Valor LM, Mulet J, Gerber S, Sala S, Sala F (2012) Expression and functional properties of alpha7 acetylcholine nicotinic receptors are modified in the presence of other receptor subunits. J Neurochem 123:504–514

Cui C, Booker TK, Allen RS, Grady SR, Whiteaker P, Marks MJ, Salminen O, Tritto T, Butt CM, Allen WR, Stitzel JA, McIntosh JM, Boulter J, Collins AC, Heinemann SF (2003) The beta3 nicotinic receptor subunit: a component of alpha-conotoxin MII-binding nicotinic acetylcholine receptors that modulate dopamine release and related behaviors. J Neurosci 23:11045–11053

Dani JA, Bertrand D (2007) Nicotinic acetylcholine receptors and nicotinic cholinergic mechanisms of the central nervous system. Annu Rev Pharmacol Toxicol 47:699–729

Drenan RM, Lester HA (2012) Insights into the neurobiology of the nicotinic cholinergic system and nicotine addiction from mice expressing nicotinic receptors harboring gain-of-function mutations. Pharmacol Rev 64:869–879

Drenan RM, Nashmi R, Imoukhuede P, Just H, McKinney S, Lester HA (2008) Subcellular trafficking, pentameric assembly, and subunit stoichiometry of neuronal nicotinic acetylcholine receptors containing fluorescently labeled alpha6 and beta3 subunits. Mol Pharmacol 73:27–41

Drisdel RC, Green WN (2000) Neuronal alpha-bungarotoxin receptors are alpha7 subunit homomers. J Neurosci 20:133–139

Fucile S (2004) Ca^{2+} permeability of nicotinic acetylcholine receptors. Cell Calcium 35:1–8

Girod R, Crabtree G, Ernstrom G, Ramirez-Latorre J, McGehee D, Turner J, Role L (1999) Heteromeric complexes of alpha 5 and/or alpha 7 subunits. Effects of calcium and potential role in nicotine-induced presynaptic facilitation. Ann N Y Acad Sci 868:578–590

Gotti C, Clementi F (2004) Neuronal nicotinic receptors: from structure to pathology. Prog Neurobiol 74:363–396

Gotti C, Clementi F, Fornari A, Gaimarri A, Guiducci S, Manfredi I, Moretti M, Pedrazzi P, Pucci L, Zoli M (2009) Structural and functional diversity of native brain neuronal nicotinic receptors. Biochem Pharmacol 78:703–711

Gotti C, Guiducci S, Tedesco V, Corbioli S, Zanetti L, Moretti M, Zanardi A, Rimondini R, Mugnaini M, Clementi F, Chiamulera C, Zoli M (2010) Nicotinic acetylcholine receptors in the mesolimbic pathway: primary role of ventral tegmental area alpha6beta2* receptors in mediating systemic nicotine effects on dopamine release, locomotion, and reinforcement. J Neurosci 30:5311–5325

Gotti C, Hanke W, Maury K, Moretti M, Ballivet M, Clementi F, Bertrand D (1994) Pharmacology and biophysical properties of alpha 7 and alpha 7-alpha 8 alpha-bungarotoxin receptor subtypes immunopurified from the chick optic lobe. Eur J Neurosci 6:1281–1291

Gotti C, Moretti M, Meinerz N, Clementi F, Gaimarri A, Collins AC, Marks MJ (2008) Partial deletion of the nicotinic cholinergic receptor {alpha}4 and (Alkondon and Albuquerque)2 subunit genes changes the acetylcholine sensitivity of receptor mediated $86Rb^+$ efflux in cortex and thalamus and alters relative expression of {alpha}4 and (Alkondon and Albuquerque)2 subunits. Mol Pharmacol 73:1796–1807

Gotti C, Zoli M, Clementi F (2006) Brain nicotinic acetylcholine receptors: native subtypes and their relevance. Trends Pharmacol Sci 27:482–491

Grady SR, Moretti M, Zoli M, Marks MJ, Zanardi A, Pucci L, Clementi F, Gotti C (2009) Rodent habenulo-interpeduncular pathway expresses a large variety of uncommon nAChR subtypes, but only the alpha3beta4* and alpha3beta3beta4* subtypes mediate acetylcholine release. J Neurosci 29:2272–2282

Groot-Kormelink PJ, Luyten WH, Colquhoun D, Sivilotti LG (1998) A reporter mutation approach shows incorporation of the "orphan" subunit beta3 into a functional nicotinic receptor. J Biol Chem 273:15317–15320

Hurst R, Rollema H, Bertrand D (2013) Nicotinic acetylcholine receptors: from basic science to therapeutics. Pharmacol Ther 137:22–54

Jensen AA, Frolund B, Liljefors T, Krogsgaard-Larsen P (2005) Neuronal nicotinic acetylcholine receptors: structural revelations, target identifications, and therapeutic inspirations. J Med Chem 48:4705–4745

Jin X, Bermudez I, Steinbach JH (2014) The nicotinic alpha5 subunit can replace either an acetylcholine-binding or nonbinding subunit in the alpha4beta2* neuronal nicotinic receptor. Mol Pharmacol 85:11–17

Khiroug SS, Harkness PC, Lamb PW, Sudweeks SN, Khiroug L, Millar NS, Yakel JL (2002) Rat nicotinic ACh receptor alpha7 and beta2 subunits co-assemble to form functional heteromeric nicotinic receptor channels. J Physiol 540:425–434

Krashia P, Moroni M, Broadbent S, Hofmann G, Kracun S, Beato M, Groot-Kormelink PJ, Sivilotti LG (2010) Human alpha3beta4 neuronal nicotinic receptors show different stoichiometry if they are expressed in Xenopus oocytes or mammalian HEK293 cells. PLoS One 5:e13611

Kuryatov A, Olale F, Cooper J, Choi C, Lindstrom J (2000) Human alpha6 AChR subtypes: subunit composition, assembly, and pharmacological responses. Neuropharmacology 13:90–2570

Kuryatov A, Onksen J, Lindstrom J (2008) Roles of accessory subunits in alpha4beta2(*) nicotinic receptors. Mol Pharmacol 74:132–143

Liu Q, Huang Y, Xue F, Simard A, DeChon J, Li G, Zhang J, Lucero L, Wang M, Sierks M, Hu G, Chang Y, Lukas RJ, Wu J (2009) A novel nicotinic acetylcholine receptor subtype in basal forebrain cholinergic neurons with high sensitivity to amyloid peptides. J Neurosci 29:918–929

Marks MJ, Collins AC (1982) Characterization of nicotine binding in mouse brain and comparison with the binding of alpha-bungarotoxin and quinuclidinyl benzilate. Mol Pharmacol 22:554–564

Marks MJ, Whiteaker P, Collins AC (2006) Deletion of the alpha7, beta2, or beta4 nicotinic receptor subunit genes identifies highly expressed subtypes with relatively low affinity for [3H] epibatidine. Mol Pharmacol 70:947–959

Mazzo F, Pistillo F, Grazioso G, Clementi F, Borgese N, Gotti C, Colombo SF (2013) Nicotine-modulated subunit stoichiometry affects stability and trafficking of alpha3beta4 nicotinic receptor. J Neurosci 33:12316–12328

Miller PS, Smart TG (2010) Binding, activation and modulation of Cys-loop receptors. Trends Pharmacol Sci 31:161–174

Moretti M, Zoli M, George AA, Lukas RJ, Pistillo F, Maskos U, Whiteaker P, and Gotti C (2014) The novel α7β2-nicotinic acetylcholine receptor subtype is expressed in mouse and human basal forebrain: biochemical and pharmacological characterisation. Mol Pharmacol 86:306–317

Moroni M, Vijayan R, Carbone A, Zwart R, Biggin PC, Bermudez I (2008) Non-agonist-binding subunit interfaces confer distinct functional signatures to the alternate stoichiometries of the alpha4beta2 nicotinic receptor: an alpha4-alpha4 interface is required for Zn^{2+} potentiation. J Neurosci 28:6884–6894

Moroni M, Zwart R, Sher E, Cassels BK, Bermudez I (2006) alpha4beta2 nicotinic receptors with high and low acetylcholine sensitivity: pharmacology, stoichiometry, and sensitivity to long-term exposure to nicotine. Mol Pharmacol 70:755–768

Moser N, Mechawar N, Jones I, Gochberg-Sarver A, Orr-Urtreger A, Plomann M, Salas R, Molles B, Marubio L, Roth U, Maskos U, Winzer-Serhan U, Bourgeois JP, Le Sourd AM, De Biasi M, Schroder H, Lindstrom J, Maelicke A, Changeux JP, Wevers A (2007) Evaluating the suitability of nicotinic acetylcholine receptor antibodies for standard immunodetection procedures. J Neurochem 102:479–492

Mukhtasimova N, Lee WY, Wang HL, Sine SM (2009) Detection and trapping of intermediate states priming nicotinic receptor channel opening. Nature 459:451–454

Nelson ME, Kuryatov A, Choi CH, Zhou Y, Lindstrom J (2003) Alternate stoichiometries of alpha4beta2 nicotinic acetylcholine receptors. Mol Pharmacol 63:332–341

Palma E, Bertrand S, Binzoni T, Bertrand D (1996) Neuronal nicotinic alpha 7 receptor expressed in Xenopus oocytes presents five putative binding sites for methyllycaconitine. J Physiol 491:151–161

Palma E, Maggi L, Barabino B, Eusebi F, Ballivet M (1999) Nicotinic acetylcholine receptors assembled from the alpha7 and beta3 subunits. J Biol Chem 274:18335–18340

Picciotto MR, Caldarone BJ, Brunzell DH, Zachariou V, Stevens TR, King SL (2001) Neuronal nicotinic acetylcholine receptor subunit knockout mice: physiological and behavioral phenotypes and possible clinical implications. Pharmacol Ther 92:89–108

Picciotto MR, Higley MJ, Mineur YS (2012) Acetylcholine as a neuromodulator: cholinergic signaling shapes nervous system function and behavior. Neuron 76:116–129

Picciotto MR, Zoli M (2008) Neuroprotection via nAChRs: the role of nAChRs in neurodegenerative disorders such as Alzheimer's and Parkinson's disease. Front Biosci 13:492–504

Quik M, Wonnacott S (2011) alpha6beta2* and alpha4beta2* nicotinic acetylcholine receptors as drug targets for Parkinson's disease. Pharmacol Rev 63:938–966

Ramirez-Latorre J, Yu CR, Qu X, Perin F, Karlin A, Role L (1996) Functional contributions of alpha5 subunit to neuronal acetylcholine receptor channels. Nature 380:347–351

Rucktooa P, Smit AB, Sixma TK (2009) Insight in nAChR subtype selectivity from AChBP crystal structures. Biochem Pharmacol 78:777–787

Taly A, Corringer PJ, Guedin D, Lestage P, Changeux JP (2009) Nicotinic receptors: allosteric transitions and therapeutic targets in the nervous system. Nat Rev Drug Discov 8:733–750

Tapia L, Kuryatov A, Lindstrom J (2007) Ca^{2+} permeability of the (alpha4)3(beta2)2 stoichiometry greatly exceeds that of (alpha4)2(beta2)3 human acetylcholine receptors. Mol Pharmacol 71:769–776

Wessler I, Kirkpatrick CJ (2008) Acetylcholine beyond neurons: the non-neuronal cholinergic system in humans. Br J Pharmacol 154:1558–1571

Zhou Y, Nelson ME, Kuryatov A, Choi C, Cooper J, Lindstrom J (2003) Human alpha4beta2 acetylcholine receptors formed from linked subunits. J Neurosci 23:9004–9015

Zwart R, Carbone AL, Moroni M, Bermudez I, Mogg AJ, Folly EA, Broad LM, Williams AC, Zhang D, Ding C, Heinz BA, Sher E (2008) Sazetidine-A is a potent and selective agonist at native and recombinant alpha 4 beta 2 nicotinic acetylcholine receptors. Mol Pharmacol 73:1838–1843

Genetics of Smoking Behaviour

Jennifer J. Ware and Marcus R. Munafò

Abstract It is now well-established that smoking-related behaviours are under a substantial degree of genetic influence. Efforts are now focused on identifying the specific genetic variants which underlie these behaviours. Within this chapter, we introduce a variety of established and emerging methods employed to identify such variants, ranging from candidate gene to whole genome sequencing approaches, and highlight what these techniques have taught us about the genetic architecture of smoking-related behaviours. Further, we discuss how phenotype refinement has developed our understanding of these relationships, affording us insight into the specific mechanisms linking genetic variants to smoking-related behaviours.

Keywords Genetics · Smoking · Heritability · Candidate gene · GWAS · Sequencing · Phenotype refinement · Recall-by-genotype

Contents

1 Introduction 20
2 Genome-Wide Association Studies 21
3 Phenotype Refinement 26
3.1 In Vitro Studies and Animal Models 26
3.2 Candidate Gene Follow-up Studies 28
3.3 Recall-by-Genotype Studies 30
4 New and Emerging Technologies 31
5 Conclusions and Future Directions 32
References 33

M.R. Munafò
MRC Integrative Epidemiology Unit, UK Centre for Tobacco and Alcohol Studies, and School of Experimental Psychology, University of Bristol, Bristol BS8 1TU, UK

J.J. Ware (✉)
MRC Integrative Epidemiology Unit and School of Social and Community Medicine, University of Bristol, Oakfield House, Oakfield Grove, Bristol BS8 2BN, UK
e-mail: jen.ware@bristol.ac.uk

D.J.K. Balfour and M.R. Munafò (eds.), *The Neurobiology and Genetics of Nicotine and Tobacco*, Current Topics in Behavioral Neurosciences 23,
DOI 10.1007/978-3-319-13665-3_2

1 Introduction

Behaviour genetics has historically relied on the use of twin, family and adoption studies to determine the proportion of variance in a phenotype that can be attributed to genetic influences. The cornerstone of this approach is the 'natural experiment' of twinning—approximately 1 in 85 live births are twin births, and of these approximately one-third will be identical (monozygotic or MZ) twins and the remaining two-thirds non-identical or fraternal (dizygotic or DZ) twins. Since MZ twins have identical genotypes, whereas DZ twins share on average only 50 % of their variegated genotype, and yet in both cases twins will be raised in (presumably) identical environments, the logic of twin studies is that if a behavioural trait is more similar in pairs of MZ twins than it is in pairs of DZ twins, then that trait must presumably be under a degree of genetic influence. Family and adoption studies provide other means by which natural variation in genotypic similarity and environmental similarity can be associated with phenotypic variability to calculate the proportion of variation in the phenotype that can be accounted for by variation in genetic and environmental influences. The proportion of variation in phenotype that is due to variation in genotype is expressed as the *heritability* of a trait (h^2)—a heritability coefficient of 0.50 means that 50 % of the variation in that trait is due to genotypic variation.

Early evidence suggested that various aspects of smoking behaviour are under partial genetic influence; concordance rates are higher in MZ compared to DZ twins for persistent smoking and for successful cessation (Carmelli et al. 1992). Further evidence has supported this, suggesting a substantial heritability of smoking initiation (Heath et al. 1993), with this effect being partly independent of genetic effects on persistent smoking behaviour (Heath and Martin 1993), which itself shows a similar heritability. Twin studies have also allowed the dissection of different tobacco use phenotypes, using a conditional approach to understand the shared and unique genetic influences on initiation of tobacco use, progression to regular use, and development of dependence. This has indicated significant overlap in the contribution of genetic factors to these distinct phenotypes, but also important unique contributions, particularly to progression to regular tobacco use and the development of dependence (Maes et al. 2004). Similar approaches have enabled the investigation of shared and unique genetic influences on tobacco use by different routes of administration (e.g., cigarette smoking versus smokeless tobacco). This indicates that genetic influences may indeed vary by route of administration (Schmitt et al. 2005). More specific tobacco use phenotypes have also been investigated using these methods, such as initial reactions to tobacco use (both positive and negative), which also appear to be under a moderate degree of genetic influence (Agrawal et al. 2014). This is of particular interest given the evidence that initial reactions to tobacco use may predict the subsequent development of regular use and dependence (Sartor et al. 2010). Nevertheless, despite the valuable insights provided by twin, family and adoption studies, they are unable to identify the specific genetic variants that influence tobacco use.

However, since the mid-1990s it has been possible to directly genotype specific variants, and therefore look for association between measured genetic variation and phenotypic variation. Genetic association studies typically use a case-control methodology to compare the prevalence of genetic variants in these two unrelated groups (e.g., tobacco dependent and non-dependent individuals). Genetic association studies typically investigate candidate genes selected on the basis of a priori evidence of relevance, based on the known or presumed neurobiology of the phenotype of interest. For example, given the acknowledged role of dopamine in addictive behaviours, candidate genes related to variation in the dopamine pathway received considerable interest (Munafò et al. 2004). However, despite over a decade of candidate gene studies, very few associations proved reproducible. It now seems that this was principally because the effects of common genetic variants on complex behavioural phenotypes, such as smoking behaviour, are very small (typically <0.1 % phenotypic variance). These studies were therefore simply underpowered to reliably detect these effects.

Replication of candidate gene studies has been uneven, at best. This may reflect the fact that initial findings were false positives (which therefore subsequently failed to replicate), or may instead reflect differences in phenotype definition across studies. Certainly some candidate gene findings have proven to be robust (notably those relating to genes encoding nicotine-metabolising enzymes, such as *CYP2A6* see chapter entitled Pharmacogenetics of Nicotine and Associated Smoking Behaviors; this volume). However, there is growing acceptance that most have not, and it remains difficult to discern from the published candidate gene literature which findings are reliable (given differences in study design and phenotypic definition, as well as uneven statistical evidence). However, in the early 2000s, candidate gene methods were rapidly supplanted by genome-wide association studies (GWAS).

2 Genome-Wide Association Studies

Genome-wide association studies adopt the same approach to sample selection and analysis as candidate gene studies (e.g., comparison of allelic frequencies between groups selected on the basis of phenotype). However, instead of focusing on a pre-specified locus, microarrays ("gene chips") are used to systematically genotype hundreds of thousands of single nucleotide polymorphisms (SNPs) across the entire genome. As such, and in contrast to candidate gene studies, they are strictly agnostic in their approach, requiring no a priori hypothesis. The multiple testing burden implicit in GWAS led to consensus that signals had to achieve a statistical significance threshold corrected for this (typically $p < 5 \times 10^{-8}$), which in turn necessitated the pooling of data across multiple studies to achieve the necessary sample size to detect the small effects associated with the common genetic variants captured by GWAS. As a direct consequence of these large-scale efforts, GWAS have been extremely successful in identifying genetic variants associated with a range of complex phenotypes, aiding identification of variants that would perhaps not have been considered previously on the basis of biological function.

The most robust finding to emerge from GWAS of smoking phenotypes is the association between a variant within the nicotinic acetylcholine receptor gene cluster *CHRNA5-A3-B4* on chromosome 15 and smoking quantity. An association between rs16969968 in *CHRNA5* and nicotine dependence (ND) was first reported in 2007 in a candidate gene study (Saccone et al. 2007), with the minor allele found to confer increased risk. The following year, the same locus (tagged by rs1051730 in *CHRNA3*, a variant highly correlated with rs16969968) was found to be associated with smoking quantity in a GWAS conducted by Thorgeirsson et al. (2008). This study also demonstrated an association between rs1051730 and nicotine dependence and two smoking-related diseases, lung cancer and peripheral arterial disease. Notably, whilst the candidate gene study was published first, it was the GWAS that made much more of an impact. This may have been because *CHRNA5* was not recognised as a particularly strong candidate at the time, given the then known neurobiology of tobacco dependence—more emphasis had been placed on genes encoding α_4 and β_2 subunits, based on evidence from animal models that these mediate the reinforcing properties of nicotine (e.g., Picciotto et al. 1998). Furthermore, the simultaneous demonstration of an association between this locus and two smoking-related diseases (i.e., lung cancer and peripheral arterial disease) lent further weight to this finding. The association between this locus and smoking quantity has since been replicated in several consequent independent studies (David et al. 2012; Liu et al. 2010; Thorgeirsson et al. 2010; Tobacco and Genetics Consortium 2010).

Early GWAS of smoking behaviours typically examined associations with directly genotyped SNPs only, which provide relatively sparse coverage of the genome, and include only common SNPs with a minor allele frequency (MAF) greater than 5 % (Caporaso et al. 2009; Thorgeirsson et al. 2008). In contrast, GWAS now commonly employ imputation to expand genomic coverage. Imputation allows us to predict ('impute') genotypes that have not been directly genotyped on a microarray. In order to do so, microarray data is matched to a genome reference panel, which consists of densely genotyped (or sequenced) genomic data from multiple individuals (see Fig. 1). The genomic reference panels typically used for imputation are HapMap (http://hapmap.ncbi.nlm.nih.gov/) and 1000 Genomes (http://www.1000genomes.org/). The most recent release of 1000 Genomes (Phase I) provides a much larger set of SNPs (~ 37 million) relative to HapMap (~ 4 million). Imputation is used in GWAS for a number of reasons. First, it can increase study power (Marchini and Howie 2010). Second, it provides us with a more detailed view of particular genomic regions which contain many associated SNPs. This fine-mapping approach may allow us to identify actual causal variants not captured by microarrays. Finally, imputation allows us to combine GWAS results across studies that have used different GWAS chips, thus enabling meta-analysis (Marchini and Howie 2010). Imputation has been used to great success in recent GWAS of smoking phenotypes, enabling the running of very large ($n > 75{,}000$), consortia-based GWAS meta-analyses, which have identified a number of new genomic regions (Liu et al. 2010; Thorgeirsson et al. 2010; Tobacco and Genetics Consortium 2010). Further, it has enabled fine-mapping of the 15q25 locus (Liu et al. 2010). These results will be discussed in turn.

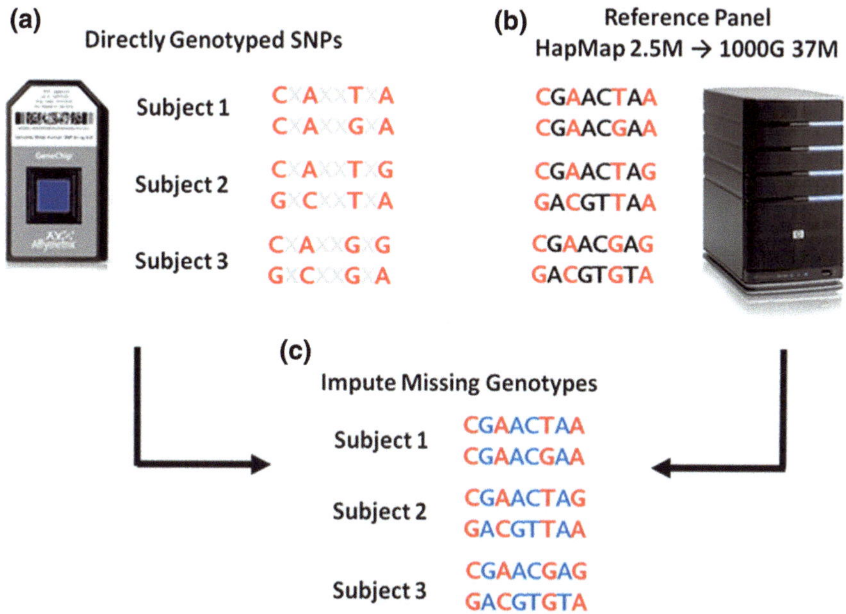

Fig. 1 Schematic overview of imputation. Imputation allows us to predict ('impute') genotypes which have not been directly genotyped on a microarray. In order to do so, microarray data is matched to a genome reference panel, which consists of densely genotyped (or sequenced) genomic data from multiple individuals. The genomic reference panels typically used for imputation are HapMap and 1000 Genomes. The most recent release of 1000 Genomes (phase I) provides a much larger set of SNPs ($\sim$37 million) relative to HapMap. This affords increased resolution of genomic regions. Imputation error increases as minor allele frequency decreases, as the imputation is based on an algorithm derived from a much smaller number of individuals (Marchini and Howie 2010). Accurate imputation of low-frequency SNPs requires genomic data from large numbers of individuals. HapMap 2 contains haplotype information on just 120 Europeans. However, HapMap 3 is based on data from 330 Europeans. Similarly, 1000 Genomes phase I includes far more haplotypes than 1000 Genomes pilot I, thus allowing for more accurate imputation of low frequency or rare SNPs. If rare variation is key to a particular trait, then 1000 Genomes Phase 1 (and later releases) may help identify these variants. Figure reproduced from Wood et al. (2013)

Consortia-based GWAS meta-analyses (typically imputed to HapMap 2) have identified a number of novel genomic regions associated with a range of smoking phenotypes, including smoking initiation and smoking cessation, as well as smoking quantity. Specifically, a nonsynonymous SNP (rs6265) in the brain-derived neurotrophic factor (*BDNF*) gene on chromosome 11 was found to be associated with smoking initiation. BDNF is involved in regulating synaptic plasticity and survival of cholinergic and dopaminergic neurons (Thorgeirsson et al. 2010; Tobacco and Genetics Consortium 2010). A variant near dopamine beta hydroxylase (*DBH*) gene on chromosome 9 was also identified in relation to smoking cessation (Thorgeirsson et al. 2010; Tobacco and Genetics Consortium 2010). DBH is an enzyme involved in the conversion of dopamine to noradrenaline. In relation to smoking quantity, genomic regions on chromosomes 8, 10 and 19

Table 1 Genomic loci identified through GWAS of smoking phenotypes

Author	Year	Phenotype	Chr	Gene	SNP	Notes
Thorgeirsson	2008	Smoking quantity	15	*CHRNA3*	rs1051730	
Liu	2009	Smoking status	4	near *IL15*	rs4956302	Marginal evidence for association ($p = 8.8 \times 10^{-8}$) in males only
Furberg	2010	Smoking quantity	15	*CHRNA3*	rs1051730	Primary signal at this locus
		Smoking quantity	15	*CHRNA5*	rs684513	Independent signal (after conditioning on rs1051730)
		Smoking quantity	10	*LOC100188947*	rs1329650	
		Smoking quantity	10	*LOC100188947*	rs1028936	
		Smoking quantity	19	*EGLN2*	rs3733829	
		Initiation	11	*BDNF*	rs6265	
		Cessation	9	*DBH*	rs3025343	
Thorgeirsson	2010	Smoking quantity	15	*CHRNA3*	rs1051730	Primary signal at this locus
		Smoking quantity	15	near *IREB2*	rs2869046	Independent signal (after conditioning on rs1051730)
		Smoking quantity	15	*AGPHD1*	rs2036534	Independent signal (after conditioning on rs1051730)
		Smoking quantity	19	*CYP2A6*	rs4105144	
		Smoking quantity	8	*CHRNB3*	rs6474412	SNP in shared LD block with *CHRNA6*
Liu	2010	Smoking quantity	15	*CHRNA3*	rs1051730	Original GWAS using HapMap (release 22)
		Smoking quantity	15	*CHRNA5*	rs55853698	1000 Genomes (Pilot 1). SNP is in high LD with rs1051730
		Smoking quantity	15	*CHRNA3*	rs6495308	Independent signal (after conditioning on rs55853698)
David	2012	Smoking quantity	15	near *CHRNA5*	rs2036527	African American sample

were identified, in addition to the 15q25 locus (Thorgeirsson et al. 2010; Tobacco and Genetics Consortium 2010). In particular, two genes with clear biological relevance were identified—two nicotinic receptor sub-unit genes (*CHRNB3/CHRNA6*) on chromosome 8 and *CYP2A6* on chromosome 19 (Thorgeirsson et al. 2010). The CYP2A6 enzyme is principally responsible for the metabolism of nicotine (see chapter entitled Pharmacogenetics of Nicotine and Associated Smoking Behaviors; this volume). However, the proportion of phenotypic variance explained by these SNPs is far less than the variance explained by rs16969968-rs1051730 at 15q25, which explains ~ 1 % of the variation in smoking quantity.

High-density imputation has afforded additional benefits; for example, it has enabled high-resolution examination of the 15q25 region in relation to smoking quantity. Liu et al. (2010) used 1000 Genomes Pilot 1 data to fine-map the 15q25 region for association with smoking quantity. This allowed for analysis of virtually all common SNPs (MAF > 5 %) in the region, and offered a fivefold increase in marker density compared to HapMap 2. Using this imputation approach, combined with meta-analysis, Liu et al. (2010) identified rs55853698 as the variant with strongest evidence for association with smoking quantity at this locus. This variant, which is absent from HapMap 2, is located within a promoter region of *CHRNA5*, and is a plausible candidate for affecting mRNA transcription. However, this variant is also in very high LD with rs16969968. Conditioning on rs55853698 revealed a second independent signal at rs6495308, located within *CHRNA3*. Conditioning on both SNPs left no residual signal, suggesting that these two variants could explain the full signal at 15q25.1 in relation to smoking quantity (Liu et al. 2010).

In summary, GWAS has dramatically advanced our understanding of the genetic underpinnings of smoking initiation, cessation, and smoking quantity (see Table 1). Despite these successes, however, the total proportion of variance explained by all variants identified through GWAS to date is far less than the heritability estimates indicated by earlier twin studies. It is possible that rare variants (MAF < 1 %), which cannot be very accurately imputed using current genomic reference panels (given the limited numbers of haplotypes on which they are based) may account for this 'missing heritability'. This is discussed in more detail in Text Box 1.

Text Box 1: The Missing Heritability Problem Genome-wide association studies (GWAS) have been extremely successful in identifying genetic variants associated with a range of complex phenotypes. As we discuss, several loci associated with various tobacco use phenotypes have been identified, through large consortium-based efforts (Tobacco and Genetics Consortium 2010). This is in stark contrast to the candidate gene literature, where few reliable signals emerged after almost two decades of effort. Despite this, variants identified to date via GWAS explain less than half the heritability of complex phenotypes such as smoking behaviour estimated by twin and family studies. This is the so-called "missing heritability" problem (Manolio et al. 2009).

It is now clear that some missing heritability will be accounted for by variants that have not yet been identified via GWAS (Yang et al. 2010). This is in part because most common variant chips have relatively poor coverage in the minor allele frequency (MAF) spectrum below 5 %. Evolutionary theory predicts that mutations strongly affecting complex phenotypes will tend to occur at low allele frequencies (Visscher et al. 2012). There is, therefore, growing interest in the potential role of low frequency and rare variants (i.e., MAF <5 and <1 %, respectively). In other words, "missing heritability" is likely to be due to two factors (Visscher et al. 2012): (i) unidentified common variants with very small effects that cannot be detected by current GWAS using existing sample sizes, and (ii) rare variants not in sufficient linkage disequilibrium with variants on currently available genotyping chips and therefore undetectable via GWAS.

3 Phenotype Refinement

As we have discussed, GWAS approaches have proven highly successful in identifying genetic variants that are associated with a variety of tobacco use phenotypes (Thorgeirsson et al. 2010; Tobacco and Genetics Consortium 2010). For example, these approaches have confirmed the importance of the nicotinic receptor subunit gene cluster *CHRNA5-A3-B4* in smoking quantity. Identification of novel genes implicated in tobacco use through genome-wide approaches (e.g., *CHRNA5*) has led to renewed interest in these genes and their products. However, these studies are limited in two important ways. First, they rely on self-report measures of smoking behaviour, often ascertained retrospectively and in different ways across different studies. This combination of self-report, retrospective recall, and the need for phenotype harmonisation, may introduce considerable measurement error. Second, these studies have not taught us what the fundamental mechanisms linking these genes to these behaviours are. Phenotype refinement provides a means by which to understand the specific mechanisms linking genetic variants to disease. Carefully designed and well-characterised phenotypes also offer greater measurement precision, in principle providing a 'cleaner' genetic signal, and improving the likelihood of effect replication. In this section, we discuss how phenotype refinement has advanced our understanding of these relationships.

3.1 *In Vitro Studies and Animal Models*

The rs16969968 variant in *CHRNA5* is non-synonymous, resulting in an amino acid change (aspartate to asparagine) in the resultant α_5 nicotinic receptor subunit protein. Research has shown that this variant is functional. In vitro studies have demonstrated that nicotinic receptor complexes containing the α_5 receptor subunit featuring the aspartic acid variant exhibit a twofold greater maximal response to a nicotine agonist compared to α_5 receptor complexes containing the asparagine variant (i.e., the risk variant robustly associated with heavier smoking) (Bierut et al. 2008).

'Knockout' mouse models have enabled us to further explore the impact of specific genes on specific phenotypes. A knockout mouse is simply a mouse in which a specific target gene has been selectively disabled. Studying the physiological characteristics or behaviours of such mice (e.g., assessing their response to nicotine), provides insight into the mechanisms which link the knocked out gene to disease. The development of α_5 knockout mouse models has illustrated the role that the *CHRNA5* gene plays in determining response to nicotine. Such mice are considered a model of individuals with reduced *CHRNA5* gene function (i.e., carriers of the rs16969968 minor allele, associated with increased heaviness of smoking). Fowler et al. (2011) conducted an elegant series of experiments using the α_5 knockout mouse model. Based on a self-administration paradigm, they observed that knockout mice responded far more vigorously than wild-type mice (i.e., mice with no inactivated genes) for nicotine infusions at high doses (see Fig. 2a). Whilst wild-type mice self-titrated delivery of nicotine dose to achieve a consistent, desired level during each test session ($\sim$1.5 mg kg^{-1}), knockout mice did not, consuming greater amounts as dosage increased (see Fig. 2b). The authors proposed that deficient α_5 signalling attenuates the negative effects of nicotine that normally serve to limit its intake, a conclusion which fits well with human research (i.e., smokers carrying the rs16969968 risk allele are likely to smoke more heavily than their counterparts without the risk allele). Fowler and colleagues further demonstrated that this effect could be 'rescued' in α_5 knockout mice through injection of a lentivirus vector into the medial habenula (MHb), an area where α_5 subunits are typically densely expressed, rescuing expression of α_5 subunits in this region. The MHb projects to the interpeduncular nucleus (IPN). Notably, the pathway between the MHb and the IPN has previously been shown to regulate avoidance of noxious substances such as quinine (Donovick et al. 1970). Fowler and colleagues further

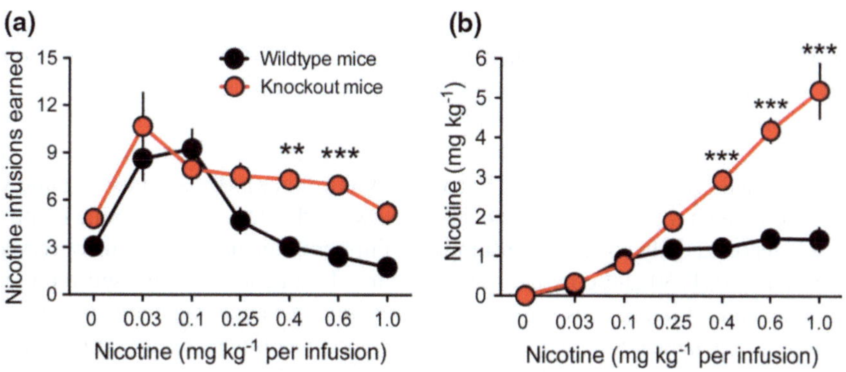

Fig. 2 Self-administration of nicotine in α_5 subunit knockout versus wild-type mice. *Panel a* Knockout and wild-type mice respond at an equivalent level for nicotine infusions at low doses. However, knockouts respond more vigorously than wild-types for high doses of nicotine. *Panel b* Wild-type mice self-titrate delivery of nicotine dose to achieve a consistent, desired level during test sessions ($\sim$1.5 mg kg^{-1}), whereas knockout mice do not, consuming greater amounts as dosage increases. Figure reprinted by permission from Macmillan Publishers Ltd: Nature (Fowler et al. 2011), copyright 2011

observed reduced activity in the IPN in response to nicotine in knockout animals. Additionally, disruption of IPN activity was found to increase nicotine self-administration. However, both knockout and wild-type mice were indistinguishable in terms of responding for low doses of nicotine (see Fig. 2), indicative of shared experience of the rewarding, stimulatory effects of nicotine at low doses in both groups. These observations are complemented by a previous study by Jackson (2010), which also illustrated the initially shared and later differential experience of reward in response to an increasing dose of nicotine between α_5 knockouts and wild-types, in this instance illustrated using a conditioned place preference task.

Fowler et al. (2011) concluded that high doses of nicotine stimulate the pathway between the MHb and the IPN pathway via nicotinic receptor complexes containing α_5 subunits. This results in the relay of an inhibitory motivational signal that serves to limit further nicotine intake. It is perhaps easiest to consider this pathway as a negative feedback loop, gated by α_5 containing nicotinic receptors, which is inadequately activated by high doses of nicotine in those with the rs16969968 risk allele.

3.2 Candidate Gene Follow-up Studies

Despite the success of GWAS, the number of individual genes identified, and in particular the proportion of phenotypic variance explained, remains relatively modest (typically <1 %). However, as we have discussed, the phenotypes employed in studies of smoking behaviour and often quite crude, for example using self-reported daily cigarette consumption as a measure of total tobacco exposure (Thorgeirsson et al. 2008). Subsequent studies have sought to follow-up on associations identified through GWAS techniques, employing a candidate gene approach in conjunction with refined phenotypes.

Munafò et al. (2012) examined the association of rs1051730 with both self-reported smoking quantity, and circulating levels of cotinine (see Fig. 3). Cotinine is the primary metabolite of nicotine, and is a much more precise and objective measure of total tobacco exposure (see chapter entitled Pharmacogenetics of Nicotine and Associated Smoking Behaviors; this volume). Perhaps unsurprisingly, they observed a much stronger association between this locus and cotinine relative to self-reported daily cigarette consumption. Furthermore, the relationship observed between rs1051730 and cotinine was robust to adjustments made for self-reported cigarette consumption. This suggests that even amongst equal cigarette consumers, there is still genetically influenced variation in total nicotine exposure. This finding is supported by a previous study conducted by Le Marchand et al. (2008), who noted that the association between the same locus and total nicotine equivalents (an extremely accurate measure of nicotine exposure, encompassing nicotine, cotinine, related glucuronides, and 3-hydroxycotinine) was also robust for adjustments made for self-reported daily cigarette consumption. Munafò et al. (2012) propose that this is the result of differences in smoking topography (i.e., how a cigarette is smoked: number of puffs per cigarette, volume of smoke inhaled and so on).

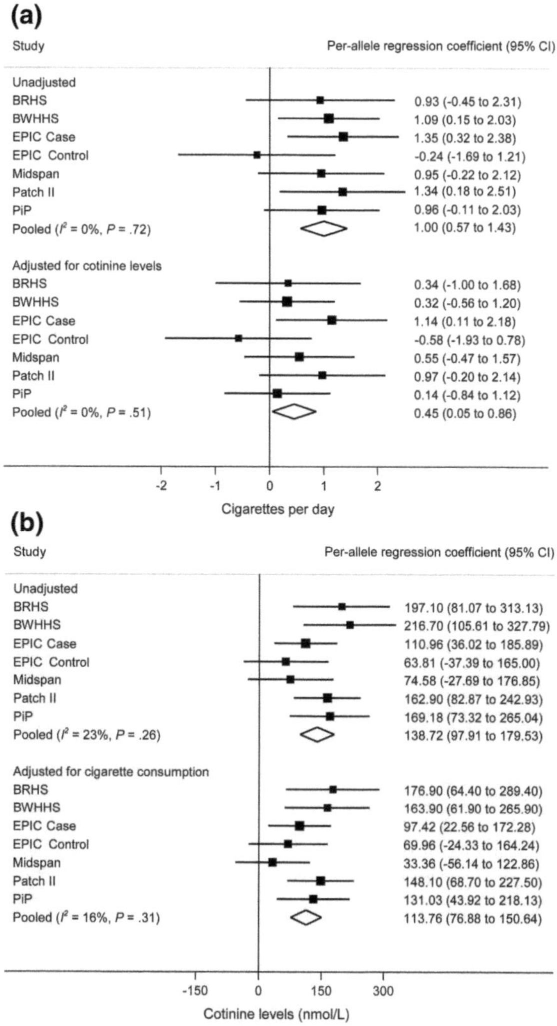

Fig. 3 Meta-analysis of association of rs1051730/rs16969968 risk allele with two measures of heaviness of smoking in current smokers. *Panel a* self-reported cigarette consumption. *Panel b* biochemically-assessed cotinine levels. Munafò et al. (2012), by permission of Oxford University Press

These findings replicate and extend previous work by Keskitalo et al. (2009), who found that rs16969968 accounts for ~1 % of the variation in self-report measures of smoking quantity, compared to ~4 % of the variation in levels of cotinine. The key observation in these studies is that carefully designed and well-characterised phenotypes offer greater measurement precision. They provide a 'cleaner' genetic signal, and improve the likelihood of effect replication. While it may not be feasible to conduct GWAS on the same scale using these biochemical phenotypes as it is using self-report measures, these findings suggest two possibilities. First, variants identified through conventional GWAS can be followed up, exploring the candidate gene so identified with greater phenotype precision. Second, smaller GWAS which use these more precise phenotypes may still have greater statistical power, and identify novel variants, compared to larger studies using less-precise phenotypes. We have recently demonstrated the utility of this approach, applying GWAS techniques to examine genetic determinants of cotinine level in current smokers (Ware et al. in preparation). A clear extension of this line of research would be to repeat this approach using total nicotine equivalents as a phenotype.

3.3 Recall-by-Genotype Studies

As we have seen, follow-up candidate gene studies of variants identified through GWAS can prove valuable in further refining the nature of the observed association. Together with the use of precisely-assessed phenotypes, they can provide important insights into underlying mechanisms of genetic action. One drawback to this approach, however, is that the measurement of detailed, precise phenotypes in large, adequately powered samples can be costly and time consuming. One approach which circumvents these issues is the 'recall-by-genotype' method, which we outline below.

The recall-by-genotype approach involves the selection of participants (or their biological samples) on the basis of a specific genotype, for further detailed, clinically-relevant, phenotype assessment. This approach has two key strengths. First, it is efficient. For example, statistical power can be improved by using genotype-specific recall instead of random sampling; it is far more powerful to selectively phenotype 100 minor homozygotes and 100 major homozygotes (at opposite ends of a biological gradient) than to phenotype 200 individuals at random, because the latter design would recruit far fewer homozygotes. Consider rs16969968 as an example. Only 16 % of Europeans are minor homozygotes at this locus. Therefore, in an unselected sample of 200 individuals, we would only expect 32 minor homozygotes. A second, related advantage of the recall-by-genotype approach is that it enables the collection of extremely detailed phenotypic data that would be impractical to collect in a much larger sample for reasons of time and expense.

We have recently adopted a recall-by-genotype approach to further investigate the association between rs16969968 and total nicotine exposure (Ware et al. 2014). We saw in the previous section that the relationship between this locus and cotinine levels was robust to adjustments made for self-reported daily cigarette consumption, suggesting that even among equal cigarette consumers there is genetically influenced variation in total nicotine exposure. Munafò et al. (2012) proposed that this is likely due to differences in the way a cigarette is smoked (e.g., number of puffs taken per cigarette, volume of smoke inhaled per puff). Smoking topography is the term used to describe these nuances in smoking behaviour, and it can be assessed precisely using a small hand-held device. It has previously been established that cigarette smokers modify their smoking behaviour to self-titrate nicotine to a level appropriate to individual need (McNeill and Munafò 2013; Strasser et al. 2007). What remains to be established, however, is whether differences in smoking topography mediate the relationship between the rs16969968 and nicotine metabolite levels. We are currently investigating this using a recall-by-genotype approach, recalling current smokers from a large cohort study on the basis of minor and major homozygote status at rs1051730-rs16969968, assessing smoking topography and cotinine levels in both groups.

4 New and Emerging Technologies

As we discussed at the beginning of the chapter, smoking-related behaviours are highly heritable. Evidence from twin studies suggests that up to 75 % of the variation in nicotine dependence, for example, can be accounted for by genetic factors. However, findings from GWAS account for only a tiny proportion of such estimates. Why might this be? Have twin studies overestimated trait heritability? Do gene × gene and gene × environment interactions account for the shortfall? These are possible explanations. However, it is likely that a substantial proportion of 'missing heritability' remains undetected by GWAS because (a) the individual effects of common SNPs studied are too small to reach strict thresholds for genome-wide significance, and (b) rare and novel genetic variants, which are not captured by GWAS approaches, also contribute to the genetic architecture of these traits. Evidence in support of the former point is provided by genome-wide complex trait analysis (GCTA), a novel method for calculating the proportion of phenotypic variance explained by genome-wide SNPs for complex traits. Using this approach, Yang et al. (2010) illustrated that *common* SNPs explain 45 % of the variance in human height. This stands in stark contrast to the SNPs identified through height GWAS, which explain ~5 % variation, a far cry from the 80 % heritability estimates calculated from twin studies. As a more relevant example, Lubke et al. (2012) illustrated that 19 % of the variation in smoking initiation could be explained by ~1.1 M common SNPs. This is a substantially larger proportion of heritability than that explained by SNPs identified in GWAS of this phenotype.

It is also likely that rare and novel genetic variants, which are not captured by GWAS approaches, account for some of this additional unexplained heritability. In contrast to GWAS, genome sequencing, involving the use of high-throughput sequencing technologies to determine the order of nucleotide bases present in a particular individual, provides high-quality rare variant genotype calls. Given the size of the human genome (which consists of ~3 billion nucleotide base pairs), and the formerly prohibitive costs of sequencing whole genomes, much research in this field has focused on exome sequencing. Exome analysis involves the study of genetic variants within the exome only (i.e., protein-coding regions of the genome), which comprises ~1 % of the human genome (Ng et al. 2009). This approach allows for the identification of rare, non-synonymous variants associated with specific phenotypes. Of note, however, is that non-coding regions of the genome (e.g., introns) may also be of importance in determining disease. Indeed, the majority of trait-associated variants identified by GWAS fall in non-coding genome regions (Hindorff et al. 2009). For this reason, whole genome sequencing (covering ~95 %

of the genome) is preferable. Expense has, up until very recently, prohibited widespread use of this technique (the first sequenced human genome cost ~$3 billion); however, recent advances in technology have resulted in a huge drop in cost in recent years, and the $1,000 genome is now a reality (Hayden 2014).

Sequencing techniques are now being employed to further our understanding of tobacco dependence and other smoking-related phenotypes. An example of such a project is the GWAS and Sequencing Consortium of Alcohol and Nicotine use (GSCAN; https://gscan.sph.umich.edu/), led by Scott Vrieze at the University of Michigan. This is a collaborative effort comprising around 30 individual studies, with the aim of identifying genes associated with various tobacco- and alcohol-related phenotypes, including cigarettes per day, pack years, smoking initiation, and age of smoking initiation. The GSCAN project comprises three complimentary approaches: GWAS meta-analysis, exome meta-analysis, and whole genome sequencing meta-analysis. All contributing studies are to be imputed using a new haplotype reference panel, affording relatively high-density genotype imputation and accurate imputation of low frequency variants. Further, given the size of the collaboration effort, the sample size promises to be extremely large (affording increased power to detect small genetic effects). The exome meta-analysis should theoretically allow for the identification of rare, non-synonymous, exonic variants associated with nicotine-related phenotypes. Finally, the sequencing arm of the GSCAN project will be focused on following up variants identified through GWAS analyses, allowing evaluation at greatly increased genomic resolution. The complimentary use of GWAS and sequencing techniques has recently been used to great success (Holm et al. 2011). In short, this large, collaborative project holds the potential to substantially advance our understanding of the genetic basis of smoking-related behaviours.

5 Conclusions and Future Directions

We have known that smoking behaviours are heritable for some time, but the candidate gene era generated findings that proved difficult or impossible to replicate reliably, and taught us little about the specific genetic underpinnings of smoking behaviour. However, the introduction of genome-wide technologies in the mid-2000s has transformed our understanding of the genetic architecture of smoking behaviour, and identified a number of common variants associated with a range of smoking-related phenotypes. In particular, these approaches have implicated five nicotinic receptor subunit genes (*CHRNA5*, *CHRNA3*, *CHRNA6*, *CHRNB3* and *CHRNB4* see chapter entitled Structure of Neuronal Nicotinic Receptors; this volume), all specifically in relation to smoking quantity (Thorgeirsson et al. 2010; Tobacco and Genetics Consortium 2010). The genomic region encoding the nicotine- and cotinine-metabolising enzyme CYP2A6 has further been implicated in smoking quantity (Thorgeirsson et al. 2010). *BDNF* has been highlighted in relation to smoking initiation and *DBH* in relation to smoking cessation (Thorgeirsson et al. 2010). These exciting initial findings have led to in vitro and preclinical research focused on determining the mechanisms underlying these associations. Follow-up studies in humans employing carefully refined phenotypes have further developed our understanding of these relationships.

However, despite these successes, we can still only account for relatively little of the heritability of smoking behaviours. New sequencing approaches, which are increasingly becoming more affordable, hold promise and may offer the potential to discover rare or novel variants associated with smoking-related behaviours. However, as Loukola et al. (2014) stress, identification is "merely the first step in the discovery process". Investigating the function of newly identified genetic variants must follow, through careful employment of the techniques discussed within this chapter (amongst others), which may ultimately lead to the development of novel treatments for tobacco dependence, and, as is the goal, an improvement in health outcomes.

References

Agrawal A, Madden PA, Bucholz KK, Heath AC, Lynskey MT (2014) Initial reactions to tobacco and cannabis smoking: a twin study. Addiction 109:663–671

Bierut LJ, Stitzel JA, Wang JC, Hinrichs AL, Grucza RA, Xuei X, Saccone NL, Saccone SF, Bertelsen S, Fox L, Horton WJ, Breslau N, Budde J, Cloninger CR, Dick DM, Foroud T, Hatsukami D, Hesselbrock V, Johnson EO, Kramer J, Kuperman S, Madden PAF, Mayo K, Nurnberger J Jr, Pomerleau O, Porjesz B, Reyes O, Schuckit M, Swan G, Tischfield JA, Edenberg HJ, Rice JP, Goate AM (2008) Variants in the nicotinic receptors alter risk for nicotine dependence. Am J Psychiatry 165:1163–1171

Caporaso N, Gu F, Chatterjee N, Sheng-Chih J, Yu K, Yeager M, Chen C, Jacobs K, Wheeler W, Landi MT, Ziegler RG, Hunter DJ, Chanock S, Hankinson S, Kraft P, Bergen AW (2009) Genome-wide and candidate gene association study of cigarette smoking behaviors. PLoS ONE 4:e4653

Carmelli D, Swan GE, Robinette D, Fabsitz R (1992) Genetic influence on smoking—a study of male twins. N Engl J Med 327:829–833

David SP, Hamidovic A, Chen GK, Bergen AW, Wessel J, Kasberger JL, Brown WM, Petruzella S, Thacker EL, Kim Y, Nalls MA, Tranah GJ, Sung YJ, Ambrosone CB, Arnett D, Bandera EV, Becker DM, Becker L, Berndt SI, Bernstein L, Blot WJ, Broeckel U, Buxbaum SG, Caporaso N, Casey G, Chanock SJ, Deming SL, Diver WR, Eaton CB, Evans DS, Evans MK, Fornage M, Franceschini N, Harris TB, Henderson BE, Hernandez DG, Hitsman B, Hu JJ, Hunt SC, Ingles SA, John EM, Kittles R, Kolb S, Kolonel LN, Le Marchand L, Liu Y, Lohman KK, McKnight B, Millikan RC, Murphy A, Neslund-Dudas C, Nyante S, Press M, Psaty BM, Rao DC, Redline S, Rodriguez-Gil JL, Rybicki BA, Signorello LB, Singleton AB, Smoller J, Snively B, Spring B, Stanford JL, Strom SS, Swan GE, Taylor KD, Thun MJ, Wilson AF, Witte JS, Yamamura Y, Yanek LR, Yu K, Zheng W, Ziegler RG, Zonderman AB, Jorgenson E, Haiman CA, Furberg H (2012) Genome-wide meta-analyses of smoking behaviors in African Americans. Transl Psychiatry 2:e119

Donovick PJ, Burright RG, Zuromski E (1970) Localization of quinine aversion within the septum, habenula, and interpeduncular nucleus of the rat. J Comp Physiol Psychol 71:376–383

Fowler CD, Lu Q, Johnson PM, Marks MJ, Kenny PJ (2011) Habenular alpha5 nicotinic receptor subunit signalling controls nicotine intake. Nature 471:597–601

Hayden EC (2014) Is the $1,000 genome for real? Nature doi:10.1038/nature.2014.14530

Heath AC, Martin NG (1993) Genetic models for the natural history of smoking: evidence for a genetic influence on smoking persistence. Addict Behav 18:19–34

Heath AC, Cates R, Martin NG, Meyer J, Hewitt JK, Neale MC, Eaves LJ (1993) Genetic contribution to risk of smoking initiation: comparisons across birth cohorts and across cultures. J Subst Abuse 5:221–246

Hindorff LA, Sethupathy P, Junkins HA, Ramos EM, Mehta JP, Collins FS, Manolio TA (2009) Potential etiologic and functional implications of genome-wide association loci for human diseases and traits. Proc Natl Acad Sci USA 106:9362–9367

Holm H, Gudbjartsson DF, Sulem P, Masson G, Helgadottir HT, Zanon C, Magnusson OT, Helgason A, Saemundsdottir J, Gylfason A, Stefansdottir H, Gretarsdottir S, Matthiasson SE, Thorgeirsson GM, Jonasdottir A, Sigurdsson A, Stefansson H, Werge T, Rafnar T, Kiemeney LA, Parvez B, Muhammad R, Roden DM, Darbar D, Thorleifsson G, Walters GB, Kong A, Thorsteinsdottir U, Arnar DO, Stefansson K (2011) A rare variant in MYH6 is associated with high risk of sick sinus syndrome. Nat Genet 43:316–320

Jackson KJ, Marks MJ, Vann RE, Chen X, Gamage TF, Warner JA, Damaj MI (2010) Role of alpha5 nicotinic acetylcholine receptors in pharmacological and behavioral effects of nicotine in mice. J Pharmacol Exp Ther 334:137–146

Keskitalo K, Broms U, Heloidie, vaara M, Ripatti S, Surakka I, Perola M, PitkÃ¤niemi J, Peltonen L, Aromaa A, Kaprio J (2009) Association of serum cotinine level with a cluster of three nicotinic acetylcholine receptor genes (CHRNA3/CHRNA5/CHRNB4) on chromosome 15. Hum Mol Genet 18:4007–4012

Le Marchand L, Derby KS, Murphy SE, Hecht SS, Hatsukami D, Carmella SG, Tiirikainen M, Wang H (2008) Smokers with the CHRNA lung cancer-associated variants are exposed to higher levels of nicotine equivalents and a carcinogenic tobacco-specific nitrosamine. Cancer Res 68:9137–9140

Liu JZ, Tozzi F, Waterworth DM, Pillai SG, Muglia P, Middleton L, Berrettini W, Knouff CW, Yuan X, Waeber G, Vollenweider P, Preisig M, Wareham NJ, Zhao JH, Loos RJ, Barroso I, Khaw KT, Grundy S, Barter P, Mahley R, Kesaniemi A, McPherson R, Vincent JB, Strauss J, Kennedy JL, Farmer A, McGuffin P, Day R, Matthews K, Bakke P, Gulsvik A, Lucae S, Ising M, Brueckl T, Horstmann S, Wichmann HE, Rawal R, Dahmen N, Lamina C, Polasek O, Zgaga L, Huffman J, Campbell S, Kooner J, Chambers JC, Burnett MS, Devaney JM, Pichard AD, Kent KM, Satler L, Lindsay JM, Waksman R, Epstein S, Wilson JF, Wild SH, Campbell H, Vitart V, Reilly MP, Li M, Qu L, Wilensky R, Matthai W, Hakonarson HH, Rader DJ, Franke A, Wittig M, Schafer A, Uda M, Terracciano A, Xiao X, Busonero F, Scheet P, Schlessinger D, St Clair D, Rujescu D, Abecasis GR, Grabe HJ, Teumer A, Volzke H, Petersmann A, John U, Rudan I, Hayward C, Wright AF, Kolcic I, Wright BJ, Thompson JR, Balmforth AJ, Hall AS, Samani NJ, Anderson CA, Ahmad T, Mathew CG, Parkes M, Satsangi J, Caulfield M, Munroe PB, Farrall M, Dominiczak A, Worthington J, Thomson W, Eyre S, Barton A, Wellcome Trust Case Control C, Mooser V, Francks C, Marchini J (2010) Meta-analysis and imputation refines the association of 15q25 with smoking quantity. Nat Genet 42:436–440

Loukola A, Hallfors J, Korhonen T, Kaprio J (2014) Genetics and smoking. Curr Addict Rep 1:75–82

Lubke GH, Hottenga JJ, Walters R, Laurin C, de Geus EJ, Willemsen G, Smit JH, Middeldorp CM, Penninx BW, Vink JM, Boomsma DI (2012) Estimating the genetic variance of major depressive disorder due to all single nucleotide polymorphisms. Biol Psychiatry 72:707–709

Maes HH, Sullivan PF, Bulik CM, Neale MC, Prescott CA, Eaves LJ, Kendler KS (2004) A twin study of genetic and environmental influences on tobacco initiation, regular tobacco use and nicotine dependence. Psychol Med 34:1251–1261

Manolio TA, Collins FS, Cox NJ, Goldstein DB, Hindorff LA, Hunter DJ, McCarthy MI, Ramos EM, Cardon LR, Chakravarti A, Cho JH, Guttmacher AE, Kong A, Kruglyak L, Mardis E, Rotimi CN, Slatkin M, Valle D, Whittemore AS, Boehnke M, Clark AG, Eichler EE, Gibson G, Haines JL, Mackay TF, McCarroll SA, Visscher PM (2009) Finding the missing heritability of complex diseases. Nature 461:747–753

Marchini J, Howie B (2010) Genotype imputation for genome-wide association studies. Nat Rev Genet 11:499–511

McNeill A, Munafò MR (2013) Reducing harm from tobacco use. J Psychopharmacol 27:13–18
Munafò MR, Clark T, Johnstone E, Murphy M, Walton R (2004) The genetic basis for smoking behavior: a systematic review and meta-analysis. Nicotine Tob Res 6:583–597
Munafò MR, Timofeeva MN, Morris RW, Prieto-Merino D, Sattar N, Brennan P, Johnstone EC, Relton C, Johnson PC, Walther D, Whincup PH, Casas JP, Uhl GR, Vineis P, Padmanabhan S, Jefferis BJ, Amuzu A, Riboli E, Upton MN, Aveyard P, Ebrahim S, Hingorani AD, Watt G, Palmer TM, Timpson NJ, Davey Smith G (2012) Association between genetic variants on chromosome 15q25 locus and objective measures of tobacco exposure. J Natl Cancer Inst 104:740–748
Ng SB, Turner E, Robertson PD, Flygare SD, Bigham AW, Lee C, Shaffer T, Wong M, Bhattacharjee A, Eichler EE, Bamshad M, Nickerson DA, Shendure J (2009) Targeted capture and massively parallel sequencing of twelve human exomes. Nature 461:272–276
Picciotto MR, Zoli M, Rimondini R, Lena C, Marubio LM, Pich EM, Fuxe K, Changeux JP (1998) Acetylcholine receptors containing the beta2 subunit are involved in the reinforcing properties of nicotine. Nature 391:173–177
Saccone SF, Hinrichs AL, Saccone NL, Chase GA, Konvicka K, Madden PAF, Breslau N, Johnson EO, Hatsukami D, Pomerleau O, Swan GE, Goate AM, Rutter J, Bertelsen S, Fox L, Fugman D, Martin NG, Montgomery GW, Wang JC, Ballinger DG, Rice JP, Bierut LJ (2007) Cholinergic nicotinic receptor genes implicated in a nicotine dependence association study targeting 348 candidate genes with 3713 SNPs. Hum Mol Genet 16:36–49
Sartor CE, Lessov-Schlaggar CN, Scherrer JF, Bucholz KK, Madden PA, Pergadia ML, Grant JD, Jacob T, Xian H (2010) Initial response to cigarettes predicts rate of progression to regular smoking: findings from an offspring-of-twins design. Addict Behav 35:771–778
Schmitt JE, Prescott CA, Gardner CO, Neale MC, Kendler KS (2005) The differential heritability of regular tobacco use based on method of administration. Twin Res Hum Genet 8:60–62
Strasser AA, Lerman C, Sanborn PM, Pickworth WB, Feldman EA (2007) New lower nicotine cigarettes can produce compensatory smoking and increased carbon monoxide exposure. Drug Alcohol Depend 86:294–300
Thorgeirsson TE, Geller F, Sulem P, Rafnar T, Wiste A, Magnusson KP, Manolescu A, Thorleifsson G, Stefansson H, Ingason A, Stacey SN, Bergthorsson JT, Thorlacius S, Gudmundsson J, Jonsson T, Jakobsdottir M, Saemundsdottir J, Olafsdottir O, Gudmundsson LJ, Bjornsdottir G, Kristjansson K, Skuladottir H, Isaksson HJ, Gudbjartsson T, Jones GT, Mueller T, Gottsater A, Flex A, Aben KK, de Vegt F, Mulders PF, Isla D, Vidal MJ, Asin L, Saez B, Murillo L, Blondal T, Kolbeinsson H, Stefansson JG, Hansdottir I, Runarsdottir V, Pola R, Lindblad B, van Rij AM, Dieplinger B, Haltmayer M, Mayordomo JI, Kiemeney LA, Matthiasson SE, Oskarsson H, Tyrfingsson T, Gudbjartsson DF, Gulcher JR, Jonsson S, Thorsteinsdottir U, Kong A, Stefansson K (2008) A variant associated with nicotine dependence, lung cancer and peripheral arterial disease. Nature 452:638–642
Thorgeirsson TE, Gudbjartsson DF, Surakka I, Vink JM, Amin N, Geller F, Sulem P, Rafnar T, Esko T, Walter S, Gieger C, Rawal R, Mangino M, Prokopenko I, Magi R, Keskitalo K, Gudjonsdottir IH, Gretarsdottir S, Stefansson H, Thompson JR, Aulchenko YS, Nelis M, Aben KK, den Heijer M, Dirksen A, Ashraf H, Soranzo N, Valdes AM, Steves C, Uitterlinden AG, Hofman A, Tonjes A, Kovacs P, Hottenga JJ, Willemsen G, Vogelzangs N, Doring A, Dahmen N, Nitz B, Pergadia ML, Saez B, De Diego V, Lezcano V, Garcia-Prats MD, Ripatti S, Perola M, Kettunen J, Hartikainen AL, Pouta A, Laitinen J, Isohanni M, Huei-Yi S, Allen M, Krestyaninova M, Hall AS, Jones GT, van Rij AM, Mueller T, Dieplinger B, Haltmayer M, Jonsson S, Matthiasson SE, Oskarsson H, Tyrfingsson T, Kiemeney LA, Mayordomo JI, Lindholt JS, Pedersen JH, Franklin WA, Wolf H, Montgomery GW, Heath AC, Martin NG, Madden PA, Giegling I, Rujescu D, Jarvelin MR, Salomaa V, Stumvoll M, Spector TD, Wichmann HE, Metspalu A, Samani NJ, Penninx BW, Oostra BA, Boomsma DI, Tiemeier H, van Duijn CM, Kaprio J, Gulcher JR, Consortium E, McCarthy MI, Peltonen L, Thorsteinsdottir U, Stefansson K (2010) Sequence variants at CHRNB3-CHRNA6 and CYP2A6 affect smoking behavior. Nat Genet 42:448–453

Tobacco-and-Genetics-Consortium (2010) Genome-wide meta-analyses identify multiple loci associated with smoking behavior. Nat Genet 42:441–447

Visscher PM, Goddard ME, Derks EM, Wray NR (2012) Evidence-based psychiatric genetics, AKA the false dichotomy between common and rare variant hypotheses. Mol Psychiatry 17:474–485

Ware JJ, Timpson N, Davey Smith G, Munafò MR (2014) A recall-by-genotype study of CHRNA5-A3-B4 genotype, cotinine and smoking topography: study protocol. BMC Med Genet 15:13

Ware JJ, Chen X, Vink J, Loukola A, Minica C, Pool R, Milaneschi Y, Mangino M, Menni C, Chen J, Peterson R, Auro K, Lyytikäinen L, Wedenoja J, Stiby AI, Hemani G, Willemsen G, Hottenga JJ, Korhonen T, Heliövaara M, Perola M, Rose R, Paternoster L, Timpson N, Wassenaar CA, Zhu AZX, Davey Smith G, Raitakari O, Lehtimäki T, Kähönen M, Koskinen S, Spector T, Penninx BWJH, Salomaa V, Boomsma D, Tyndale RF, Kaprio J, Munafò MR (in preparation) Genome-wide meta-analysis of cotinine levels in cigarette smokers identifies locus at 4q13.2.

Wood AR, Perry JR, Tanaka T, Hernandez DG, Zheng HF, Melzer D, Gibbs JR, Nalls MA, Weedon MN, Spector TD, Richards JB, Bandinelli S, Ferrucci L, Singleton AB, Frayling TM (2013) Imputation of variants from the 1000 Genomes Project modestly improves known associations and can identify low-frequency variant-phenotype associations undetected by HapMap based imputation. PLoS ONE 8:e64343

Yang J, Benyamin B, McEvoy BP, Gordon S, Henders AK, Nyholt DR, Madden PA, Heath AC, Martin NG, Montgomery GW, Goddard ME, Visscher PM (2010) Common SNPs explain a large proportion of the heritability for human height. Nat Genet 42:565–569

Pharmacogenetics of Nicotine and Associated Smoking Behaviors

Julie-Anne Tanner, Meghan J. Chenoweth and Rachel F. Tyndale

Abstract This chapter summarizes genetic factors that contribute to variation in nicotine pharmacokinetics and nicotine's pharmacological action in the central nervous system (CNS), and how this in turn influences smoking behaviors. Nicotine, the major psychoactive compound in cigarette smoke, is metabolized by a number of enzymes, including CYP2A6, CYP2B6, FMOs, and UGTs, among others. Variation in the genes encoding these enzymes, in particular *CYP2A6*, can alter the rate of nicotine metabolism and smoking behaviors. Faster nicotine metabolism is associated with higher cigarette consumption and nicotine dependence, as well as lower quit rates. Variation in nicotine's CNS targets and downstream signaling pathways can also contribute to interindividual differences in smoking patterns. Binding of nicotine to neuronal nicotinic acetylcholine receptors (nAChRs) mediates the release of several neurotransmitters including dopamine and serotonin. Genetic variation in nAChRs, and in transporter and enzyme systems that leads to altered CNS levels of dopamine and serotonin, is associated with a number of smoking behaviors. To date, the precise mechanism underpinning many of these findings remains unknown. Considering the complex etiology of nicotine addiction, a more comprehensive approach that assesses the contribution of multiple gene variants, and their interaction with environmental factors, will likely improve personalized therapeutic approaches and increase smoking cessation rates.

Keywords Nicotine · Genetic variation · Smoking · CYP2A6 · nAChRs · Dopamine · Serotonin

J.-A. Tanner · M.J. Chenoweth · R.F. Tyndale (✉)
Departments of Pharmacology and Toxicology and Psychiatry,
University of Toronto, Toronto, ON, Canada
e-mail: r.tyndale@utoronto.ca

J.-A. Tanner · M.J. Chenoweth · R.F. Tyndale
Campbell Family Mental Health Research Institute, Centre for Addiction and Mental Health, Toronto, ON, Canada

D.J.K. Balfour and M.R. Munafò (eds.), *The Neurobiology and Genetics of Nicotine and Tobacco*, Current Topics in Behavioral Neurosciences 23,
DOI 10.1007/978-3-319-13665-3_3

Contents

1 Introduction ... 38
2 Pharmacogenetics of Nicotine Pharmacokinetics ... 39
2.1 Genetic Variation in Nicotine Metabolism ... 39
2.2 Genetic Variation in the Renal Elimination of Nicotine ... 52
3 Pharmacogenetics of Nicotine's Central Nervous System Targets ... 52
3.1 Genetic Variation in Nicotinic Acetylcholine Receptors ... 53
3.2 Genetic Variation in the Dopaminergic System ... 58
3.3 Genetic Variation in the Serotonergic System ... 67
4 Conclusions ... 71
References ... 71

1 Introduction

The characterization of nicotine as an addictive drug is widely accepted, but the degree of tobacco dependence and patterns of smoking behaviors varies substantially between individuals. Genetic and environmental factors contribute to this variation by influencing the pharmacological actions of nicotine both in the central nervous system (CNS) and in the periphery. During smoking, nicotine is inhaled as a component of cigarette smoke into the lungs where it is absorbed into the bloodstream, facilitating rapid transport to the brain where nicotine easily crosses the blood–brain barrier (Benowitz 2010). Individuals adjust their smoking behaviors in order to maintain desirable plasma nicotine levels and avoid symptoms of tobacco withdrawal. However, alterations in both the nicotine metabolic pathway (nicotine pharmacokinetics) and the response of the brain to nicotine (nicotine pharmacodynamics) can lead to variation in nicotine plasma levels and CNS response, respectively, in turn altering smoking behaviors (Liu et al. 2010; Malaiyandi et al. 2005; Pianezza et al. 1998; Picciotto et al. 1998).

Nicotine is primarily metabolized in the liver (Benowitz et al. 2009); however, extrahepatic enzymes also contribute to nicotine biotransformation (Su et al. 2000). Variation in nicotine metabolism leads to different rates of nicotine inactivation and clearance (Benowitz and Jacob 1994; Nakajima and Yokoi 2005). This, in turn, influences the duration of the neurobiological effects of nicotine within several brain regions, including the midbrain dopamine systems, part of the brain reward system, altering smoking behaviors (Grenhoff et al. 1986; Ikemoto et al. 2006). Nicotine binds to neuronal nicotinic acetylcholine receptors (nAChRs) (Liu et al. 2012; Picciotto et al. 1998), which are ligand-gated ion channels, leading to a cascade of effects that include the release of neurotransmitters such as dopamine, serotonin, norepinephrine, acetylcholine, γ-aminobutyric acid, glutamate, and endorphins (Benowitz 2008). The nicotine-evoked dopamine release in the nucleus accumbens is critical for the reinforcing effects of tobacco use, and can lead to the development of nicotine dependence (Dani and Heinemann 1996).

Several twin studies have investigated the heritability of smoking behaviors, highlighting the role of genetics in smoking. Substantial genetic influence has been demonstrated for smoking initiation (Koopmans et al. 1999; Vink et al. 2005), amount smoked (Koopmans et al. 1999), persistence of smoking (True et al. 1997), nicotine dependence (Vink et al. 2005), nicotine withdrawal and cessation (Xian et al. 2003; see also chapter entitled Genetics of Smoking Behaviour; this volume). In contrast to studies that provide total heritability estimates, genome-wide association studies (GWAS) and candidate gene studies are designed to focus on the influences of specific genes. Such studies have been used to identify genetic risk and/or protective factors for smoking behaviors.

The present chapter will outline sources of variation in smoking behaviors stemming from genetic variation in nicotine pharmacokinetic and pharmacodynamic pathways. Gene variants that alter the metabolism and renal clearance of nicotine, as well as CNS response to nicotine, including variation in the cholinergic, Dopaminergic, and serotonergic systems, will be discussed.

2 Pharmacogenetics of Nicotine Pharmacokinetics

Nicotine is the major psychoactive compound present in cigarette smoke. The duration of nicotine's psychoactive effects depends, to a large extent, on the rate of removal of nicotine from the body. However, there is wide interindividual variation in nicotine clearance rates (Benowitz and Jacob 1994; Nakajima and Yokoi 2005), which in turn is associated with altered smoking behaviors (Pianezza et al. 1998; Tyndale and Sellers 2002), warranting investigation of the sources of genetic variation in the pathways mediating nicotine clearance. Polymorphic genes encode the enzymes responsible for nicotine clearance via metabolism and renal elimination. The impact of this genetic variation on smoking behaviors will be described in the following section.

2.1 Genetic Variation in Nicotine Metabolism

2.1.1 Cytochrome P450 2A6 (CYP2A6)

The vast majority of nicotine clearance ($\sim$90 %) occurs via metabolic (i.e., non-renal) clearance (Benowitz et al. 2009) (Fig. 1). CYP2A6 is the enzyme responsible for $\sim$90 % of the metabolic inactivation of nicotine to cotinine (Messina et al. 1997), and this pathway accounts for 70–90 % of the total clearance of nicotine (Benowitz and Jacob 1994). Cotinine's metabolism to *trans*-3′-hydroxycotinine (3HC) is entirely catalyzed by CYP2A6 (Nakajima et al. 1996). The rate of conversion of nicotine to its metabolites can vary substantially between individuals due primarily to genetic variation in CYP2A6 enzymatic function, although the environment can

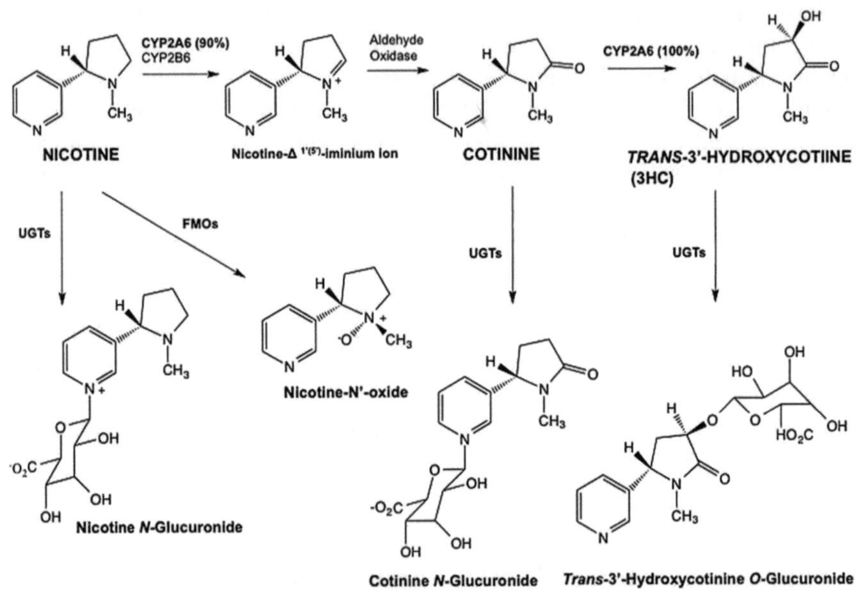

Fig. 1 Main hepatic pathways of nicotine metabolism

also influence this. The gene encoding CYP2A6 is highly polymorphic, with more than 35 variant alleles, resulting in substantial interindividual variation in nicotine metabolism (variants characterized to date found at http://www.cypalleles.ki.se/cyp2a6.htm). These *CYP2A6* allelic variants have different degrees of impact on CYP2A6 function, with some variants resulting in complete loss-of-enzyme function (**2, *4*), a reduction in enzyme expression or function (**5, *6, *7, *9, *10, *11, *12, *17, *18, *19, *20, *21, *23–28, *35, *39–40, *42–45*), no established effect on enzyme function (**8, *13, *14, *15, *16, *22, *29–34, *41*), or an increase in enzyme activity (**1B, *1X2A, *1X2B*) (Table 1). Variability in CYP2A6 activity, measured by *CYP2A6* genotype or phenotype, leads to variation in the rate of nicotine metabolism and is associated with altered smoking behaviors.

2.1.2 3HC/COT as a Biomarker of CYP2A6 Enzymatic Activity

The ratio of 3HC to cotinine (3HC/COT; also known as the nicotine metabolite ratio (NMR)) is a phenotypic biomarker of CYP2A6 enzymatic activity and, like *CYP2A6* genotype, is associated with smoking behaviors. Before discussing the impact of *CYP2A6* genetic variation on smoking phenotypes, it is useful to understand the utility and characteristics of this biomarker when investigating genetic associations. The utility of NMR as a biomarker of CYP2A6 activity among smokers derives from the exclusive metabolism of COT to 3HC by CYP2A6, as well as the relatively long half-life of COT ($\sim$16 h) and the formation dependence

Table 1 *CYP2A6* gene variant frequencies and associated changes in enzyme function across different ethnic groups. Alleles included have a frequency >1 % in one or more ethnic groups

CYP2A6 Allele	Genetic change (Amino acid change)	CYP2A6 activity	Allele frequency (%)					
			European Ancestry	African American	Japanese	Chinese	Korean	Alaska Native
*1B	58 bp gene conversion with *CYP2A7*	Increased	27.6–35.0	11.2–18.2	25.6–54.6	40.6–51.3	37.1–57.0	65.3
*1X2A and B	Crossover with *CYP2A7*	Increased	0–1.7	0	0.4	0	0.2	0
*2	1799T>A (L160H)	Inactive	1.1–5.3	0–1.1	0	0	0	0.4
*4	Gene deletion	Inactive	0.13–4.2	0.5–2.7	17.0–24.2	4.9–15.1	10.8–11.0	14.5
*5	6582G>T (G479V)	Decreased	0–0.3	0	0	0.5–1.2	0.5	–
*7	6558T>C (I471T)	Decreased	0–0.3	0	6.3–12.6	2.2–9.8	3.6–9.8	0
*9	-48T>G	Decreased	5.2–8.0	5.7–9.6	19.0–20.7	15.6–15.7	19.6–22.3	8.9
*10	6558T>C (I471T), 600G>T (R485L)	Decreased	0	0	1.1–4.3	0.4–4.3	0.5–4.1	1.9
*12	Exons 1-2 from *CYP2A7*, exons 3-9 from *CYP2A6*, 10 amino acid substitution	Decreased	0–3.0	0–0.4	0–0.8	0	0	0.4
*17	5065G>A (V365M)	Decreased	0	7.1–10.5	0	0	0	0
*18	5668A>T (Y392F)	Decreased	1.1–2.1	0	0	0	0.5	–
*20	Deletions at nucleotides 2140 and 2141, frame-shift at codon 196	Decreased	0	1.1–1.7	0	–	0	–
*21	6573A>G (K476R),	Decreased	0–2.3	0–0.6	0	3.4	0	–
*23	2161C>T (R203C)	Decreased	0	1.1–2.0	0	0	0	–
*24	594G>C (V110L), 6458A>T (N438Y)	Decreased	0	0.7–2.3	0	0	0	–
*25	1672T>C (F118L)	Decreased	0	0.5–1.2	0	0	0	–
*28	5745A>G (N418D), 5750G>C (E419D)	Decreased	–	0.9–2.4	–	–	–	–
*35	6458A>T (N438Y)	Decreased	0	2.5–2.9	0.8	0.5	–	0

of 3HC (Benowitz et al. 2009; Benowitz and Jacob 2001; Nakajima et al. 1996). The long half-life of COT promotes stability of the relative levels of both COT and 3HC in regular smokers, irrespective of the level of cigarette consumption, making this a useful biomarker for assessing the rate of CYP2A6 activity and nicotine metabolism in both heavy and light smokers. In addition, the NMR is stable over time in regular smokers (Lea et al. 2006; Mooney et al. 2008; St Helen et al. 2012, 2013). NMR is highly correlated with the rate of nicotine clearance, due to the major role of CYP2A6-mediated inactivation of nicotine in nicotine clearance (Dempsey et al. 2004; Levi et al. 2007). NMR is also strongly associated with *CYP2A6* genotype, as decrease- and loss-of-function *CYP2A6* variants result in slower CYP2A6 activity and corresponding decreases in NMR (Malaiyandi et al. 2006b). In addition to its association with *CYP2A6* genotype, NMR also captures environmental sources of variability in CYP2A6 activity, and is often used together with *CYP2A6* genotype to investigate contributions of nicotine metabolism variation to smoking behaviors.

2.1.3 Effect of CYP2A6 Genetic Variation on Smoking Behaviors

Blood nicotine concentration is negatively correlated with cigarette craving (Jarvik et al. 2000), and smokers are known to titrate their smoking levels and intensity to maintain desirable nicotine levels in the body (McMorrow and Foxx 1983). Variation in nicotine pharmacokinetics can contribute to variation in smoking acquisition, the level of consumption (cigarettes smoked per day), puff volume, nicotine dependence, cravings and withdrawal, and cessation success between smokers. A significant proportion of variation in these behaviors is captured by variation in *CYP2A6* genotype and phenotype via altered nicotine metabolism rates.

2.1.4 Acquisition of Smoking Behaviors

The vast majority of smoking initiation occurs in adolescence (Giovino 1999), and *CYP2A6* variation influences smoking acquisition during this developmental period. Relative to faster *CYP2A6* nicotine metabolizers, slower *CYP2A6* metabolizers are at increased risk of acquiring nicotine dependence, described below. The mechanism(s) underpinning this elevated risk is not known, but may relate to longer durations of action of nicotine in the CNS among those with slower nicotine clearance, and/or genotype group differences in initial smoking experiences. A longer duration of action of nicotine in the CNS may in turn lead to prolongation of the reinforcing properties of nicotine, thus increasing the risk for the development of dependence. Adolescents with one or two copies of the loss-of-function *CYP2A6* alleles **2* and **4*, classified as slow nicotine metabolizers (Table 1), are at significantly greater risk of developing nicotine dependence defined by the International Classification of Diseases (ICD-10) and the modified Fagerström Tolerance Questionnaire (mFTQ) after initiating smoking compared to normal metabolizers

(Al Koudsi and Tyndale 2010; Al Koudsi 2010; Karp et al. 2006; O'Loughlin et al. 2004). Although initial smoking experiences have been shown to be predictive of nicotine dependence (Pomerleau et al. 1998), the proportion of *CYP2A6* faster and slower metabolizers experiencing early smoking symptoms (e.g., rush, buzz, dizziness, nausea, relaxation) appears similar among adolescents (Audrain-McGovern et al. 2007; O'Loughlin et al. 2004). Whether the *intensity* of early smoking symptoms differs according to *CYP2A6* genotype group remains to be determined and, if differences exist, may help to explain these findings.

CYP2A6 also influences the progression in nicotine dependence among adolescent smokers. Once dependent, those smokers with slower *CYP2A6* activity demonstrate slower progression in nicotine dependence scores (on mFTQ) over time, relative to those with faster *CYP2A6* activity (Audrain-McGovern et al. 2007). Thus, while their initial risk for nicotine dependence is greater, those with slower nicotine metabolism do not appear to escalate in nicotine dependence as rapidly as faster metabolizers (Audrain-McGovern et al. 2007). When looking at a single measure of nicotine dependence, rather than progression in nicotine dependence, a cross-sectional study demonstrated that adolescent smokers with slower nicotine metabolism (lower NMR) displayed greater mFTQ dependence relative to those with higher NMR (Rubinstein et al. 2013). In contrast, a separate study showed a trend toward higher scores on the Hooked on Nicotine Checklist (HONC), a measure of nicotine dependence, among adolescent light smokers (smoking 1–6 cigarettes per day; CPD) with higher NMR (Rubinstein et al. 2008). The different patterns of dependence scores across NMR groups in the separate studies may reflect, in part, that the specific study populations were at different stages of smoking onset, as well as differences in the aspects of dependence captured by the two scales.

2.1.5 Cigarette Consumption and Smoking Topography

To achieve desirable levels of nicotine in the brain and bloodstream, smokers regulate the number of cigarettes smoked per day, as well as the intensity (i.e., puff volume) with which they smoke each cigarette (McMorrow and Foxx 1983). Several biomarkers of smoking consumption have been established, including the levels of exhaled carbon monoxide (CO), plasma and urinary cotinine, and urinary total nicotine equivalents (TNE) (Hartz et al. 2012). Slower *CYP2A6* nicotine metabolizers smoke fewer CPD than faster *CYP2A6* metabolizers, as indicated by lower breath CO, TNE, and/or depth of inhalation (Ariyoshi et al. 2002; Benowitz et al. 2002; Liu et al. 2011; Malaiyandi et al. 2006b; Rao et al. 2000; Schoedel et al. 2004; Zhu et al. 2013a). The lower levels of smoking exhibited by slower nicotine metabolizers likely stems from their slower elimination of nicotine. It appears, however, that the titration of nicotine levels may be more pronounced in nicotine-dependent smokers as Schoedel et al. (2004), in adults, showed that slow metabolizers of European descent smoke fewer CPD only when they meet criteria for tobacco dependence.

In a separate study of smokers of European ancestry, slower *CYP2A6* metabolizers smoked fewer cigarettes while maintaining a level of smoking intensity (determined by the CO to cotinine ratio) similar to that of individuals with a wild-type (*CYP2A6*1/*1*) genotype (Rao et al. 2000). Individuals with the *CYP2A6* gene duplication (associated with an increased rate of nicotine metabolism, Table 1) smoked a similar number of CPD as individuals with the *CYP2A6*1/*1* genotype, but displayed increased smoking intensity (higher CO to cigarette and nicotine to cigarette ratios) (Rao et al. 2000). Different titration patterns are also seen among adolescent smokers. Although adolescents who are slower *CYP2A6* nicotine metabolizers are faster to acquire nicotine dependence (O'Loughlin et al. 2004), dependent slower *CYP2A6* metabolizers smoke *fewer* CPD than dependent faster *CYP2A6* metabolizers (Al Koudsi 2010; Audrain-McGovern et al. 2007; O'Loughlin et al. 2004). Similar findings were obtained in studies of Chinese and Japanese adult smokers (Ariyoshi et al. 2002; Benowitz et al. 2002; Liu et al. 2011). Chinese and Japanese smokers living in their native countries smoke relatively high quantities of cigarettes overall, but a similar impact of *CYP2A6* variability on CPD remains apparent. For example, Japanese slow metabolizers smoke ~15 CPD compared to ~20 CPD smoked by intermediate metabolizers, and ~25–30 CPD smoked by normal metabolizers (Ariyoshi et al. 2002; Fujieda et al. 2004). In addition to the association of *CYP2A6* genotype with cigarette consumption, variation in NMR also influences daily tobacco consumption, with lower NMR, indicative of slower nicotine metabolism, corresponding to lower levels of smoking (Benowitz et al. 2003). This has been demonstrated in Europeans, African Americans, and Alaska Natives (Benowitz et al. 2003; Mwenifumbo et al. 2007; Zhu et al. 2013a).

In contrast to smokers of European descent, African American and Alaska Native smokers (predominantly reporting smoking less than 10 CPD) and smokeless tobacco users do not display *CYP2A6* genotype associations with reported levels of use (CPD, chews per day) (Ho et al. 2009a; Zhu et al. 2013a). For example, the number of cigarettes smoked per day by these two populations does not differ between slow and normal nicotine metabolizers (Ho et al. 2009a; Zhu et al. 2013a). However, when using a more robust measure of consumption, levels of urinary TNE, light smokers, and smokeless tobacco users who are slow metabolizers have lower levels of tobacco consumption (Zhu et al. 2013a). Light smokers with low NMR may achieve lower doses of tobacco by smoking less intensely. This suggests that self-reported CPD may be a relatively poor biomarker of total tobacco exposure, especially among lighter smokers.

Cotinine, the primary metabolite of nicotine, has been used as a biomarker of tobacco exposure in a large number of studies, including those focused on *CYP2A6* genetic influences on smoking. In heavy smokers, plasma cotinine is correlated with tobacco consumption (Benowitz et al. 2011). However, cotinine levels can overestimate tobacco exposure in *CYP2A6* slower metabolizers because slower CYP2A6 activity influences cotinine clearance more than cotinine formation (Zhu et al. 2013b). In support of this, slow metabolizers exhibit greater plasma cotinine levels than normal metabolizers, despite smoking fewer cigarettes and having lower TNE (Ho et al. 2009a; Rao et al. 2000; Zhu et al. 2013a, b). The higher plasma cotinine levels in slower versus normal metabolizers does not accurately reflect their differences in smoking. Therefore, other biomarkers should be used to determine the level of tobacco consumption, especially when contrasting groups with differing CYP2A6 activity, such as those with differing genotypes (e.g., ethnic groups with different *CYP2A6* genotype frequencies) and those of different gender (females are faster nicotine metabolizers than males) (Zhu et al. 2013b).

The measurement of TNE, as opposed to plasma cotinine, self-reported CPD, and breath CO, is likely the most accurate biomarker of nicotine exposure available thus far. Urinary TNE is advantageous over these other markers of dose because TNE is associated with total nicotine exposure (Benowitz et al. 2010; Scherer et al. 2007), and accounts for multiple metabolites of nicotine ($\sim$90 % of nicotine dose) (Benowitz et al. 1994), thus making it a measure of dose which is independent of variation in these metabolic pathways. *CYP2A6* genotype is associated with differences in TNE, even among light smokers, whereas the other relatively weaker markers of smoking are not (Zhu et al. 2013a).

2.1.6 Nicotine Dependence

Several assessments are available to evaluate nicotine dependence in smokers. The Fagerström Test of Nicotine Dependence (FTND) (Heatherton et al. 1991), HONC, and DSM-IV criteria are common examples of scales that measure the level of nicotine dependence. *CYP2A6* genotype or NMR variability have not consistently been shown to be associated with nicotine dependence, with one assessment measure in particular yielding differing results—FTND. Several studies have shown higher FTND scores, indicative of greater nicotine dependence, in faster nicotine metabolizers compared to slower nicotine metabolizers when characterized by *CYP2A6* genotype (Kubota et al. 2006; Wassenaar et al. 2011) and NMR (Sofuoglu et al. 2012). In contrast, other studies do not show a significant association between the rate of nicotine metabolism and the level of dependence (Benowitz et al. 2009; Ho et al. 2009b; Johnstone et al. 2006; Malaiyandi et al. 2006a; Schnoll et al. 2009b). When measuring dependence in those with lower cigarette consumption, however, HONC may be a more appropriate measure to ascertain nicotine dependence (Wellman et al. 2006).

Potential explanations for the different findings with FTND include the heavy weighting toward CPD and the lack of assessment of withdrawal or craving, which can be influenced by the rate of nicotine metabolism (Jarvik et al. 2000; Kubota et al. 2006). Another potential explanation is the possibility of increased selection bias in clinical smoking cessation trials compared to observational studies. In clinical trials, one may be less likely to see differences in dependence between slow and normal metabolizers as this particular group of slow metabolizers may differ as they are smokers who have not been successful in quitting smoking. This could manifest as greater dependence than would be observed in a random sample of smokers who are not seeking help for cessation. Discordant findings involving FTND may also result from sampling a mixed gender population, as no association between NMR and FTND was seen in females (Schnoll et al. 2014). In contrast, male smokers with higher NMR have higher FTND scores than males with lower NMR (Schnoll et al. 2014). The lack of association of NMR and nicotine dependence in females suggests that non-nicotinic factors may have a relatively stronger influence on dependence in females than in males (Schnoll et al. 2014).

2.1.7 Cessation and Response to Pharmacotherapies

Several studies have directly assessed relationships between *CYP2A6* genotype and NMR with cessation, where slower nicotine metabolism, as measured by genotype or phenotype, predicts increased quit rates in adult (Chen et al. 2014; Gu et al. 2000; Ho et al. 2009b; Lerman et al. 2006; Patterson et al. 2008; Schnoll et al. 2009) and adolescent (Chenoweth et al. 2013) smokers. Variation in CYP2A6 activity can influence spontaneous smoking cessation success as well as response to smoking cessation pharmacotherapies.

Relatively higher quit rates observed among slower nicotine metabolizers compared to faster nicotine metabolizers may be due in part to their less severe withdrawal symptoms (Kubota et al. 2006), lower numbers of cigarettes smoked (resulting in fewer learning trials and possibly less ingrained smoking behaviors) (Ariyoshi et al. 2002; Audrain-McGovern et al. 2007; Benowitz et al. 2002; Liu et al. 2011; O'Loughlin et al. 2004; Rao et al. 2000), as well as reduced brain response to smoking cues as demonstrated through functional magnetic resonance imaging (Tang et al. 2012). Relative to faster nicotine metabolizers, slower nicotine metabolizers likely experience lower fluctuations in blood nicotine concentrations throughout the day. These latter imaging findings regarding decreased response to smoking cues among slower *CYP2A6* metabolizers suggest slower metabolizers likely form weaker conditioned responses to smoking cues, perhaps due to relatively lower surges in brain nicotine concentration, and resulting dopamine levels, achieved during ad lib smoking (Tang et al. 2012).

Slower nicotine metabolizers have higher spontaneous quit rates compared to normal nicotine metabolizers (Gu et al. 2000) as well as increased quitting on certain smoking cessation pharmacotherapies. Cessation trials involving nicotine replacement therapies (NRTs) and bupropion indicate that the rate of nicotine metabolism can predict cessation success (retrospective analyses). NRTs include nicotine gum, patch, nasal spray, and lozenge, which act to replace nicotine that would normally be consumed from cigarettes, with the intention of helping to reduce a smoker's cravings and likelihood to smoke. Smokers with lower NMR display greater cessation success when treated with transdermal nicotine (nicotine patch) compared to faster metabolizers, but the same effect is not seen when using nicotine nasal spray (Lerman et al. 2006; Schnoll et al. 2009). The lack of effect observed for subjects using nasal spray is likely due to the smokers' differential titration of the dose, as usage in this trial differed by their rate of nicotine metabolism (Malaiyandi et al. 2006a). In contrast, Chen et al. (2014) showed that NRT increases cessation success compared to placebo in fast, but not slow, nicotine metabolizers as defined by

CYP2A6 genotype. When treated with bupropion, a dopamine and norepinephrine reuptake inhibitor and weak nAChR antagonist (Warner and Shoaib 2005), which is not metabolized by CYP2A6, there was no difference in quit rates by NMR in those receiving bupropion therapy. In contrast, in the placebo arm, slow nicotine metabolizers experienced higher quit rates relative to faster nicotine metabolizers (Patterson et al. 2008). A more recent study, however, demonstrated that bupropion was effective in prolonging abstinence in both slower and faster nicotine metabolizers relative to placebo, suggesting no effect of *CYP2A6* genotype on response to bupropion (Chen et al. 2014). Varenicline, a newer pharmacotherapy for smoking cessation that acts as a partial agonist at α4β2 nAChRs, competes with nicotine for α4β2 binding sites, blocking nicotine-evoked dopamine release (Garrison and Dugan 2009). King et al. (2012) assessed whether certain genes play an important role in cessation success with this drug as well as with bupropion. Continuous abstinence when taking varenicline was not associated with *CYP2A6* genotype.

The associations between CYP2A6 and NMR and smoking cessation outcomes were elucidated from retrospective analyses of smoking cessation clinical trial data. The utility of NMR as a predictive biomarker of smoking cessation outcomes is being prospectively investigated in a phase III clinical trial (NCT0131001) involving varenicline, nicotine patch, and placebo. Randomization to treatment group was stratified prospectively according to participant NMR. Varenicline, compared to nicotine patch, was associated with greater quitting among normal nicotine metabolizers, whereas for slow metabolizers patch worked well and had fewer side effects than Varenicline for slow metabolizers (Lerman et al. 2014).

2.1.8 Interethnic Variation in CYP2A6 and Smoking Behaviors

In addition to interindividual differences in smoking behavior *within* ethnicities, smoking patterns also vary *between* ethnicities. The variable smoking patterns between different ethnic groups are associated with differences in the rate of nicotine metabolism, which are consistent with population differences in *CYP2A6* allele frequencies. Multiple studies have shown that *CYP2A6* allele frequencies vary significantly between European, African American, Asian, and Alaska Native populations (Table 1). In ethnic groups that have a higher frequency of *CYP2A6* decrease- and loss-of-function variants, a lower overall NMR and slower rate of nicotine metabolism is observed. For example, relative to those of European descent, African Americans have a higher frequency of several *CYP2A6* alleles that result in a decrease or loss of CYP2A6 enzyme function (Al Koudsi et al. 2009; Fukami et al. 2005; Mwenifumbo et al. 2008, 2010; Nakajima et al. 2006). Consistent with the higher frequency of decrease- and loss-of-function of *CYP2A6* alleles in African Americans, African American adolescent smokers display slower nicotine metabolism, as measured by NMR, relative to adolescent smokers of European ancestry (Rubinstein et al. 2013).

Similarly, Asian smokers metabolize nicotine more slowly than European smokers, due in part to the higher frequency of specific *CYP2A6* slower metabolism variants in Asian populations, relative to Europeans (Nakajima et al. 2006; Schoedel et al. 2004). Japanese subjects exhibited significantly lower cotinine/nicotine ratios compared to Europeans (Nakajima et al. 2006), and Chinese-American smokers had significantly lower nonrenal clearance of nicotine and cotinine compared to European smokers, which was associated with lower nicotine intake from smoking in Chinese-Americans than Europeans (Benowitz et al. 2002). Consistent with these findings, Asian adolescent smokers also possess a lower average population NMR, relative to adolescent smokers of European descent (Rubinstein et al. 2013).

Unlike African American and Asian populations, Alaska Natives metabolize nicotine more rapidly than Europeans as shown by a greater NMR in the overall population and among the wild-type (*CYP2A6*1/*1*) subgroup (Binnington et al. 2012). This is consistent with higher levels of nicotine intake during tobacco use in Alaska Native smokers compared to the general United States population (Benowitz et al. 2012). The observed higher NMR among Alaska Natives does not correspond to a lower frequency of *CYP2A6* decrease- and loss-of-function variants compared to Europeans; the reason(s) for this higher NMR has not yet been determined, but may relate to uncharacterized genetic variation or to dietary inducers.

2.1.9 Cytochrome P450 2B6 (CYP2B6)

Similar to CYP2A6, CYP2B6 is involved in the metabolism of nicotine (Yamazaki et al. 1999), however its role in hepatic nicotine metabolism to cotinine is minor ($\sim$10 %) relative to CYP2A6 ($\sim$90 %) (Al Koudsi and Tyndale 2010). Similar to *CYP2A6*, the *CYP2B6* gene is highly polymorphic (variants characterized to date found at http://www.cypalleles.ki.se/cyp2b6.htm). However, consistent with its minor role in hepatic nicotine metabolism there is no detectable impact of *CYP2B6* gene variants on the modulation of peripheral nicotine metabolism (Lee et al. 2007b). Conversely, when CYP2A6 expression or activity is reduced, CYP2B6 may play a relatively more prominent role in nicotine metabolism, and *CYP2B6* genetic variation may represent a source of variation in nicotine metabolism in those with reduced CYP2A6 activity (Ring et al. 2007). CYP2B6 is expressed in the liver, but it is also expressed in the brains of nonhuman primates and humans (Ferguson et al. 2013; Miksys et al. 2003), thus potentially modulating CNS nicotine metabolism and the duration of action of nicotine in the brain.

2.1.10 Effect of CYP2B6 Genetic Variation on Smoking Behaviors

The impact of *CYP2B6* variation on nicotine metabolism appears negligible (Al Koudsi and Tyndale 2010; Lee et al. 2007b), suggesting that the role of CYP2B6 in the modulation of smoking behaviors is also minor. However, since CYP2B6 metabolizes the smoking cessation drug bupropion (Faucette et al. 2000), the association between variation in *CYP2B6* and cessation success has been investigated. There is inconsistency regarding whether slower CYP2B6 activity promotes smoking cessation when taking bupropion. The unpredictability of treatment outcome based on *CYP2B6* genotype may stem from the fact that both bupropion and the CYP2B6-mediated metabolite hydroxybupropion are therapeutically active (Carroll et al. 2014; Damaj et al. 2004; Zhu et al. 2012).

In a placebo-controlled smoking cessation clinical trial involving bupropion, subjects taking placebo who possessed the *CYP2B6*5* variant, resulting in decreased *CYP2B6* expression (Lang et al. 2001), reported more cravings post-quit and lower quit rates compared to those without the *CYP2B6*5* allele (Lerman et al. 2002). In the bupropion arm, male smokers possessing the *CYP2B6*5* variant also exhibited decreased abstinence, but the association with *CYP2B6* genotype was not observed in females. However, interpretation of the *CYP2B6*5* data is unclear as the impact of this variant in vivo is not well established (Burger et al. 2006; Kirchheiner et al. 2003; Wyen et al. 2008). If this allele does reduce activity, the slower bupropion metabolism may result in decreased cessation success, potentially due to lower levels of bupropion's pharmacologically active metabolite hydroxybupropion, which similarly inhibits dopamine and norepinephrine transporters (Carroll et al. 2014; Damaj et al. 2004). In support of this, smokers with higher levels of hydroxybupropion, suggestive of greater CYP2B6 activity, have a greater likelihood of being abstinent than smokers with low levels of hydroxybupropion (Zhu et al. 2012). It is worth noting several things about this study. The first is that only about 60 % of the participants were actually taking bupropion at week 3—thus without biomarkers it is difficult to assess adherence and thus the effect of a gene on the drug effect. Also, while *CYP2B6* genotype altered hydroxybupropion levels, there was no direct association between *CYP2B6* genotype and cessation rates, suggesting that the genotype effects are not large enough to alter smoking cessation, that the study was insufficiently powered for this comparison, or that other undetected alleles or environmental influences on CYP2B6 may have muted the association. A separate placebo-controlled bupropion clinical trial found that smokers with one or two copies of the *CYP2B6*6* haplotype, which is associated with decreased bupropion metabolism (Zhu et al. 2012), had similar rates of abstinence on bupropion (Lee et al. 2007a). Conversely, King et al. (2012) showed an association between continuous smoking abstinence at weeks 9–12 of bupropion treatment and through nondrug follow-up from weeks 12–52 and several *CYP2B6* SNPs. However, these SNPs are not yet functionally described, and a direct impact on treatment outcome could not be determined. Thus it remains to be clarified whether *CYP2B6* genetic variation has a substantial impact on quitting, in both placebo and bupropion treatment arms.

Lee et al. (2007b) assessed the association between *CYP2B6*6* and abstinence rates when subjects were administered different forms of NRT (nicotine patch and nasal spray), and found no difference in levels of nicotine, or abstinence on nicotine patch or nasal spray, between wild-type and *CYP2B6*6* groups. This suggests a negligible effect of *CYP2B6* on nicotine kinetics or cessation using NRTs.

2.1.11 Cytochrome P450 2A13 (CYP2A13)

CYP2A13 is also capable of metabolizing nicotine to cotinine in humans (Bao et al. 2005). However, the role that CYP2A13 plays in nicotine clearance is relatively minor as its low levels and extrahepatic localization to the nasal mucosa, lung, and trachea limits its overall contribution to systemic nicotine metabolism (Su et al. 2000). Although *CYP2A13* genetic variation does not substantially affect the rate of nicotine metabolism, *CYP2A13* variants are implicated in lung cancer risk in smokers, likely via CYP2A13-mediated metabolic activation of the tobacco-specific nitrosamine NNK (Su et al. 2000). CYP2A13 has the highest affinity for NNK of all P450 enzymes investigated (Jalas et al. 2005) and is located within the lung where it may mediate this activation. Subjects having one or two copies of a *CYP2A13* variant that result in markedly reduced metabolic activity toward NNK are associated with a lower risk of developing lung adenocarcinoma (Wang et al. 2003).

2.1.12 Flavin-Containing Monooxygenases (FMOs)

FMOs are involved in the metabolism of nicotine, although the relative contribution depends on the subtype. FMO3, the most common form of FMO present in human liver (Hines et al. 1994), metabolically inactivates nicotine to nicotine *N*′-oxide (Cashman et al. 1992). FMO1 is predominantly expressed in the human kidneys and potentially in the brain, and has been shown to metabolize nicotine more efficiently than FMO3 in vitro (Hinrichs et al. 2011). However, the extrahepatic expression of FMO1 likely limits its overall impact on nicotine *N*′-oxidation. The genes encoding both FMO1 and FMO3 are polymorphic, indicating a potential role of *FMO* variation in interindividual differences in nicotine metabolism.

2.1.13 Association of FMO1 and FMO3 Genetic Variation with Nicotine Metabolism and Smoking Behaviors

FMO3 may play a relatively larger role in nicotine metabolism in subjects with reduced CYP2A6 expression or activity compared to those with normal CYP2A6 function (Yamanaka et al. 2004). A larger proportion of an oral nicotine dose is excreted as nicotine *N*′-oxide in urine among those homozygous for the *CYP2A6* gene deletion (**4*) (∼30 %) relative to those with wild-type *CYP2A6* genotype (∼5–7 %) (Yamanaka et al. 2004). In addition, the *FMO3* SNP rs2266782 (G>A, E158K), a decrease-of-function variant, was associated with slower nicotine metabolism in slower *CYP2A6* metabolizers, but not in faster *CYP2A6* metabolizers, in African Canadian nonsmokers receiving oral nicotine (Chenoweth et al. 2014). Despite the association between *FMO3* rs2266782 and modestly slower nicotine metabolism among those with slower CYP2A6 activity, *FMO3* rs2266782 did not alter cigarette consumption or TNE, a measure of daily nicotine intake and smoke exposure, examined among African North American light smokers (<10 CPD) (Chenoweth et al. 2014).

Bloom et al. (2013) investigated several *FMO3* haplotypes prevalent in subjects of European descent, which included the following coding SNPs: rs1800822 (C>T), rs2266782 (G>A, E158K, previously discussed), rs1736557 (G>A, V257M), rs909530 (T>C), and rs2266780 (A>G, E308G). Variation in *FMO3* haplotype was associated with differences in cigarette consumption among European American heavier smokers ($\sim$20–25 CPD), with the *FMO3* haplotypes CGGCA and CAGTG corresponding to an increase in consumption of $\sim$3 cigarettes per day, relative to the *FMO3* haplotypes CAGCA, CGACA, and TGGTA (Bloom et al. 2013). However, this association was apparent only in those with faster, rather than slower, CYP2A6 activity, and it is unclear which SNPs comprising the haplotypes are causal (Bloom et al. 2013).

The role of *FMO1* genetic variation in smoking behaviors has not been widely investigated. In European American smokers, two *FMO1* SNPs, rs10912765 and rs7877, were associated with nicotine dependence (Hinrichs et al. 2011). These SNPs are present in the 5′ and 3′ regions of *FMO1* and are thought to play a role in *FMO1* regulation (Hinrichs et al. 2011). Together these findings suggest that variation in FMO activity may influence the rate of nicotine metabolism. However, it is not yet clear if the changes are sufficiently large to alter smoking behaviors, including cigarette consumption and nicotine dependence.

2.1.14 UDP-Glucuronyltransferases (UGTs)

Nicotine and cotinine are metabolized by UGTs through the process of *N*-glucuronidation, which results in the excretion of nicotine- and cotinine-glucuronide in the urine. Nicotine-glucuronide and cotinine-glucuronide account for 4 and 17 % of the total nicotine metabolites excreted in the urine, respectively (Benowitz et al. 2009; Benowitz and Jacob 1994; Byrd et al. 1992). Variation in *UGT* genes may result in altered rates of nicotine metabolism, potentially impacting smoking behaviors.

2.1.15 Association of UGT1A and UGT2B Genetic Variation with Nicotine Metabolism and Smoking Behaviors

Interindividual variation in nicotine glucuronidation rates has been observed, with genetic factors contributing to this variation (Benowitz et al. 1999; Lessov-Schlaggar et al. 2009). Multiple UGTs may be responsible for nicotine glucuronidation, in particular UGT1A4 and UGT2B10. Variation in *UGT1A4* alters the glucuronidation of non-nicotinic substrates, but potential effects on nicotine metabolism and resulting changes in smoking behavior appear to be negligible (Kaivosaari et al. 2007). Two SNPs in *UGT1A4* (rs6755571 C>A, P24T and rs2011425 T>G, L48V), known as *UGT1A4*2* and *UGT1A4*3,* respectively, exhibit decreased glucuronidation activity toward the steroid dihydrotestosterone. *UGT1A4*3* is also associated with increased glucuronidation efficiency (V_{max}/K_m) for the substrate clozapine (Ehmer et al. 2004; Mori et al. 2005). Considering that UGT1A4 does not appear to play a significant role in hepatic nicotine glucuronidation (Kaivosaari et al. 2007), it is unlikely that *UGT1A4* variants would alter smoking behaviors, although this requires confirmation.

A SNP occurring in the *UGT2B10* gene (rs61750900 G>T, D67R), referred to as *UGT2B10*2*, exhibits significantly reduced *N*-glucuronidation of nicotine and cotinine (Chen et al. 2007, 2010). One study has found that individuals with the *UGT2B10*2* allele consume less nicotine, exemplified by lower urinary TNE relative to those with the *UGT2B10*1/*1* genotype (Berg et al. 2010), while another did not (Zhu et al. 2013c). This suggests that the *UGT2B10*2* allele may lower smoking levels in some circumstances or populations.

2.2 Genetic Variation in the Renal Elimination of Nicotine

Renal elimination is responsible for approximately 5 % of total nicotine clearance (Benowitz et al. 2009; Benowitz and Jacob 1985). A twin study provides evidence of substantial genetic influence (approximately 40 % for nicotine, 60 % for cotinine) on variation in nicotine and cotinine renal clearance (Benowitz et al. 2008). The gene(s) contributing to the observed genetic variation remain to be characterized. Organic cation transporters (OCTs), which have been shown to transport nicotine in a human carcinoma cell line (Takami et al. 1998; Urakami et al. 2002; Zevin et al. 1998), may play a role in nicotine elimination through active transport of nicotine across kidney cells. The gene encoding OCT2 is polymorphic, and *OCT2* variation alters the renal clearance of metformin, a substrate of OCT2 (Yoon et al. 2013). Increased renal clearance of nicotine in smokers, achieved through urine acidification (results in ionic trapping of nicotine, a weak base), was previously associated with an increase in daily nicotine intake (Benowitz and Jacob 1985). Thus, if a role for OCT2 in nicotine transport in vivo is elucidated, *OCT2* variation may influence the renal clearance of nicotine, and lead to compensatory changes in smoking behaviors. OCT2 is also expressed at the blood–brain barrier; however, it is not clear if variation in OCT2 at this site would be sufficient to alter smoking behaviors. Thus genetic variation in nicotine transport systems, while a very minor pathway for nicotine clearance, may represent another source of variation in smoking behaviors.

3 Pharmacogenetics of Nicotine's Central Nervous System Targets

In addition to variation in the rate of nicotine clearance, alteration of nicotine's CNS targets and downstream signaling pathways can modify nicotine's psychoactive effects. As previously mentioned, the binding of nicotine to neuronal nAChRs

(Liu et al. 2012) leads to a cascade of downstream signaling events involving the release of dopamine, serotonin, norepinephrine, acetylcholine, γ-aminobutyric acid, glutamate, and endorphins (Benowitz 2008). Dopamine release, which mediates the primary reinforcing effects of nicotine intake, contributes to the development of nicotine dependence (Dani and Heinemann 1996), and variation in the neurobiological pathways that regulate response to nicotine may contribute to differences in multiple smoking behaviors, including the level of smoking and dependence. Polymorphic genes encode nAChR subunits, dopamine transporters, dopamine receptors, and dopamine-metabolizing enzymes, as well as enzymes responsible for serotonin synthesis and transport. The impact of genetic variation in the CNS on smoking behaviors will be described in the following section.

3.1 Genetic Variation in Nicotinic Acetylcholine Receptors

The nAChRs are pentameric ligand-gated ion channels, comprised of a combination of nine α subunits (α2–10) and three β subunits (β2–4); these are encoded by genes in which polymorphisms can functionally alter receptor response to nicotine binding (Ho and Tyndale 2007; see also chapter entitled Structure of Neuronal Nicotinic Receptors; this volume). Considering that downstream responses to nAChR activation modulate reward from nicotine administration (Corrigall et al. 1994), nAChR variation is a biologically plausible source of interindividual differences in smoking patterns. The specific nAChR subunit gene possessing the variant allele is an important determinant of the functional outcome, as nAChR subtypes exert differential effects in response to nicotine binding (Marks 2013). In the absence of α4 and β2 subunits, for example, mice display a lack of dopamine release upon nicotine challenge and will not self-administer nicotine (Marubio et al. 2003; Picciotto et al. 1998) suggesting α4 and β2 subunits are required for the reinforcing properties of nicotine. The role of genetic variation in nAChR subunits in cigarette smoking behaviors is discussed in detail below.

3.1.1 Acquisition of Smoking Behaviors

Genetic variation in nAChR subunits is associated with variation in both the risk for smoking initiation, as well as the age of onset of smoking. For example, the β2 nAChR subunit has been implicated in smoking initiation. The β2 subunit mediates nicotine-stimulated dopamine release (Picciotto et al. 1998), thus influencing reinforcement from smoking. *CHRNB2* (the gene encoding the β2 subunit) variants that decrease β2 nAChR subunit function may protect against the initiation of regular smoking by decreasing nicotine-evoked dopamine release during early smoking experiences. A *CHRNB2* haplotype, consisting of five SNPs (CACTA—rs2280781 (T>C), rs12072348 (C>A), rs3766927 (A>G), rs2072659 (C>G), and rs2072660 (T>C)), was associated with a decreased risk of being a regular smoker (defined as smoking daily for ≥1 year) in a group of Israeli women (Greenbaum et al. 2006). The functional impacts of these SNPs are still unclear, but based on this observed protective effect of the CACTA haplotype on smoking initiation (Greenbaum et al. 2006), the causal SNP(s) within this haplotype may act to reduce β2 expression and/or function, in turn reducing dopamine release. The rs2072659 and rs2072660 SNPs from the CACTA haplotype are located in the 3′UTR of *CHRNB2*, a putative site involved in mRNA stability, and therefore these SNPs potentially influence β2 subunit expression in the brain. However, in a separate study that included both women and men of European ancestry, no associations between several different *CHRNB2* SNPs and smoking initiation were observed (Silverman et al. 2000).

Genetic variation located upstream of nAChR subunit genes can also influence smoking initiation. Homozygosity for rs1996371 (T>C), a functionally uncharacterized SNP-located upstream of *CHRNB4*, was associated with a younger age of onset of daily smoking in smokers of European ancestry (Kapoor et al. 2012). To date, a potential effect of rs1996371 on nicotinic receptor expression and/or function has not been described, and the reason for the increased risk of daily smoking at younger ages remains unclear. In mice, *CHRNA5/A3/B4* overexpression increases functional α3β4 receptors in the brain, and enhances nicotine self-administration (Gallego et al. 2012). Perhaps rs1996371, given its location upstream of *CHRNB4*, increases β4-containing nAChR expression in the brain, promoting earlier onset of regular daily smoking. In support of the role of α3β4 receptors in promoting nicotine intake, nicotine self-administration by rats is attenuated by α3β4 receptor antagonists (Glick et al. 2002).

An interaction between the age of smoking initiation and nAChR genetic variation on risk for regular smoking has also been observed. This suggests that the impact of nAChR genetic variation on risk for regular smoking may be stronger in those who start smoking at younger ages (Hartz et al. 2012). Subjects heterozygous for the nonsynonymous SNP rs16969968 (G>A, D398N) in *CHRNA5*, who initiated smoking at 16 years or younger, displayed an increased risk of heavy smoking in adulthood compared to later-onset smokers who also possessed this variant (Hartz et al. 2012). Rs16969968 has been shown to reduce nAChR response to the nicotinic agonist epibatidine in vitro, as indicated by a decrease in intracellular calcium levels following epibatidine binding (Bierut et al. 2008). Thus, while lower responses to nicotine (e.g., via lower β2 and β4 expression and/or function) may be protective against smoking initiation, lower CNS responses to nicotine (e.g., via rs16969968) may promote greater risk for regular smoking, especially among early initiators.

3.1.2 Cigarette Consumption and Smoking Topography

In addition to its association with increased risk of regular smoking among early initiators, the rs16969968 A allele has been associated with greater cigarette consumption (Breetvelt et al. 2012; Liu et al. 2010). The reduced nAChR response to nicotinic agonists observed for the rs16969968 minor allele (A) (Bierut et al. 2008) suggests that individuals with this allele may experience lower levels of nicotinic receptor activation during ad lib smoking. Smokers possessing the rs16969968 A allele may compensate for the lower level of nicotinic receptor activation by smoking more cigarettes per day to increase the level of nicotine acting at nAChRs, in order to maintain a desirable level of nicotine reinforcement and to avoid withdrawal. In a population of smokers of European descent, rs16969968 AA homozygotes smoked significantly more cigarettes per day than GA heterozygotes and GG homozygotes (Breetvelt et al. 2012). Rs1051730 (C>T), in perfect linkage disequilibrium (LD) with rs16969968, has similarly been associated with smoking quantity. The rs1051730 minor allele (T) was associated with higher levels of smoking, consistent with the increased smoking observed among those with the rs16969968 A allele (Breetvelt et al. 2012; Thorgeirsson et al. 2008; Wassenaar et al. 2011). Although rs16969968 and rs1051730 have mostly been investigated separately, more recently they have been referred to as a single-risk allele, rs1051730–rs16969968 (Munafo et al. 2012; Tobacco and Genetics Consortium 2010). Among smokers of European descent, those with the rs1051730–rs16969968 risk allele (either heterozygous or homozygous for the minor alleles in rs1051730 and rs16969968) reported higher cigarette consumption (Munafo et al. 2012). Among African American smokers, a GWAS showed that cigarette consumption was associated with a different *CHRNA5* SNP, rs2036527 (G>A), located upstream of *CHRNA5* (David et al. 2012).

In addition to SNPs in *CHRNA3* and *CHRNA5* genes being implicated in smoking, rs2236196 (G>A) in *CHRNA4* was significantly associated with the number of cigarettes smoked per day in European American and African American populations, with the G allele having been more prevalent in heavier smokers (Han et al. 2011). As previously mentioned, mice lacking the α4 nAChR subunit do not demonstrate increased striatal dopamine release in response to nicotine stimulation (Marubio et al. 2003), supporting the concept that genetic variation in *CHRNA4* may be a risk factor for smoking.

3.1.3 Nicotine Dependence

Genetic variation in the α4 nAChR subunit has also been implicated in the risk for nicotine dependence. Two SNPs in exon 5 of the *CHRNA4* gene, rs1044396 (C>T) and rs1044397 (G>A) are in near complete LD with each other; these two SNPs are part of the haplotype block GCTATA (rs2273504 G>A, rs2273502 C>T, rs1044396 C>T, rs1044397 G>A, rs3827020 T>C, rs2236196 A>G) that has been associated with significantly lower FTND and Revised Tolerance Questionnaire scores in a Chinese population (Feng et al. 2004). The mechanism of protection against nicotine dependence afforded by these SNPs is currently unknown. However, given that the activation of α4-containing nAChRs is crucial in mediating nicotine reinforcement (Tapper et al. 2004), these SNPs may result in lower α4 subunit function, reducing the risk for nicotine dependence.

Consistent with the role of the α3 and α5 subunit SNPs rs1051730 and rs16969968, respectively, in smoking quantity, these SNPs were also associated with nicotine dependence in subjects of European ancestry (Chen et al. 2009; Wassenaar et al. 2011). The minor alleles of rs1051730 and rs16969968 (T and A, respectively) were found to be more prevalent in smokers with higher FTND scores, suggestive of greater nicotine dependence. However, in a functional magnetic resonance imaging study assessing brain reactivity to smoking cues, nicotine-dependent women with the rs16969968 GG genotype exhibited *greater* brain response to smoking images than women who possessed the minor rs16969968 risk allele (A) (Janes et al. 2012). As posited by Janes et al. (2012), expression of the rs16969968 risk allele in *CHRNA5* may diminish nAChR response to agonist binding, reducing intracellular calcium influx (Bierut et al. 2008) and thus inhibiting nicotine's role in facilitating long-term potentiation (Jia et al. 2010). Long-term potentiation, the primary process involved in learning and memory, relies on intracellular calcium influx in the hippocampus (Jia et al. 2010), a brain region expressing nAChR α5 subunits (Wada et al. 1990). Thus, it is possible that smokers who possess the rs16969968 risk allele may have inhibited formation of drug-cue associations due to deficits in working memory, potentially leading to their lower brain reactivity to smoking cues, even though they may be highly dependent smokers.

3.1.4 Cessation and Response to Pharmacotherapies

A smoker's ability to successfully quit smoking is determined by multiple factors, including the level of nicotine dependence, withdrawal, craving, and response to smoking cues (Norregaard et al. 1993; Zhou et al. 2009). Nicotine withdrawal results from a decrease in brain reward function that stems from a decrease in dopamine release (Benowitz 2009). One of the primary symptoms of withdrawal is cigarette craving (Baker et al. 2012). Smokers can experience withdrawal after only hours of smoking abstinence (Brown et al. 2013), for example during sleep as nicotine levels in the body decline (Herskovic et al. 1986). A state of withdrawal will continue for an extended duration if abstinence continues. It is thought that by sustaining a state of nAChR desensitization, smokers are able to avoid symptoms of withdrawal, such as irritability, anxiety, difficulty concentrating, and tobacco cravings (Benowitz 2008). This suggests that variation in the genes encoding nAChR subunits may play a role in modulating nicotine withdrawal symptoms and resulting cessation success. Genetic variation in nAChRs that modulate other factors associated with smoking cessation success, such as nicotine dependence levels and response to smoking cues, may also influence cessation.

Genetic variation in nAChRs appear to alter spontaneous smoking cessation. The *CHRNA5* SNP rs569207 (C>T) was associated with the number of quit attempts among smokers of European ancestry, with TT homozygotes having displayed a higher number of quit attempts relative to subjects with CC homozygosity, which may suggest a reduced ability to quit smoking among TT homozygotes (Budulac et al. 2012). In addition, rs569207 is part of the CTGAG haplotype (rs680244 (T>C), rs569207 (C>T), rs16969968 (G>A), rs578776 (G>A), and rs1051730

(C>T)) that has been associated with more severe withdrawal than the CCAGA haplotype, which includes the wild-type allele of rs569207 (Baker et al. 2009). Although the functional significance of rs569207 is currently unknown, it is possible that the T allele of rs569207 is associated with lower cessation success through the modulation of withdrawal severity.

Genetic variation in the α3 subunit was also associated with spontaneous cessation success. In addition to its association with higher cigarette consumption and nicotine dependence scores (Breetvelt et al. 2012; Munafo et al. 2012; Thorgeirsson et al. 2008; Wassenaar et al. 2011), the *CHRNA3/A5* SNP rs1051730 (C>T) was associated with lower smoking cessation success during spontaneous quitting (Budulac et al. 2012). In addition, the rs16969968 (in complete LD with *CHRNA3/5* SNP rs1051730) and rs680244 haplotype AC was associated with later age of self-reported cessation in smokers of European descent, compared to subjects with the GC haplotype (Chen et al. 2012). Another *CHRNA3/A5* SNP (rs660652, A>G), found in the 3′UTR region of the *CHRNA3* gene, was associated with increased quit attempts among European smokers who possessed the minor allele (G), relative to AG/AA smokers (Erlich et al. 2010). This further supports the observation that α3/α5 nAChR genetic variation alters the success of spontaneous quit attempts.

The response to smoking cessation pharmacotherapies may also be influenced by nAChR genetic variation, although findings are inconsistent. The rs1051730 T allele was weakly associated with lower abstinence at 4-week follow-up after an 8-week open-label trial involving nicotine patch in European smokers, but was not associated with abstinence at 12- and 26-weeks of follow-up, or in a separate placebo-controlled nicotine patch trial (Munafo et al. 2011). In a separate study, rs1051730 was associated with less smoking cessation success in smokers assigned to placebo when assessed at end of treatment and 6-month follow-up (Bergen et al. 2013). The A allele of the SNP rs2036527 (G>A), which is in strong LD with rs1051730, was associated with lower abstinence among African American light smokers on nicotine gum, bupropion, or either treatment during treatment and at end of treatment. However, the rs2036527 A allele was not associated with abstinence at 6-month follow-up or at any time point in those taking placebo (Zhu et al. 2014). Together, these results suggest that the minor allele of rs1051730 is a risk allele for lower cessation success among treatment-seeking smokers.

Conversely, increased abstinence on NRT (nicotine patch, lozenge, gum, nasal spray) has been demonstrated for smokers of European descent possessing one or two copies of the rs1051730 and rs588765 (*CHRNA5* SNP, C>T) minor alleles, compared to wild-type smokers on NRT, at 6-month follow-up (Bergen et al. 2013). A similar impact of NRT (nicotine gum) on abstinence was observed among African American smokers possessing the rs588765 T allele; however, the association between rs588765 genotype and abstinence was not observed at 6-month follow-up as it was in subjects of European descent (Zhu et al. 2014). Although the direction of the effect of the minor allele of rs1051730 in smoking cessation is still unclear, it appears that possession of the rs588765 T allele may benefit smokers trying to quit who are using a form of NRT.

A study assessing the impact of haplotypes formed by rs16969968 (in complete LD with rs1051730) and rs680244 showed that haplotype status was not associated with 7-day abstinence at the end of treatment among individuals receiving active pharmacotherapy (nicotine patch, nicotine lozenge, bupropion, or combination of NRT and bupropion) (Chen et al. 2012). However, active treatment was associated with less risk of smoking relapse compared to placebo among smokers with the rs16969968 and rs680244 haplotypes GT (high risk haplotype) and AC, but not GC (low risk haplotype). Associations were not dependent on the type of active treatment (bupropion vs. NRT, and combination) (Chen et al. 2012).

Varenicline, a relatively new smoking cessation drug, is a partial agonist at α4β2 nAChRs and also inhibits the ability of nicotine, a full agonist at α4β2 nAChRs, to bind to and activate these receptors (Garrison and Dugan 2009). Variation in *CHRNA4* (rs3787138, rs2236196, rs6062899), *CHRNA5* (rs518425), *CHRNA7* (rs6494121), and *CHRNB2* (rs3811450) was associated with continuous abstinence during weeks 9–12 within the varenicline treatment group (King et al. 2012). Based on the findings described, the influence of α3 and α5 genetic variants on cessation remains unclear. It remains to be elucidated whether the effects of SNPs at these loci are dependent on active treatment, treatment type, or whether their effect on cessation success is nonspecific.

3.2 Genetic Variation in the Dopaminergic System

Following the binding and activation of nAChRs by nicotine, dopamine release in the VTA is central in mediating the reinforcing properties of nicotine (Corrigall et al. 1994; see also chapter entitled The Role of Mesoaccumbens Dopamine in Nicotine Dependence; volume 24). Variation in both the levels of dopamine in the brain (via dopamine transport and metabolism) and the brain response to dopamine (via dopamine receptor binding) may influence the reinforcing properties associated with cigarette smoking. Variation in the reinforcing properties of nicotine may in turn alter smoking behaviors, including cessation. The influence of genetic variation in the dopaminergic system on smoking behaviors is outlined in the following sections.

3.2.1 Dopamine Transporters

The dopamine transporter (DAT) is a Na^+/Cl^--dependent transmembrane protein that regulates the reuptake and release of dopamine to presynaptic terminals (VanNess et al. 2005). DAT is encoded by the ~64 kb *DAT1/SLC6A3* gene that contains a common 40-bp Variable Number Tandem Repeats (VNTR) polymorphism (3–13 repeats) in the 3′-UTR of the gene. The 9- and 10-repeat alleles are the most common; however, their functional outcome remains to be characterized (Kang et al. 1999). It is plausible that this VNTR polymorphism affects mRNA stability, nuclear transport, and/or protein synthesis as it is located in the 3′-UTR of the *SLC6A3* gene (Nakamura et al. 1998). Modified dopamine reuptake, due to increased or decreased DAT activity, can alter the availability of dopamine for binding at its receptors. This, in turn, is likely to modify dopamine-mediated reinforcement from cigarette smoking and therefore impact patterns of tobacco use.

3.2.2 Likelihood of Being a Smoker

Genetic variation in the *SLC6A3* gene, characterized by a VNTR 9-repeat variant located in the 3′UTR that has an unknown functional outcome, is associated with increased or decreased DAT levels, depending on the study. In one study, the 9-repeat allele was associated with a *reduction* in DAT protein in the putamen compared to the 10-repeat allele, suggesting less reuptake and greater availability of dopamine in this brain region (Heinz et al. 2000). In contrast, van Dyck et al. (2005) demonstrated an opposite effect, with the *SLC6A3–9* allele resulting in *increased* DAT in both the putamen and caudate, and presumably increased reuptake and lower availability of dopamine in these brain regions. Dopamine release in the striatum plays a role in decision-making and reward, with learning of action-reward associations being linked to the putamen and caudate nucleus (Balleine et al. 2007). If the 9-repeat allele results in greater DAT expression and more dopamine reuptake in the putamen and caudate, individuals possessing this variant may experience less nicotine-evoked reward, in turn lowering their risk for smoking. Genetic variation in the *SLC6A3* gene is associated with differences in smoking risk, although studies addressing this relationship have not yielded consistent findings. Lerman et al. (1999) demonstrated that European and African American individuals possessing the *SLC6A3* 9-repeat allele were significantly less likely to be smokers, relative to individuals without the 9-repeat allele. However, this finding was not replicated in a subsequent investigation, where no significant difference in 9-repeat allele frequency among never, former, and current smokers in European or African American populations was observed (Vandenbergh et al. 2002).

3.2.3 Acquisition of Smoking Behaviors

The age of smoking initiation is influenced by DAT genetic variation. Smokers who possess one or two copies of the *SLC6A3* 9-repeat allele were less likely to initiate smoking at an early age in a mixed European and African American population (<16 years old) (Lerman et al. 1999), relative to those without the *SLC6A3* 9-repeat allele. In a separate study, European smokers with the *SLC6A*3 9-repeat variant displayed a lower risk of starting smoking before the age of 20 compared to subjects without the 9-repeat allele (Sieminska et al. 2009). Since an earlier age of smoking onset is associated with higher levels of tobacco consumption and increased difficulty quitting smoking (Breslau and Peterson 1996; Chen and Millar 1998; Cui et al. 2006; Taioli and Wynder 1991), the reduced risk of early smoking initiation among those with the *SLC6A3* 9-repeat allele supports the classification of the 9-repeat allele as protective against smoking. The protective role of the 9-repeat allele against early smoking initiation may stem from increased DAT levels (van Dyck et al. 2005), leading to lower reinforcement from smoking during initial smoking experiences.

In contrast to the *SLC6A3* 9-repeat allele, rs27072 (G>A), a SNP present in the 3′-UTR of the *SLC6A3* gene, was associated with an *increased* risk of early smoking initiation. In a Chinese population, the rs27072 A allele was significantly associated with smoking initiation at ages younger than 18 years, and may also increase the risk of developing nicotine dependence among early-initiating smokers (Ling et al. 2004). The function of the rs27072 variant is unknown; however, it may play a role in the regulation of *SLC6A3* mRNA expression due to its location in the 3′UTR. If the SNP (minor allele, A) exerts a negative regulatory effect leading to decreased mRNA expression and DAT levels in the brain, this may increase the duration of dopamine action (via reduced dopamine reuptake), resulting in greater sensitivity to nicotine during early smoking experiences.

3.2.4 Response to Smoking Cues

Although the *SLC6A3* 9-repeat allele plays a protective role against early smoking, once regular smoking is established, this DAT polymorphism may increase the response to smoking cues and influence cravings, which may promote nicotine dependence. Smokers with the *SLC6A3* 9-repeat allele exhibited greater brain response to smoking cues (compared to nonsmoking cues) than smokers possessing the *SLC6A3* 10-repeat allele (Franklin et al. 2009). In addition, self-reported craving was associated with increases in brain activity in reward-related regions, including the subventicular extended amygdala, insula, and post-central gyrus, in subjects possessing the *SLC6A3* 9-repeat allele (Franklin et al. 2011). In experimental animals, dysfunction in the extended amygdala has been associated with drug dependency (Di Chiara et al. 1999), supporting the observation of increased brain activity in this region during craving in smokers with the 9-repeat allele.

3.2.5 Cessation and Response to Pharmacotherapies

A smoker's ability to quit smoking spontaneously, as well as through the use of pharmacological aids, may also be altered by genetic variation in DAT. Lerman et al. (1999) showed that European and African American smokers with the *SLC6A3* 9-repeat allele exhibited a longer duration of last quit attempt compared to those with other genotypes, suggesting greater duration of smoking abstinence among those with the 9-repeat allele. In support of this, the *SLC6A3* 9-repeat allele was associated with increased smoking cessation in a meta-analysis (Stapleton et al. 2007). However, in Korean smokers receiving bupropion therapy, those with the *SLC6A3* 9-repeat allele were less likely to achieve abstinence relative to smokers with the 10-repeat allele (Han et al. 2008). As one pharmacological action of bupropion is to occupy DAT (Learned-Coughlin et al. 2003) and inhibit dopamine reuptake, it is plausible that the 9-repeat variant influences the efficacy of bupropion for smoking cessation.

3.2.6 Dopamine Metabolism

Dopamine is metabolized by multiple enzymes, including monoamine oxidases (MAOs), catechol-O-methyltransferase (COMT), and dopamine β-hydroxylase (DBH) (Meiser et al. 2013); the genes encoding these enzymes (*MAO-A* and *MAO-B, COMT,* and *DBH*, respectively) are polymorphic, with functionally significant genetic variants altering both transcriptional efficiency and enzyme activity. Differences in rates of dopamine metabolism can result in a shorter or longer duration of dopamine in the CNS, influencing reinforcement from smoking, which in turn may regulate smoking behaviors. Relevant *MAO-A*, *COMT*, and *DBH* variants, their impact on smoking behaviors, and biological explanations of these effects are described below.

3.2.7 Likelihood of Being a Smoker

A reduction in the amount of dopamine in the brain, due to greater dopamine metabolism, may influence the reinforcement from smoking. Alterations in the reinforcement from smoking may in turn increase an individual's desire to smoke. Among Japanese adults, the risk of being a current smoker versus never smoker (<100 cigarettes smoked lifetime) was lower in females possessing a 30-bp VNTR 4-repeat allele, located in the promoter region of *MAO-A*, compared to those without the 4-repeat allele (Ito et al. 2003). The 3.5 and 4-repeat variants of *MAO-A* are transcribed more efficiently than the 3-repeat or 5-repeat alleles (Sabol et al. 1998), presumably resulting in higher levels of MAO-A enzyme, leading to greater metabolic inactivation of dopamine via oxidative deamination. Similar findings were observed in Chinese males, where the 4-repeat allele was associated with a decreased risk of current smoking (vs. former smoking) compared to the 3-repeat allele (Jin et al. 2006). Thus, the 4-repeat allele may lead to lower dopamine levels via greater dopamine degradation, likely reducing the reinforcing effects from smoking, resulting in a lower risk of being a smoker.

COMT is also involved in the metabolism of dopamine, suggesting that variation in COMT activity may also influence dopamine degradation, in turn affecting smoking behaviors. The *COMT* Val158Met (rs4680, G>A) polymorphism is associated with decreased COMT enzymatic activity (Lachman et al. 1996), potentially resulting in reduced dopamine degradation, which may lead to increased brain levels of dopamine. Among Thai subjects, those with one or two copies of the *COMT* Met allele were more likely to be smokers than subjects homozygous for the Val allele (Suriyaprom et al. 2013). The increased risk of smoking among those with the Met allele may be due to reduced dopamine degradation and increased reinforcement (via increased dopamine levels) during smoking. However, among European and Japanese populations, the Met allele was associated with a *decreased* risk of being a smoker, relative to subjects homozygous for the Val allele (Colilla et al. 2005; Nedic et al. 2010; Tochigi et al. 2007). It has been proposed that the decreased risk of smoking among those with the Met allele in some populations may be due to lower novelty seeking and risk-taking behavior personality traits associated with the Met allele, relative to those with the Val/Val genotype (Tsai et al. 2004), although at present it is not clear how to reconcile these apparently discordant genotype association studies.

3.2.8 Cigarette Consumption

Smokers with greater COMT enzymatic activity (higher dopamine inactivation) may smoke more heavily in order to maintain desirable dopamine levels in the brain. Among Japanese males, *COMT* Val homozygosity was associated with a greater risk of heavy smoking relative to Val/Met and Met/Met genotypes (Tochigi et al. 2007). Greater MAO-A activity (higher dopamine inactivation) was also associated with heavier smoking in a group of tobacco- and alcohol-dependent males of European descent (Wiesbeck et al. 2006). Smokers who possessed the more efficiently transcribed allele (4-repeat) smoked an average of ~31 CPD, whereas those with the 3-repeat allele smoked an average of ~24 CPD. As discussed previously, the increased activity *MAO-A* 4-repeat allele was associated with a decreased risk of smoking (Ito et al. 2003), presumably via lower dopamine levels and lower nicotine reinforcement, in turn reducing the risk for smoking. Subjects who possess this 4-repeat allele and *do* continue smoking, however, may smoke more CPD to compensate for higher MAO-A activity, in order to maintain desirable levels of dopamine (Wiesbeck et al. 2006). Tobacco smoke also contains the MAO-A inhibitors harman and norharman (Herraiz and Chaparro 2005), which may also contribute to the reinforcing properties associated with cigarette smoking. It is possible that the higher CPD, and presumably greater exposure to MAO-A inhibitors, observed among those with the *MAO-A* 4-repeat allele occurs partially in response to their increased MAO-A expression/activity.

Another *MAO-A* SNP relevant to variation in smoking behavior is rs1137070 (T>C) in exon 14, which causes reduced MAO-A enzyme activity (Hotamisligil and Breakefield 1991). Relative to the C allele, the T allele of rs1137070 was associated with higher CPD (McKinney et al. 2000), potentially due to greater MAO enzyme activity and therefore reduced dopamine levels, among those with the rs1137070 T allele. Thus, the lower availability of dopamine in the brain may mediate the heavier smoking observed among those with the rs1137070 T allele. In this same study, the *DBH* SNP rs77905 (G>A) predicted greater cigarette consumption among smokers with the A allele (McKinney et al. 2000). When attempting to replicate these findings in a larger cohort, however, there was no significant association between either SNP with CPD (Johnstone et al. 2002). The original results stemmed from a pilot investigation that sampled a small subset

(<15 %) of the available cohort, whereas the follow-up study (Johnstone et al. 2002) sampled 73 % of the total cohort. There were no differences in demographic characteristics, including age, sex, and ethnicity, between the two subsets. However, *DBH* rs77905 homozygotes were overrepresented in the original sample, although still remaining consistent with expected frequencies as predicted by the Hardy–Weinberg equilibrium, which may have contributed to the positive association between the A allele and CPD (McKinney et al. 2000). The literature on *DBH* variation and smoking behaviors is less extensive compared to other dopamine-metabolizing enzymes, adding to the difficulty in interpreting these discordant findings.

3.2.9 Nicotine Dependence

The role of the *COMT* Met allele, associated with reduced smoking risk and lower CPD, in nicotine dependence has also been assessed. Smokers with one or two copies of the Met allele possessed higher FTND scores, suggestive of greater nicotine dependence (Guo et al. 2007). However, this association did not survive statistical correction for multiple comparisons.

The *MAO-A* 4-repeat allele, which results in greater transcriptional efficiency, was associated with significantly higher FTND scores in male smokers possessing the 4-repeat variant compared to those without (Ito et al. 2003). This suggests that greater dopamine metabolism may promote greater nicotine dependence. Smokers with the 4-repeat variant may experience a different level of reward from smoking, contributing to stronger nicotine dependence.

3.2.10 Cessation and Response to Pharmacotherapies

Multiple studies have investigated associations between the *COMT* Val158Met polymorphism and smoking cessation with and without pharmacotherapy. In terms of spontaneous quitting, the *COMT* Val allele, associated with faster COMT activity, was associated with greater smoking cessation compared to the Val/Met and Met/Met genotypes based on combined retrospective and prospective analyses of elderly current, former, and never smokers of European ancestry (Omidvar et al. 2009). In a separate study, however, *COMT* Val158Met variation was not associated with smoking cessation in elderly subjects from Saarland, Germany (Breitling et al. 2009). Neither study, however, assessed the subjects' level of nicotine dependence. It is possible that differences in nicotine dependence levels across the studies may have led to discordant findings regarding *COMT* Val158Met genotype and cessation success.

When assessing the impact of the *COMT* Val158Met genotype on cessation success with NRT (nicotine patch and nasal spray), the faster activity Val allele was associated with reduced cessation success in smokers of European ancestry at end of treatment compared to smokers possessing the Met allele who were also using

NRT (Colilla et al. 2005; David et al. 2011; Munafo et al. 2008). Collectively, these results suggest that while those with the Val allele may be more likely to quit smoking in the absence of pharmacotherapy, those with the Met allele may benefit from NRT; however, this requires further validation.

3.2.11 Dopamine Receptors

Dopamine receptors, of which there are five subtypes (D_1–D_5), are G-protein coupled receptors (GPCRs). The D_1–D_5 receptor subtypes are encoded by the genes *DRD1–5*. Variation in *DRD* genes is associated with alterations in smoking behaviors, with genetic variation in *DRD2* and *DRD4* being most commonly investigated. Different receptor subtypes modulate emotion, motivation, and behaviors, suggesting a potentially important role of dopamine receptor variation in differential smoking patterns.

3.2.12 Initiation

Variation in the gene encoding the dopamine receptor subtype 4 (*DRD4*) has been associated with age of smoking initiation in multiple studies. Within *DRD4*, there is a common 48-bp VNTR in exon 3 in which the "long" allele contains 6–10 repeats, and the "short" allele has 2–5 repeats. Relative to *DRD4* VNTR short alleles, the *DRD4* 7-repeat allele has two to threefold lower potency for coupling with adenylyl cyclase, an important effector in GPCR signaling (Oak et al. 2000). In a study assessing associations between dopamine receptor genetic variation and smoking behaviors at age 15, adolescents of European descent possessing the 7-repeat form of the 48-bp VNTR in exon 3 of *DRD4* began smoking at a younger age compared to other genotypes (Laucht et al. 2008). These findings are supported by data from Shields et al. (1998) showing that African American smokers with one or two copies of the long allele (6-repeat–8-repeat) were more likely to start smoking before the age of 16 compared to smokers homozygous for the short allele (2-repeat to 5-repeat). The earlier smoking initiation observed among those with the *DRD4* long allele may stem from greater novelty seeking in adolescence in this group; however, this genotype-novelty seeking association may be restricted to males (Laucht et al. 2005), and is not consistently replicated (Vandenbergh et al. 1997).

3.2.13 Cigarette Consumption

As previously mentioned, the *DRD4* VNTR long allele results in lower potency for receptor coupling with adenylyl cyclase than the shorter alleles (Oak et al. 2000), which presumably decreases downstream signaling following dopamine binding. Smokers possessing the long allele may smoke more in order to compensate for the reduced adenylyl cyclase activation and downstream signaling. In a study consisting of predominantly smokers of European descent, males possessing the *DRD4* long allele (7-repeat) smoked significantly more CPD relative to smokers with other genotypes (Laucht et al. 2005), suggesting that lower adenylyl cyclase activation may promote higher levels of smoking. However, no association was observed between the *DRD4* long or short alleles with CPD in European or African American smokers (Shields et al. 1998).

Variation in additional dopamine receptor subtype genes was also associated with CPD and the heaviness of smoking index (HSI), which assesses how soon after waking a subject smokes and number of CPD. Among smokers of European descent, those with the most common allele (84-bp allele) for the *DRD2* intron 2 simple tandem repeat polymorphism (*Taq*I C) smoked fewer CPD and scored lower on the HSI, compared to smokers without this allele (Vandenbergh et al. 2007). In the same study, variation in *DRD3* was also associated with CPD and the HSI. The Ser9Gly (S9G) mutation in *DRD3*, which alters the function of D_3 receptors by increasing their affinity for dopamine (Lundstrom and Turpin 1996), was associated with higher CPD and HSI scores (Vandenbergh et al. 2007). The strengthened interaction between dopamine and D_3 dopamine receptors containing the S9G polymorphism may enhance the reinforcing effects associated with smoking, in turn promoting greater levels of smoking.

A smaller number of studies have assessed associations between *DRD1* genetic variation and smoking behaviors. The *DRD1* rs686 A allele was associated with greater D_1 dopamine receptor expression (Huang et al. 2008), suggesting that greater receptor availability (and presumably more dopamine binding) may promote heavier smoking and nicotine dependence in current smokers. Huang et al. (2008) showed that the A allele of the rs686 SNP (G>A) in the 3′-UTR of the *DRD1* gene was associated with heavier smoking as assessed by self-reports of smoking quantity in African American smokers. In contrast, no genotype associations with smoking behaviors were observed in European American smokers (Huang et al. 2008).

3.2.14 Nicotine Dependence

Consistent with its effects on the quantity of smoking, the A allele of the rs686 SNP (G>A) in the 3′-UTR of the *DRD1* gene was associated with greater FTND- and HSI-assessed nicotine dependence in African American smokers (Huang et al. 2008). A reduction in D_2 receptor density has been associated with greater dependence (Huang et al. 2009). The A allele of rs2734849 (G>A, R490H), a SNP located in the C-terminal of the ankyrin repeat and kinase domain containing 1 (*ANKK1*) gene, is associated with reduced expression of the transcription factor nuclear factor-κB (NF-κB). This, in turn, leads to lower expression of NF-κB-regulated genes, including *DRD2*, resulting in reduced D_2 receptor levels in the brain (Bontempi et al. 2007; Huang et al. 2009). Among African American smokers, the rs2734849 A allele was associated with an increased risk for nicotine dependence (Huang et al. 2009). Animal studies using positron emission tomography (PET) demonstrate that low levels of *DRD2* expression in the striatum increases impulsivity and the self-administration of cocaine, supporting the notion that lower D_2 receptor density may increase susceptibility to drug dependence (Dalley et al. 2007; Nader et al. 2006). In addition, antagonism of D_2 dopamine receptors by haloperidol increases smoking behavior in heavy smokers, whereas activation of these receptors by bromocriptine was associated with a decrease in nicotine intake and smoking behavior (Caskey et al. 1999, 2002).

3.2.15 Cessation and Response to Pharmacotherapies

The dopamine receptor D_2 SNP *Taq*I A1 (rs1800497, C>T, Q713K) may influence cessation success, as demonstrated by several studies. The *Taq*I A1 variant, present at high frequency in a number of populations, is located approximately 10 kb downstream of *DRD2* in the coding region of *ANKK1* (Neville et al. 2004). There is some evidence that this variant reduces striatal D_2 receptor density (Jonsson et al. 1999); however, this has not been replicated in all studies (Laruelle et al. 1998). Reduced dopamine receptor density is implicated in addiction (Volkow et al. 2011), and a lower prevalence of D_2 receptors in the brain is associated with decreased motivation in mice (Trifilieff et al. 2013). Together these findings suggest that smokers possessing the *Taq*I A1 variant, resulting in lower D_2 receptor density, may experience lower motivation to quit smoking. Consistent with this notion, a higher prevalence of this *Taq*I A1 SNP (C>T) was observed in European smokers exhibiting shorter quit attempts (<6 months) compared to smokers who were able to quit for longer periods of time (>6 months), suggesting lower cessation success among those with the rs1800497 T allele (Comings et al. 1996). However, results from meta-analyses indicate that there is a lack of strong evidence for an association between this *Taq*I A1 polymorphism and smoking persistence or cessation success (Munafo et al. 2004, 2009b).

Results are also inconsistent for the association between the TaqI genotype and effectiveness of NRT. In smokers of mixed ethnicity, women possessing the *Taq*I A1 T allele exhibited significantly higher cessation rates on nicotine patch, while women with two copies of the C allele did not (Yudkin et al. 2004). In men, however, there were no genotype associations with nicotine patch effectiveness (Yudkin et al. 2004). Contradictory to these findings, it was shown that possessing one or more copies of the *Taq*I A1 T allele (vs. C allele) was associated with lower abstinence on nicotine patch in women of European descent (Munafo et al. 2009a). In a combined sample of men and women, there was no significant association between *Taq*I A1 genotype and abstinence (Munafo et al. 2009a). Similarly, in a predominantly European population, there was no association between the *Taq*I A1 polymorphism and overall cessation success on NRT (all formulations) (Stapleton et al. 2011). However, there was an interaction between depression and cessation success in subjects possessing the *Taq*I A1 T allele, such that individuals with a previous diagnosis of depression displayed lower quit rates, even with the help of NRT (Stapleton et al. 2011). This suggests that subjects meeting criteria for depression may not benefit from NRT therapy, and may be candidates for other cessation pharmacotherapies, such as bupropion or varenicline.

Genetic variation that causes decreased dopamine receptor density (*Taq*I A1 T allele) (Jonsson et al. 1999) may increase an individual's likelihood to relapse to smoking when using bupropion, a dopamine reuptake inhibitor (Warner and Shoaib 2005). Studies of bupropion efficacy and *Taq*I A1 genotype have demonstrated that smokers of European ancestry taking bupropion and possessing two copies of the C allele were more likely to remain abstinent at end of treatment and at 6-month follow-up, and have a greater reduction in craving, compared to subjects taking placebo (also C/C) (David et al. 2003, 2007a, c). This highlights a subgroup with potential for targeted treatment with bupropion.

As previously discussed, there is evidence for the role of dopaminergic Genetic variation in modulating smoking cessation success, including the dopamine transporter *SLC6A3* 9-repeat versus 10-repeat (Han et al. 2008; Lerman et al. 1999), the Val allele of the *COMT* Val158Met polymorphism (David et al. 2011; Omidvar et al. 2009), the *DRD2 Taq*I A1 (C>T) (David et al. 2007c), and the *DRD4* VNTR long allele (Leventhal et al. 2012). Interestingly, an additive genetic efficacy score (AGES) based on the combined influence of functional polymorphisms in the dopamine system (i.e., *DRD2 Taq*I SNP rs1800497, *COMT* Val158Met polymorphism, *DRD4* exon 3 VNTR, and *SLC6A3* 3′ VNTR) was predictive of relapse after a quit attempt in European-ancestry smokers (David et al. 2013). A higher AGES increased the risk for relapse in smokers on placebo, whereas this effect was not observed in those taking bupropion, suggesting that bupropion may be a beneficial treatment for smoking cessation in those with higher genetic risk of relapse (higher AGES) (David et al. 2013). Considering the variability in genetic studies to date, and the interrelatedness of the dopamine system, the use of AGES to predict smoking behaviors and cessation may be clinically useful. AGES considers a more cumulative effect of genetic influence on smoking compared with single gene association studies.

3.3 Genetic Variation in the Serotonergic System

Serotonin in the CNS is involved in the regulation of appetite, mood, and sleep (Halford et al. 2005; Ursin 2002; Young and Leyton 2002), and also plays an important role in nicotine-mediated reward. Nicotine has been shown to stimulate the release of serotonin in the frontal cortex of rats (Ribeiro et al. 1993). The activation of serotonin receptors in rats modulates dopamine release, leading to the activation of midbrain dopaminergic neurons (Porras et al. 2002). In addition, serotonin is involved in the regulation of processes such as emotional memory and decision-making (Merens et al. 2007), which may play a role in addiction. Variation in the expression or function of the enzymes responsible for serotonin synthesis and transport likely alters the regulation of serotonin levels and serotonin-mediated downstream effects, including reward from smoking.

3.3.1 Serotonin Synthesis

Tryptophan hydroxylase, encoded by the genes *TPH1* and *TPH2*, is the rate-limiting enzyme responsible for the biosynthesis of serotonin. Polymorphisms in *TPH* may lead to variation in the formation levels of serotonin, which may in turn influence the reinforcing properties of nicotine and alter smoking behaviors. In particular, associations between smoking status and smoking initiation with *TPH* variation have been investigated, described below.

3.3.2 Likelihood of Being a Smoker and Initiation

Impulsivity is associated with the initiation of drug use, including smoking (Mitchell 2004). In mice, depletion of serotonin in the brain (via TPH knock-out) is associated with intense compulsive and impulsive behaviors (Angoa-Perez et al. 2012). In addition, both A alleles of the *TPH1* SNPs rs1800532 (A>C) and rs1799913 (A>C), which are located in intron 7 and are in complete LD (Nielsen et al. 1997), have been associated with conditions relating to poor impulse control (Mann et al. 1997; Manuck et al. 1999). Although the functional impact of the *TPH1* SNPs is unknown, these results suggest that the A alleles may be associated with decreased *TPH1* expression or enzyme activity, resulting in lower serotonin synthesis. In subjects of European ancestry, these A alleles were more commonly found in regular smokers versus lifetime nonsmokers (Sullivan et al. 2001), suggesting increased smoking risk. In a separate study, the rs1799913 A/A genotype was associated with earlier age of smoking initiation among European smokers, relative to rs1799913 AC/CC genotypes (Lerman et al. 2001). The rs1800532 A/A genotype was also associated with earlier smoking among Japanese smokers (Mizuno et al. 2004). Although the mechanism underpinning these findings remains to be elucidated, it is possible that the increased risk of smoking and early initiation among subjects with the rs1800532 and rs1799913 A alleles may stem from reduced serotonin levels and greater impulsivity.

3.3.3 Serotonin Transporter

The serotonin transporter, encoded by the polymorphic gene *SLC6A4/5-HTT*, is an influx transporter that transports serotonin from the synapse to presynaptic neurons. Lower levels of serotonin reuptake in the CNS may increase the duration of action of serotonin in the brain (Mathews et al. 2004). This in turn may modulate the level and/or duration of serotonin receptor binding and resulting downstream signaling effects. Thus, genetic variation in the serotonin transporter that functionally impacts serotonin levels may modulate smoking behaviors.

3.3.4 Likelihood of Being a Smoker and Initiation

A common polymorphism in the *5-HTT* serotonin transporter gene has been associated with altered *5-HTT* transcription (Heils et al. 1996). This 44-bp insertion (L allele) or deletion (S allele) in the promoter region of *5-HTT* is referred to as the *5-HTT*-linked polymorphic region (*5-HTTLPR*), with the S allele reducing transcription of the gene and leading to lower serotonin reuptake (Heils et al. 1996). Relative to the *5-HTTLPR* S allele, the L allele was associated with an increased risk of smoking as shown by the greater prevalence of the L allele in Chinese and Japanese male smokers relative to nonsmokers (Chu et al. 2009; Ishikawa et al. 1999). Due to greater serotonin reuptake activity, those with the L allele may have lower brain serotonin levels. The increased risk of smoking among these individuals may stem from the nicotine-mediated boost in serotonin levels that occurs via smoking, which may compensate for the normally lower levels of synaptic serotonin among individuals with the L allele due to greater synaptic serotonin reuptake.

While this association was seen in two Asian populations, several studies have not replicated this in European and African American populations (Lerman et al. 1998; Sieminska et al. 2008). In adolescents of European descent, the *5-HTTLPR* S allele appears to *increase* the risk for smoking. The S/S genotype was more common in adolescent current smokers relative to nonsmokers, and in adolescents who had an early onset of smoking initiation (Gerra et al. 2005). However, this genetic association may interact with contextual factors, including peer smoking. The positive association between the S allele and smoking was observed only among adolescents in schools with high levels of smoking. In contrast, in schools with lower levels of smoking, the S allele was associated with a lower level of smoking (Daw et al. 2013). This indicates that decreased serotonin reuptake conferred by the S allele (via reduced *5-HTT* transcription) may interact with environmental factors to modulate smoking risk in adolescents.

3.3.5 Cigarette Consumption and Nicotine Dependence

As previously mentioned, the L allele has been associated with greater *5-HTT* transcription compared to the S allele (Heils et al. 1996), suggesting greater serotonin transporter levels in subjects possessing the L versus S allele. Higher serotonin transporter density, and thus potentially lower synaptic levels of serotonin, may motivate regular smokers to smoke more CPD in order to increase brain serotonin levels, which may in turn increase the risk for nicotine dependence. Chu et al. (2009) demonstrated the presence of a *5-HTTLPR* allele-dose effect on cigarette consumption and nicotine dependence among Chinese male smokers, where L/L individuals displayed the highest cigarette consumption and FTND scores, followed by heterozygotes (S/L) and S/S homozygotes. In European smokers, however, there was no association between *5-HTTLPR* genotype and CPD (Sieminska et al. 2008; Trummer et al. 2006). Potential reasons for these discordant findings include differences in ethnicity (Chinese vs. European ancestry), *5-HTTLPR* allele frequencies, gender (male vs. male and female smokers), and the noisiness/weakness of the phenotype measures (Chu et al. 2009; Sieminska et al. 2008). Gender selection in studies of smoking quantity may be an important factor considering that a larger proportion of men smoke compared to women (24 vs. 18 % in the United States in 2012) (Agaku et al. 2014), therefore there may be a different probability of observing significant outcomes with one gender compared to another. In addition, environmental factors interact with genotype to mediate behavioral outcomes, thus it is possible that environmental influences and/or personality traits unique to each study population may modulate the associations between *5-HTTLPR* genotype and cigarette consumption and nicotine dependence.

3.3.6 Cessation and Response to Pharmacotherapies

Among Japanese males, the *5-HTTLPR* S/S genotype was more prevalent in former versus current smokers (Sekiguchi et al. 2012) suggesting that lower levels of *5-HTT* transcription and thus less serotonin reuptake may help to increase quitting among smokers. In studies using cessation pharmacotherapy, however, *5-HTTLPR* genotype did not predict cessation success on placebo, and there was no observed interaction between *5-HTTLPR* genotype and nicotine patch treatment on cessation in smokers of European ancestry (David et al. 2007b, 2008). There was a pattern of increased cessation success in the nicotine patch group compared to the placebo group, although this was not significant (David et al. 2007b).

While cessation success on NRT does not appear to vary according to *5-HTT* genotype group, an association between *5-HTTLPR* genotype and cessation rates on bupropion has been observed. Subjects of European ancestry receiving bupropion therapy to aid smoking cessation exhibited prolonged abstinence on bupropion (vs. placebo) when they possessed one or two copies of the L allele, relative to subjects with the S/S genotype (Quaak et al. 2012). Bupropion does not exhibit strong serotonergic effects (Gobbi et al. 2003; Kugaya et al. 2003), thus the specific mechanism for this association is currently unclear.

Some of the observed associations between *5-HTTLPR* genetic variation and smoking behaviors may result from modulation of amygdala activity. The amygdala plays a major role in the processing of memory and emotional reactions (McGaugh 2004; Phelps 2006), which can influence craving and response to emotionally-charged drug cues. For example, when using functional magnetic resonance imaging to assess amygdala activation in response to neutral, pleasant, or unpleasant stimuli, subjects with one or two copies of the S allele exhibited increased amygdala activation in response to unpleasant stimuli compared to L/L subjects (Kobiella et al. 2011). As stronger amygdala response to smoking-cessation messages predicts an increased likelihood of cessation success (Jasinska et al. 2012), this may explain, in part, the higher frequencies of the *5-HTTLPR* S allele among former compared to current smokers (Sekiguchi et al. 2012). Considering, however, that the *5-HTT* genotype does not always predict cessation success (e.g., in clinical trials of cessation pharmacotherapies), it seems likely that there are additional predispositions or brain regions in which variability in the serotonin system helps modulate smoking behavior and the ability to quit smoking.

4 Conclusions

Smoking behaviors are mediated by multiple different CNS processes, which in turn are influenced by the pharmacokinetic parameters that regulate nicotine levels in the body. Genetic variation in nicotine metabolism and CNS response to nicotine contributes to interindividual differences in smoking behaviors, such as the initiation of smoking, smoking quantity, nicotine dependence, and cessation success with or without pharmacotherapy. As discussed in this chapter, there are a broad range of genetic factors that can contribute to the modulation of smoking behavior. This suggests that a more comprehensive assessment of a smoker's genetic profile, considering the combination of genes described here, may be necessary in order to form associations between genetic variation and behavior, or to predict a smoker's ability to quit smoking using pharmacotherapy. Identification of novel genetic variants, as well as functional characterization of known and novel variants, will provide a better understanding of the underlying mechanisms and assist in establishing genetic predictors of smoking behaviors and response to cessation pharmacotherapy. This information can be used to identify novel potential therapeutic targets as well as in personalizing therapy to increase smoking cessation rates.

Acknowledgments We acknowledge the support of the Endowed Chair in Addiction for the Department of Psychiatry, CIHR grant TMH109787, NIH grant DA 020830, Centre for Addiction and Mental Health and the Canada Foundation for Innovation (#20289 and #16014), the CAMH Foundation, and the Ontario Ministry of Research and Innovation.

References

Agaku IT, King BA, Dube SR (2014) Centers for disease control, prevention. Current cigarette smoking among adults—United States, 2005–2012. MMWR Morb Mortal Wkly Rep 63:29–34

Al Koudsi N, Tyndale RF (2010) Hepatic CYP2B6 is altered by genetic, physiologic, and environmental factors but plays little role in nicotine metabolism. Xenobiotica 40:381–392

Al Koudsi N, Ahluwalia JS, Lin SK, Sellers EM, Tyndale RF (2009) A novel CYP2A6 allele (CYP2A6*35) resulting in an amino-acid substitution (Asn438Tyr) is associated with lower CYP2A6 activity in vivo. Pharmacogenomics J 9:274–282

Al Koudsi N, O'Loughlin J, Rodriguez D, Audrain-McGovern J, Tyndale RF (2010) The genetic aspects of nicotine metabolism and their impact on adolescent nicotine dependence. J Pediatr Biochem 1:105–123

Angoa-Perez M, Kane MJ, Briggs DI, Sykes CE, Shah MM, Francescutti DM, Rosenberg DR, Thomas DM, Kuhn DM (2012) Genetic depletion of brain 5HT reveals a common molecular pathway mediating compulsivity and impulsivity. J Neurochem 121:974–984

Ariyoshi N, Miyamoto M, Umetsu Y, Kunitoh H, Dosaka-Akita H, Sawamura Y, Yokota J, Nemoto N, Sato K, Kamataki T (2002) Genetic polymorphism of CYP2A6 gene and tobacco-induced lung cancer risk in male smokers. Cancer Epidemiol Biomark Prev 11:890–894

Audrain-McGovern J, Al Koudsi N, Rodriguez D, Wileyto EP, Shields PG, Tyndale RF (2007) The role of CYP2A6 in the emergence of nicotine dependence in adolescents. Pediatrics 119: e264–e274

Baker TB, Weiss RB, Bolt D, von Niederhausern A, Fiore MC, Dunn DM, Piper ME, Matsunami N, Smith SS, Coon H, McMahon WM, Scholand MB, Singh N, Hoidal JR, Kim SY, Leppert MF, Cannon DS (2009) Human neuronal acetylcholine receptor A5-A3-B4 haplotypes are associated with multiple nicotine dependence phenotypes. Nicotine Tob Res 11:785–796

Baker TB, Breslau N, Covey L, Shiffman S (2012) DSM criteria for tobacco use disorder and tobacco withdrawal: a critique and proposed revisions for DSM-5. Addiction 107:263–275

Balleine BW, Delgado MR, Hikosaka O (2007) The role of the dorsal striatum in reward and decision-making. J Neurosci 27:5–8161

Bao Z, He XY, Ding X, Prabhu S, Hong JY (2005) Metabolism of nicotine and cotinine by human cytochrome P450 2A13. Drug Metab Dispos 33:61–258

Benowitz NL (2008) Clinical pharmacology of nicotine: implications for understanding, preventing, and treating tobacco addiction. Clin Pharmacol Ther 83:41–531

Benowitz NL (2009) Pharmacology of nicotine: addiction, smoking-induced disease, and therapeutics. Annu Rev Pharmacol Toxicol 49:57–71

Benowitz NL (2010) Nicotine addiction. N Engl J Med 362:303–2295

Benowitz NL, Jacob P 3rd (1985) Nicotine renal excretion rate influences nicotine intake during cigarette smoking. J Pharmacol Exp Ther 234:5–153

Benowitz NL, Jacob P 3rd (1994) Metabolism of nicotine to cotinine studied by a dual stable isotope method. Clin Pharmacol Ther 56:93–483

Benowitz NL, Jacob P 3rd (2001) Trans-3′-hydroxycotinine: disposition kinetics, effects and plasma levels during cigarette smoking. Br J Clin Pharmacol 51:9–53

Benowitz NL, Jacob P 3rd, Fong I, Gupta S (1994) Nicotine metabolic profile in man: comparison of cigarette smoking and transdermal nicotine. J Pharmacol Exp Ther 268:296–303

Benowitz NL, Perez-Stable EJ, Fong I, Modin G, Herrera B, Jacob P 3rd (1999) Ethnic differences in N-glucuronidation of nicotine and cotinine. J Pharmacol Exp Ther 291:203–1196

Benowitz NL, Perez-Stable EJ, Herrera B, Jacob P 3rd (2002) Slower metabolism and reduced intake of nicotine from cigarette smoking in Chinese-Americans. J Natl Cancer Inst 94:15–108

Benowitz NL, Pomerleau OF, Pomerleau CS, Jacob P 3rd (2003) Nicotine metabolite ratio as a predictor of cigarette consumption. Nicotine Tob Res 5:621–624

Benowitz NL, Lessov-Schlaggar CN, Swan GE (2008) Genetic influences in the variation in renal clearance of nicotine and cotinine. Clin Pharmacol Ther 84:243–247

Benowitz NL, Hukkanen J, Jacob P, 3rd (2009) Nicotine chemistry, metabolism, kinetics and biomarkers. Handb Exp Pharmacol 192:29–60

Benowitz NL, Dains KM, Dempsey D, Havel C, Wilson M, Jacob P 3rd (2010) Urine menthol as a biomarker of mentholated cigarette smoking. Cancer Epidemiol Biomark Prev 19:9–3013

Benowitz NL, Dains KM, Dempsey D, Wilson M, Jacob P (2011) Racial differences in the relationship between number of cigarettes smoked and nicotine and carcinogen exposure. Nicotine Tob Res 13:83–772

Benowitz NL, Renner CC, Lanier AP, Tyndale RF, Hatsukami DK, Lindgren B, Stepanov I, Watson CH, Sosnoff CS, Jacob P 3rd (2012) Exposure to nicotine and carcinogens among Southwestern Alaskan Native cigarette smokers and smokeless tobacco users. Cancer Epidemiol Biomark Prev 21:42–934

Berg JZ, von Weymarn LB, Thompson EA, Wickham KM, Weisensel NA, Hatsukami DK, Murphy SE (2010) UGT2B10 genotype influences nicotine glucuronidation, oxidation, and consumption. Cancer Epidemiol Biomark Prev 19:31–1423

Bergen AW, Javitz HS, Krasnow R, Nishita D, Michel M, Conti DV, Liu J, Lee W, Edlund CK, Hall S, Kwok PY, Benowitz NL, Baker TB, Tyndale RF, Lerman C, Swan GE (2013)

Nicotinic acetylcholine receptor variation and response to smoking cessation therapies. Pharmacogenet Genomics 23:94–103

Bierut LJ, Stitzel JA, Wang JC, Hinrichs AL, Grucza RA, Xuei X, Saccone NL, Saccone SF, Bertelsen S, Fox L, Horton WJ, Breslau N, Budde J, Cloninger CR, Dick DM, Foroud T, Hatsukami D, Hesselbrock V, Johnson EO, Kramer J, Kuperman S, Madden PA, Mayo K, Nurnberger J Jr, Pomerleau O, Porjesz B, Reyes O, Schuckit M, Swan G, Tischfield JA, Edenberg HJ, Rice JP, Goate AM (2008) Variants in nicotinic receptors and risk for nicotine dependence. Am J Psychiatry 165:1163–1171

Binnington MJ, Zhu AZ, Renner CC, Lanier AP, Hatsukami DK, Benowitz NL, Tyndale RF (2012) CYP2A6 and CYP2B6 genetic variation and its association with nicotine metabolism in South Western Alaska Native people. Pharmacogenet Genomics 22:429–440

Bloom AJ, Murphy SE, Martinez M, von Weymarn LB, Bierut LJ, Goate A (2013) Effects upon in-vivo nicotine metabolism reveal functional variation in FMO3 associated with cigarette consumption. Pharmacogenet Genomics 23:62–68

Bontempi S, Fiorentini C, Busi C, Guerra N, Spano P, Missale C (2007) Identification and characterization of two nuclear factor-kappaB sites in the regulatory region of the dopamine D2 receptor. Endocrinology 148:2563–2570

Breetvelt EJ, Numans ME, Aukes MF, Hoeben W, Strengman E, Luykx JJ, Bakker SC, Kahn RS, Ophoff RA, Boks MP (2012) The association of the alpha-5 subunit of the nicotinic acetylcholine receptor gene and the brain-derived neurotrophic factor gene with different aspects of smoking behavior. Psychiatr Genet 22:96–98

Breitling LP, Dahmen N, Illig T, Rujescu D, Nitz B, Raum E, Winterer G, Rothenbacher D, Brenner H (2009) Variants in COMT and spontaneous smoking cessation: retrospective cohort analysis of 925 cessation events. Pharmacogenet Genomics 19:657–659

Breslau N, Peterson EL (1996) Smoking cessation in young adults: age at initiation of cigarette smoking and other suspected influences. Am J Public Health 86:214–220

Brown J, Hajek P, McRobbie H, Locker J, Gillison F, McEwen A, Beard E, West R (2013) Cigarette craving and withdrawal symptoms during temporary abstinence and the effect of nicotine gum. Psychopharmacology 229:209–218

Budulac SE, Vonk JM, Postma DS, Siedlinski M, Timens W, Boezen MH (2012) Nicotinic acetylcholine receptor variants are related to smoking habits, but not directly to COPD. PLoS ONE 7:e33386

Burger D, van der Heiden I, la Porte C, van der Ende M, Groeneveld P, Richter C, Koopmans P, Kroon F, Sprenger H, Lindemans J, Schenk P, van Schaik R (2006) Interpatient variability in the pharmacokinetics of the HIV non-nucleoside reverse transcriptase inhibitor efavirenz: the effect of gender, race, and CYP2B6 polymorphism. Br J Clin Pharmacol 61:148–154

Byrd GD, Chang KM, Greene JM, deBethizy JD (1992) Evidence for urinary excretion of glucuronide conjugates of nicotine, cotinine, and trans-3′-hydroxycotinine in smokers. Drug Metabol Dispos 20:192–197

Carroll FI, Blough BE, Mascarella SW, Navarro HA, Lukas RJ, Damaj MI (2014) Bupropion and bupropion analogs as treatments for CNS disorders. Adv Pharmacol 69:177–216

Cashman JR, Park SB, Yang ZC, Wrighton SA, Jacob P 3rd, Benowitz NL (1992) Metabolism of nicotine by human liver microsomes: stereoselective formation of trans-nicotine N′-oxide. Chem Res Toxicol 5:639–646

Caskey NH, Jarvik ME, Wirshing WC (1999) The effects of dopaminergic D2 stimulation and blockade on smoking behavior. Exp Clin Psychopharmacol 7:72–78

Caskey NH, Jarvik ME, Wirshing WC, Madsen DC, Iwamoto-Schaap PN, Eisenberger NI, Huerta L, Terrace SM, Olmstead RE (2002) Modulating tobacco smoking rates by dopaminergic stimulation and blockade. Nicotine Tob Res 4:259–266

Chen J, Millar WJ (1998) Age of smoking initiation: implications for quitting. Health Rep 9:39–46 (Eng); 39–48(Fre)

Chen G, Blevins-Primeau AS, Dellinger RW, Muscat JE, Lazarus P (2007) Glucuronidation of nicotine and cotinine by UGT2B10: loss of function by the UGT2B10 Codon 67 (Asp>Tyr) polymorphism. Cancer Res 67:9024–9029

Chen X, Chen J, Williamson VS, An SS, Hettema JM, Aggen SH, Neale MC, Kendler KS (2009) Variants in nicotinic acetylcholine receptors alpha5 and alpha3 increase risks to nicotine dependence. Am J Med Genet Part B Neuropsychiatr Genet 150B:926–933

Chen G, Giambrone NE Jr, Dluzen DF, Muscat JE, Berg A, Gallagher CJ, Lazarus P (2010) Glucuronidation genotypes and nicotine metabolic phenotypes: importance of functional UGT2B10 and UGT2B17 polymorphisms. Cancer Res 70:7543–7552

Chen LS, Baker TB, Piper ME, Breslau N, Cannon DS, Doheny KF, Gogarten SM, Johnson EO, Saccone NL, Wang JC, Weiss RB, Goate AM, Bierut LJ (2012) Interplay of genetic risk factors (CHRNA5-CHRNA3-CHRNB4) and cessation treatments in smoking cessation success. Am J Psychiatry 169:735–742

Chen LS, Bloom AJ, Baker TB, Smith SS, Piper ME, Martinez M, Saccone N, Hatsukami D, Goate A, Bierut L (2014) Pharmacotherapy effects on smoking cessation vary with nicotine metabolism gene (CYP2A6). Addiction 109:128–137

Chenoweth MJ, O'Loughlin J, Sylvestre MP, Tyndale RF (2013) CYP2A6 slow nicotine metabolism is associated with increased quitting by adolescent smokers. Pharmacogenet Genomics 23:232–235

Chenoweth MJ, Zhu AZ, Sanderson Cox L, Ahluwalia JS, Benowitz NL, Tyndale RF (2014) Variation in P450 oxidoreductase (POR) A503V and flavin-containing monooxygenase (FMO)-3 E158K is associated with minor alterations in nicotine metabolism, but does not alter cigarette consumption. Pharmacogenet Genomics 24:172–176

Chu SL, Xiao D, Wang C, Jing H (2009) Association between 5-hydroxytryptamine transporter gene-linked polymorphic region and smoking behavior in Chinese males. Chin Med J 122:1365–1368

Colilla S, Lerman C, Shields PG, Jepson C, Rukstalis M, Berlin J, DeMichele A, Bunin G, Strom BL, Rebbeck TR (2005) Association of catechol-O-methyltransferase with smoking cessation in two independent studies of women. Pharmacogenet Genomics 15:393–398

Comings DE, Ferry L, Bradshaw-Robinson S, Burchette R, Chiu C, Muhleman D (1996) The dopamine D2 receptor (DRD2) gene: a genetic risk factor in smoking. Pharmacogenetics 6:73–79

Corrigall WA, Coen KM, Adamson KL (1994) Self-administered nicotine activates the mesolimbic dopamine system through the ventral tegmental area. Brain Res 653:278–284

Cui Y, Wen W, Moriarty CJ, Levine RS (2006) Risk factors and their effects on the dynamic process of smoking relapse among veteran smokers. Behav Res Ther 44:967–981

Dalley JW, Fryer TD, Brichard L, Robinson ES, Theobald DE, Laane K, Pena Y, Murphy ER, Shah Y, Probst K, Abakumova I, Aigbirhio FI, Richards HK, Hong Y, Baron JC, Everitt BJ, Robbins TW (2007) Nucleus accumbens D2/3 receptors predict trait impulsivity and cocaine reinforcement. Science 315:1267–1270

Damaj MI, Carroll FI, Eaton JB, Navarro HA, Blough BE, Mirza S, Lukas RJ, Martin BR (2004) Enantioselective effects of hydroxy metabolites of bupropion on behavior and on function of monoamine transporters and nicotinic receptors. Mol Pharmacol 66:675–682

Dani JA, Heinemann S (1996) Molecular and cellular aspects of nicotine abuse. Neuron 16:905–908

David SP, Niaura R, Papandonatos GD, Shadel WG, Burkholder GJ, Britt DM, Day A, Stumpff J, Hutchison K, Murphy M, Johnstone E, Griffiths SE, Walton RT (2003) Does the DRD2-Taq1 A polymorphism influence treatment response to bupropion hydrochloride for reduction of the nicotine withdrawal syndrome? Nicotine Tob Res 5:935–942

David SP, Brown RA, Papandonatos GD, Kahler CW, Lloyd-Richardson EE, Munafo MR, Shields PG, Lerman C, Strong D, McCaffery J, Niaura R (2007a) Pharmacogenetic clinical trial of sustained-release bupropion for smoking cessation. Nicotine Tob Res 9:821–833

David SP, Munafo MR, Murphy MF, Walton RT, Johnstone EC (2007b) The serotonin transporter 5-HTTLPR polymorphism and treatment response to nicotine patch: follow-up of a randomized controlled trial. Nicotine Tob Res 9:225–231

David SP, Strong DR, Munafo MR, Brown RA, Lloyd-Richardson EE, Wileyto PE, Evins EA, Shields PG, Lerman C, Niaura R (2007c) Bupropion efficacy for smoking cessation is influenced by the DRD2 Taq1A polymorphism: analysis of pooled data from two clinical trials. Nicotine Tob Res 9:1251–1257

David SP, Johnstone EC, Murphy MF, Aveyard P, Guo B, Lerman C, Munafo MR (2008) Genetic variation in the serotonin pathway and smoking cessation with nicotine replacement therapy: new data from the Patch in Practice trial and pooled analyses. Drug Alcohol Depend 98:77–85

David SP, Johnstone EC, Churchman M, Aveyard P, Murphy MF, Munafo MR (2011) Pharmacogenetics of smoking cessation in general practice: results from the patch II and patch in practice trials. Nicotine Tob Res 13:157–167

David SP, Hamidovic A, Chen GK, Bergen AW, Wessel J, Kasberger JL, Brown WM, Petruzella S, Thacker EL, Kim Y, Nalls MA, Tranah GJ, Sung YJ, Ambrosone CB, Arnett D, Bandera EV, Becker DM, Becker L, Berndt SI, Bernstein L, Blot WJ, Broeckel U, Buxbaum SG, Caporaso N, Casey G, Chanock SJ, Deming SL, Diver WR, Eaton CB, Evans DS, Evans MK, Fornage M, Franceschini N, Harris TB, Henderson BE, Hernandez DG, Hitsman B, Hu JJ, Hunt SC, Ingles SA, John EM, Kittles R, Kolb S, Kolonel LN, Le Marchand L, Liu Y, Lohman KK, McKnight B, Millikan RC, Murphy A, Neslund-Dudas C, Nyante S, Press M, Psaty BM, Rao DC, Redline S, Rodriguez-Gil JL, Rybicki BA, Signorello LB, Singleton AB, Smoller J, Snively B, Spring B, Stanford JL, Strom SS, Swan GE, Taylor KD, Thun MJ, Wilson AF, Witte JS, Yamamura Y, Yanek LR, Yu K, Zheng W, Ziegler RG, Zonderman AB, Jorgenson E, Haiman CA, Furberg H (2012) Genome-wide meta-analyses of smoking behaviors in African Americans. Transl Psychiatry 2:e119

David SP, Strong DR, Leventhal AM, Lancaster MA, McGeary JE, Munafo MR, Bergen AW, Swan GE, Benowitz NL, Tyndale RF, Conti DV, Brown RA, Lerman C, Niaura R (2013) Influence of a dopamine pathway additive genetic efficacy score on smoking cessation: results from two randomized clinical trials of bupropion. Addiction 108:2202–2211

Daw J, Shanahan M, Harris KM, Smolen A, Haberstick B, Boardman JD (2013) Genetic sensitivity to peer behaviors: 5HTTLPR, smoking, and alcohol consumption. J Health Soc Behav 54:92–108

Dempsey D, Tutka P, Jacob P 3rd, Allen F, Schoedel K, Tyndale RF, Benowitz NL (2004) Nicotine metabolite ratio as an index of cytochrome P450 2A6 metabolic activity. Clin Pharmacol Ther 76:64–72

Di Chiara G, Tanda G, Bassareo V, Pontieri F, Acquas E, Fenu S, Cadoni C, Carboni E (1999) Drug addiction as a disorder of associative learning. Role of nucleus accumbens shell/extended amygdala dopamine. Ann N Y Acad Sci 877:461–485

Ehmer U, Vogel A, Schutte JK, Krone B, Manns MP, Strassburg CP (2004) Variation of hepatic glucuronidation: novel functional polymorphisms of the UDP-glucuronosyltransferase UGT1A4. Hepatology 39:970–977

Erlich PM, Hoffman SN, Rukstalis M, Han JJ, Chu X, Linda Kao WH, Gerhard GS, Stewart WF, Boscarino JA (2010) Nicotinic acetylcholine receptor genes on chromosome 15q25.1 are associated with nicotine and opioid dependence severity. Hum Genet 128:491–499

Faucette SR, Hawke RL, Lecluyse EL, Shord SS, Yan B, Laethem RM, Lindley CM (2000) Validation of bupropion hydroxylation as a selective marker of human cytochrome P450 2B6 catalytic activity. Drug Metab Dispos 28:1222–1230

Feng Y, Niu T, Xing H, Xu X, Chen C, Peng S, Wang L, Laird N, Xu X (2004) A common haplotype of the nicotine acetylcholine receptor alpha 4 subunit gene is associated with vulnerability to nicotine addiction in men. Am J Hum Genet 75:112–121

Ferguson CS, Miksys S, Palmour RM, Tyndale RF (2013) Ethanol self-administration and nicotine treatment induce brain levels of CYP2B6 and CYP2E1 in African green monkeys. Neuropharmacology 72:74–81

Franklin TR, Lohoff FW, Wang Z, Sciortino N, Harper D, Li Y, Jens W, Cruz J, Kampman K, Ehrman R, Berrettini W, Detre JA, O'Brien CP, Childress AR (2009) DAT genotype modulates brain and behavioral responses elicited by cigarette cues. Neuropsychopharmacology 34:717–728

Franklin TR, Wang Z, Li Y, Suh JJ, Goldman M, Lohoff FW, Cruz J, Hazan R, Jens W, Detre JA, Berrettini W, O'Brien CP, Childress AR (2011) Dopamine transporter genotype modulation of neural responses to smoking cues: confirmation in a new cohort. Addict Biol 16:308–322

Fujieda M, Yamazaki H, Saito T, Kiyotani K, Gyamfi MA, Sakurai M, Dosaka-Akita H, Sawamura Y, Yokota J, Kunitoh H, Kamataki T (2004) Evaluation of CYP2A6 genetic polymorphisms as determinants of smoking behavior and tobacco-related lung cancer risk in male Japanese smokers. Carcinogenesis 25:2451–2458

Fukami T, Nakajima M, Higashi E, Yamanaka H, McLeod HL, Yokoi T (2005) A novel CYP2A6*20 allele found in African-American population produces a truncated protein lacking enzymatic activity. Biochem Pharmacol 70:801–808

Gallego X, Molas S, Amador-Arjona A, Marks MJ, Robles N, Murtra P, Armengol L, Fernandez-Montes RD, Gratacos M, Pumarola M, Cabrera R, Maldonado R, Sabria J, Estivill X, Dierssen M (2012) Overexpression of the CHRNA5/A3/B4 genomic cluster in mice increases the sensitivity to nicotine and modifies its reinforcing effects. Amino Acids 43:897–909

Garrison GD, Dugan SE (2009) Varenicline: a first-line treatment option for smoking cessation. Clin Ther 31:463–491

Gerra G, Garofano L, Zaimovic A, Moi G, Branchi B, Bussandri M, Brambilla F, Donnini C (2005) Association of the serotonin transporter promoter polymorphism with smoking behavior among adolescents. Am J Med Genet Part B Neuropsychiatr Genet 135B:73–78

Giovino GA (1999) Epidemiology of tobacco use among US adolescents. Nicotine Tob Res 1:S31–S40

Glick SD, Maisonneuve IM, Kitchen BA (2002) Modulation of nicotine self-administration in rats by combination therapy with agents blocking alpha 3 beta 4 nicotinic receptors. Eur J Pharmacol 448:185–191

Gobbi G, Slater S, Boucher N, Debonnel G, Blier P (2003) Neurochemical and psychotropic effects of bupropion in healthy male subjects. J Clin Psychopharmacol 23:233–239

Greenbaum L, Kanyas K, Karni O, Merbl Y, Olender T, Horowitz A, Yakir A, Lancet D, Ben-Asher E, Lerer B (2006) Why do young women smoke? I. Direct and interactive effects of environment, psychological characteristics and nicotinic cholinergic receptor genes. Mol Psychiatry 11:223

Grenhoff J, Aston-Jones G, Svensson TH (1986) Nicotinic effects on the firing pattern of midbrain dopamine neurons. Acta Physiol Scand 128:351–358

Gu DF, Hinks LJ, Morton NE, Day IN (2000) The use of long PCR to confirm three common alleles at the CYP2A6 locus and the relationship between genotype and smoking habit. Ann Hum Genet 64:383–390

Guo S, da Chen F, Zhou DF, Sun HQ, Wu GY, Haile CN, Kosten TA, Kosten TR, Zhang XY (2007) Association of functional catechol O-methyl transferase (COMT) Val108Met polymorphism with smoking severity and age of smoking initiation in Chinese male smokers. Psychopharmacology 190:449–456

Halford JC, Harrold JA, Lawton CL, Blundell JE (2005) Serotonin (5-HT) drugs: effects on appetite expression and use for the treatment of obesity. Curr Drug Targets 6:201–213

Han DH, Joe KH, Na C, Lee YS (2008) Effect of genetic polymorphisms on smoking cessation: a trial of bupropion in Korean male smokers. Psychiatr Genet 18:11–16

Han S, Yang BZ, Kranzler HR, Oslin D, Anton R, Gelernter J (2011) Association of CHRNA4 polymorphisms with smoking behavior in two populations. Am J Med GenetPart B Neuropsychiatr Genet 156B:421–429

Hartz SM, Short SE, Saccone NL, Culverhouse R, Chen L, Schwantes-An TH, Coon H, Han Y, Stephens SH, Sun J, Chen X, Ducci F, Dueker N, Franceschini N, Frank J, Geller F, Gubjartsson D, Hansel NN, Jiang C, Keskitalo-Vuokko K, Liu Z, Lyytikainen LP, Michel M, Rawal R, Rosenberger A, Scheet P, Shaffer JR, Teumer A, Thompson JR, Vink JM, Vogelzangs N, Wenzlaff AS, Wheeler W, Xiao X, Yang BZ, Aggen SH, Balmforth AJ, Baumeister SE, Beaty T, Bennett S, Bergen AW, Boyd HA, Broms U, Campbell H, Chatterjee N, Chen J, Cheng YC, Cichon S, Couper D, Cucca F, Dick DM, Foroud T, Furberg H, Giegling I, Gu F, Hall AS, Hallfors J, Han S, Hartmann AM, Hayward C, Heikkila K, Hewitt JK, Hottenga JJ, Jensen MK, Jousilahti P, Kaakinen M, Kittner SJ, Konte B, Korhonen T, Landi MT, Laatikainen T, Leppert M, Levy SM, Mathias RA, McNeil DW, Medland SE, Montgomery GW, Muley T, Murray T, Nauck M, North K, Pergadia M, Polasek O, Ramos

EM, Ripatti S, Risch A, Ruczinski I, Rudan I, Salomaa V, Schlessinger D, Styrkarsdottir U, Terracciano A, Uda M, Willemsen G, Wu X, Abecasis G, Barnes K, Bickeboller H, Boerwinkle E, Boomsma DI, Caporaso N, Duan J, Edenberg HJ, Francks C, Gejman PV, Gelernter J, Grabe HJ, Hops H, Jarvelin MR, Viikari J, Kahonen M, Kendler KS, Lehtimaki T, Levinson DF, Marazita ML, Marchini J, Melbye M, Mitchell BD, Murray JC, Nothen MM, Penninx BW, Raitakari O, Rietschel M, Rujescu D, Samani NJ, Sanders AR, Schwartz AG, Shete S, Shi J, Spitz M, Stefansson K, Swan GE, Thorgeirsson T, Volzke H, Wei Q, Wichmann HE, Amos CI, Breslau N, Cannon DS, Ehringer M, Grucza R, Hatsukami D, Heath A, Johnson EO, Kaprio J, Madden P, Martin NG, Stevens VL, Stitzel JA, Weiss RB, Kraft P, Bierut LJ (2012) Increased genetic vulnerability to smoking at CHRNA5 in early-onset smokers. Arch Gen Psychiatry 69:854–860

Heatherton TF, Kozlowski LT, Frecker RC, Fagerstrom KO (1991) The fagerstrom test for nicotine dependence: a revision of the fagerstrom tolerance questionnaire. Br J Addict 86:1119–1127

Heils A, Teufel A, Petri S, Stober G, Riederer P, Bengel D, Lesch KP (1996) Allelic variation of human serotonin transporter gene expression. J Neurochem 66:2621–2624

Heinz A, Goldman D, Jones DW, Palmour R, Hommer D, Gorey JG, Lee KS, Linnoila M, Weinberger DR (2000) Genotype influences in vivo dopamine transporter availability in human striatum. Neuropsychopharmacology 22:133–139

Herraiz T, Chaparro C (2005) Human monoamine oxidase is inhibited by tobacco smoke: beta-carboline alkaloids act as potent and reversible inhibitors. Biochem Biophys Res Commun 326:378–386

Herskovic JE, Rose JE, Jarvik ME (1986) Cigarette desirability and nicotine preference in smokers. Pharmacol Biochem Behav 24:171–175

Hines RN, Cashman JR, Philpot RM, Williams DE, Ziegler DM (1994) The mammalian flavin-containing monooxygenases: molecular characterization and regulation of expression. Toxicol Appl Pharmacol 125:1–6

Hinrichs AL, Murphy SE, Wang JC, Saccone S, Saccone N, Steinbach JH, Goate A, Stevens VL, Bierut LJ (2011) Common polymorphisms in FMO1 are associated with nicotine dependence. Pharmacogenet Genomics 21:397–402

Ho MK, Tyndale RF (2007) Overview of the pharmacogenomics of cigarette smoking. Pharmacogenomics J 7:81–98

Ho MK, Faseru B, Choi WS, Nollen NL, Mayo MS, Thomas JL, Okuyemi KS, Ahluwalia JS, Benowitz NL, Tyndale RF (2009a) Utility and relationships of biomarkers of smoking in African-American light smokers. Cancer Epidemiol Biomark Prev 18:3426–3434

Ho MK, Mwenifumbo JC, Al Koudsi N, Okuyemi KS, Ahluwalia JS, Benowitz NL, Tyndale RF (2009b) Association of nicotine metabolite ratio and CYP2A6 genotype with smoking cessation treatment in African-American light smokers. Clin Pharmacol Ther 85:635–643

Hotamisligil GS, Breakefield XO (1991) Human monoamine oxidase A gene determines levels of enzyme activity. Am J Hum Genet 49:383–392

Huang W, Ma JZ, Payne TJ, Beuten J, Dupont RT, Li MD (2008) Significant association of DRD1 with nicotine dependence. Hum Genet 123:133–140

Huang W, Payne TJ, Ma JZ, Beuten J, Dupont RT, Inohara N, Li MD (2009) Significant association of ANKK1 and detection of a functional polymorphism with nicotine dependence in an African-American sample. Neuropsychopharmacology 34:319–330

Ikemoto S, Qin M, Liu ZH (2006) Primary reinforcing effects of nicotine are triggered from multiple regions both inside and outside the ventral tegmental area. J Neurosci 26:723–730

Ishikawa H, Ohtsuki T, Ishiguro H, Yamakawa-Kobayashi K, Endo K, Lin YL, Yanagi H, Tsuchiya S, Kawata K, Hamaguchi H, Arinami T (1999) Association between serotonin transporter gene polymorphism and smoking among Japanese males. Cancer Epidemiol Biomark Prev 8:831–833

Ito H, Hamajima N, Matsuo K, Okuma K, Sato S, Ueda R, Tajima K (2003) Monoamine oxidase polymorphisms and smoking behaviour in Japanese. Pharmacogenetics 13:73–79

Jalas JR, Hecht SS, Murphy SE (2005) Cytochrome P450 enzymes as catalysts of metabolism of 4-(methylnitrosamino)-1-(3-pyridyl)-1-butanone, a tobacco specific carcinogen. Chem Res Toxicol 18:95–110

Janes AC, Smoller JW, David SP, Frederick BD, Haddad S, Basu A, Fava M, Evins AE, Kaufman MJ (2012) Association between CHRNA5 genetic variation at rs16969968 and brain reactivity to smoking images in nicotine dependent women. Drug Alcohol Depend 120:7–13

Jarvik ME, Madsen DC, Olmstead RE, Iwamoto-Schaap PN, Elins JL, Benowitz NL (2000) Nicotine blood levels and subjective craving for cigarettes. Pharmacol Biochem Behav 66:553–558

Jasinska AJ, Chua HF, Ho SS, Polk TA, Rozek LS, Strecher VJ (2012) Amygdala response to smoking-cessation messages mediates the effects of serotonin transporter gene variation on quitting. NeuroImage 60:766–773

Jia Y, Yamazaki Y, Nakauchi S, Ito K, Sumikawa K (2010) Nicotine facilitates long-term potentiation induction in oriens-lacunosum moleculare cells via Ca2+ entry through non-alpha7 nicotinic acetylcholine receptors. Eur J Neurosci 31:463–476

Jin Y, Chen D, Hu Y, Guo S, Sun H, Lu A, Zhang X, Li L (2006) Association between monoamine oxidase gene polymorphisms and smoking behaviour in Chinese males. Int J Neuropsychopharmacol 9:557–564

Johnstone EC, Clark TG, Griffiths SE, Murphy MF, Walton RT (2002) Polymorphisms in dopamine metabolic enzymes and tobacco consumption in smokers: seeking confirmation of the association in a follow-up study. Pharmacogenetics 12:585–587

Johnstone E, Benowitz N, Cargill A, Jacob R, Hinks L, Day I, Murphy M, Walton R (2006) Determinants of the rate of nicotine metabolism and effects on smoking behavior. Clin Pharmacol Ther 80:319–330

Jonsson EG, Nothen MM, Grunhage F, Farde L, Nakashima Y, Propping P, Sedvall GC (1999) Polymorphisms in the dopamine D2 receptor gene and their relationships to striatal dopamine receptor density of healthy volunteers. Mol Psychiatry 4:290–296

Kaivosaari S, Toivonen P, Hesse LM, Koskinen M, Court MH, Finel M (2007) Nicotine glucuronidation and the human UDP-glucuronosyltransferase UGT2B10. Mol Pharmacol 72:761–768

Kang AM, Palmatier MA, Kidd KK (1999) Global variation of a 40-bp VNTR in the 3′-untranslated region of the dopamine transporter gene (SLC6A3). Biol Psychiatry 46:151–160

Kapoor M, Wang JC, Bertelsen S, Bucholz K, Budde JP, Hinrichs A, Agrawal A, Brooks A, Chorlian D, Dick D, Hesselbrock V, Foroud T, Kramer J, Kuperman S, Manz N, Nurnberger J Jr, Porjesz B, Rice J, Tischfield J, Xuei X, Schuckit M, Edenberg HJ, Bierut LJ, Goate AM (2012) Variants located upstream of CHRNB4 on chromosome 15q25.1 are associated with age at onset of daily smoking and habitual smoking. PLoS ONE 7:e33513

Karp I, O'Loughlin J, Hanley J, Tyndale RF, Paradis G (2006) Risk factors for tobacco dependence in adolescent smokers. Tob Control 15:199–204

King DP, Paciga S, Pickering E, Benowitz NL, Bierut LJ, Conti DV, Kaprio J, Lerman C, Park PW (2012) Smoking cessation pharmacogenetics: analysis of varenicline and bupropion in placebo-controlled clinical trials. Neuropsychopharmacology 37:641–650

Kirchheiner J, Klein C, Meineke I, Sasse J, Zanger UM, Murdter TE, Roots I, Brockmoller J (2003) Bupropion and 4-OH-bupropion pharmacokinetics in relation to genetic polymorphisms in CYP2B6. Pharmacogenetics 13:619–626

Kobiella A, Reimold M, Ulshofer DE, Ikonomidou VN, Vollmert C, Vollstadt-Klein S, Rietschel M, Reischl G, Heinz A, Smolka MN (2011) How the serotonin transporter 5-HTTLPR polymorphism influences amygdala function: the roles of in vivo serotonin transporter expression and amygdala structure. Transl Psychiatry 1:e37

Koopmans JR, Slutske WS, Heath AC, Neale MC, Boomsma DI (1999) The genetics of smoking initiation and quantity smoked in Dutch adolescent and young adult twins. Behav Genet 29:383–393

Kubota T, Nakajima-Taniguchi C, Fukuda T, Funamoto M, Maeda M, Tange E, Ueki R, Kawashima K, Hara H, Fujio Y, Azuma J (2006) CYP2A6 polymorphisms are associated with nicotine dependence and influence withdrawal symptoms in smoking cessation. Pharmacogenomics J 6:115–119

Kugaya A, Seneca NM, Snyder PJ, Williams SA, Malison RT, Baldwin RM, Seibyl JP, Innis RB (2003) Changes in human in vivo serotonin and dopamine transporter availabilities during chronic antidepressant administration. Neuropsychopharmacology 28:413–420

Lachman HM, Papolos DF, Saito T, Yu YM, Szumlanski CL, Weinshilboum RM (1996) Human catechol-O-methyltransferase pharmacogenetics: description of a functional polymorphism and its potential application to neuropsychiatric disorders. Pharmacogenetics 6:243–250

Lang T, Klein K, Fischer J, Nussler AK, Neuhaus P, Hofmann U, Eichelbaum M, Schwab M, Zanger UM (2001) Extensive genetic polymorphism in the human CYP2B6 gene with impact on expression and function in human liver. Pharmacogenetics 11:399–415

Laruelle M, Gelernter J, Innis RB (1998) D2 receptors binding potential is not affected by Taq1 polymorphism at the D2 receptor gene. Mol psychiatry 3:261–265

Laucht M, Becker K, El-Faddagh M, Hohm E, Schmidt MH (2005) Association of the DRD4 exon III polymorphism with smoking in fifteen-year-olds: a mediating role for novelty seeking? J Am Acad Child Adolesc Psychiatry 44:477–484

Laucht M, Becker K, Frank J, Schmidt MH, Esser G, Treutlein J, Skowronek MH, Schumann G (2008) Genetic variation in dopamine pathways differentially associated with smoking progression in adolescence. J Am Acad Child Adolesc Psychiatry 47:673–681

Lea RA, Dickson S, Benowitz NL (2006) Within-subject variation of the salivary 3HC/COT ratio in regular daily smokers: prospects for estimating CYP2A6 enzyme activity in large-scale surveys of nicotine metabolic rate. J Anal Toxicol 30:386–389

Learned-Coughlin SM, Bergstrom M, Savitcheva I, Ascher J, Schmith VD, Langstrom B (2003) In vivo activity of bupropion at the human dopamine transporter as measured by positron emission tomography. Biol Psychiatry 54:800–805

Lee AM, Jepson C, Hoffmann E, Epstein L, Hawk LW, Lerman C, Tyndale RF (2007a) CYP2B6 genotype alters abstinence rates in a bupropion smoking cessation trial. Biol Psychiatry 62:635–641

Lee AM, Jepson C, Shields PG, Benowitz N, Lerman C, Tyndale RF (2007b) CYP2B6 genotype does not alter nicotine metabolism, plasma levels, or abstinence with nicotine replacement therapy. Cancer Epidemiol Biomark Prev 16:1312–1314

Lerman C, Shields PG, Audrain J, Main D, Cobb B, Boyd NR, Caporaso N (1998) The role of the serotonin transporter gene in cigarette smoking. Cancer Epidemiol Biomark Prev 7:253–255

Lerman C, Caporaso NE, Audrain J, Main D, Bowman ED, Lockshin B, Boyd NR, Shields PG (1999) Evidence suggesting the role of specific genetic factors in cigarette smoking. Health Psychol 18:14–20

Lerman C, Caporaso NE, Bush A, Zheng YL, Audrain J, Main D, Shields PG (2001) Tryptophan hydroxylase gene variant and smoking behavior. Am J Med Genet 105:518–520

Lerman C, Shields PG, Wileyto EP, Audrain J, Pinto A, Hawk L, Krishnan S, Niaura R, Epstein L (2002) Pharmacogenetic investigation of smoking cessation treatment. Pharmacogenetics 12:627–634

Lerman C, Tyndale R, Patterson F, Wileyto EP, Shields PG, Pinto A, Benowitz N (2006) Nicotine metabolite ratio predicts efficacy of transdermal nicotine for smoking cessation. Clin Pharmacol Ther 79:600–608

Lerman C, Schnoll RA, Hawk LW, Cinciripini P, George TP, Wileyto EP, Swan GE, Benowitz NL, Heitjan DF, Tyndale RF (2015) A randomized placebo-controlled trial to test a genetically-informed biomarker for personalizing treatment for tobacco dependence. Lancet Respir Med (in press)

Lessov-Schlaggar CN, Benowitz NL, Jacob P, Swan GE (2009) Genetic influences on individual differences in nicotine glucuronidation. Twin Res Hum Genet 12:507–513

Leventhal AM, David SP, Brightman M, Strong D, McGeary JE, Brown RA, Lloyd-Richardson EE, Munafo M, Uhl GR, Niaura R (2012) Dopamine D4 receptor gene variation moderates the efficacy of bupropion for smoking cessation. Pharmacogenomics J 12:86–92

Levi M, Dempsey DA, Benowitz NL, Sheiner LB (2007) Population pharmacokinetics of nicotine and its metabolites I. Model development. J Pharmacokinet Pharmacodyn 34:5–21

Ling D, Niu T, Feng Y, Xing H, Xu X (2004) Association between polymorphism of the dopamine transporter gene and early smoking onset: an interaction risk on nicotine dependence. J Hum Genet 49:35–39

Liu JZ, Tozzi F, Waterworth DM, Pillai SG, Muglia P, Middleton L, Berrettini W, Knouff CW, Yuan X, Waeber G, Vollenweider P, Preisig M, Wareham NJ, Zhao JH, Loos RJ, Barroso I, Khaw KT, Grundy S, Barter P, Mahley R, Kesaniemi A, McPherson R, Vincent JB, Strauss J, Kennedy JL, Farmer A, McGuffin P, Day R, Matthews K, Bakke P, Gulsvik A, Lucae S, Ising M, Brueckl T, Horstmann S, Wichmann HE, Rawal R, Dahmen N, Lamina C, Polasek O, Zgaga L, Huffman J, Campbell S, Kooner J, Chambers JC, Burnett MS, Devaney JM, Pichard AD, Kent KM, Satler L, Lindsay JM, Waksman R, Epstein S, Wilson JF, Wild SH, Campbell H, Vitart V, Reilly MP, Li M, Qu L, Wilensky R, Matthai W, Hakonarson HH, Rader DJ, Franke A, Wittig M, Schafer A, Uda M, Terracciano A, Xiao X, Busonero F, Scheet P, Schlessinger D, St Clair D, Rujescu D, Abecasis GR, Grabe HJ, Teumer A, Volzke H, Petersmann A, John U, Rudan I, Hayward C, Wright AF, Kolcic I, Wright BJ, Thompson JR, Balmforth AJ, Hall AS, Samani NJ, Anderson CA, Ahmad T, Mathew CG, Parkes M, Satsangi J, Caulfield M, Munroe PB, Farrall M, Dominiczak A, Worthington J, Thomson W, Eyre S, Barton A, Francks C, Marchini J (2010) Meta-analysis and imputation refines the association of 15q25 with smoking quantity. Nat Genet 42:436–440

Liu T, David SP, Tyndale RF, Wang H, Zhou Q, Ding P, He YH, Yu XQ, Chen W, Crump C, Wen XZ, Chen WQ (2011) Associations of CYP2A6 genotype with smoking behaviors in southern China. Addiction 106:985–994

Liu L, Zhao-Shea R, McIntosh JM, Gardner PD, Tapper AR (2012) Nicotine persistently activates ventral tegmental area dopaminergic neurons via nicotinic acetylcholine receptors containing alpha4 and alpha6 subunits. Mol Pharmacol 81:541–548

Lundstrom K, Turpin MP (1996) Proposed schizophrenia-related gene polymorphism: expression of the Ser9Gly mutant human dopamine D3 receptor with the Semliki Forest virus system. Biochem Biophys Res Commun 225:1068–1072

Malaiyandi V, Sellers EM, Tyndale RF (2005) Implications of CYP2A6 genetic variation for smoking behaviors and nicotine dependence. Clin Pharmacol Ther 77:145–158

Malaiyandi V, Goodz SD, Sellers EM, Tyndale RF (2006a) CYP2A6 genotype, phenotype, and the use of nicotine metabolites as biomarkers during ad libitum smoking. Cancer Epidemiol Biomark Prev 15:1812–1819

Malaiyandi V, Lerman C, Benowitz NL, Jepson C, Patterson F, Tyndale RF (2006b) Impact of CYP2A6 genotype on pretreatment smoking behaviour and nicotine levels from and usage of nicotine replacement therapy. Mol Psychiatry 11:400–409

Mann JJ, Malone KM, Nielsen DA, Goldman D, Erdos J, Gelernter J (1997) Possible association of a polymorphism of the tryptophan hydroxylase gene with suicidal behavior in depressed patients. Am J Psychiatry 154:1451–1453

Manuck SB, Flory JD, Ferrell RE, Dent KM, Mann JJ, Muldoon MF (1999) Aggression and anger-related traits associated with a polymorphism of the tryptophan hydroxylase gene. Biol Psychiatry 45:603–614

Marks MJ (2013) Genetic matters: thirty years of progress using mouse models in nicotinic research. Biochem Pharmacol 86:1105–1113

Marubio LM, Gardier AM, Durier S, David D, Klink R, Arroyo-Jimenez MM, McIntosh JM, Rossi F, Champtiaux N, Zoli M, Changeux JP (2003) Effects of nicotine in the dopaminergic system of mice lacking the alpha4 subunit of neuronal nicotinic acetylcholine receptors. Eur J Neurosci 17:1329–1337

Mathews TA, Fedele DE, Coppelli FM, Avila AM, Murphy DL, Andrews AM (2004) Gene dose-dependent alterations in extraneuronal serotonin but not dopamine in mice with reduced serotonin transporter expression. J Neurosci Methods 140:169–181

McGaugh JL (2004) The amygdala modulates the consolidation of memories of emotionally arousing experiences. Annu Rev Neurosci 27:1–28

McKinney EF, Walton RT, Yudkin P, Fuller A, Haldar NA, Mant D, Murphy M, Welsh KI, Marshall SE (2000) Association between polymorphisms in dopamine metabolic enzymes and tobacco consumption in smokers. Pharmacogenetics 10:483–491

McMorrow MJ, Foxx RM (1983) Nicotine's role in smoking: an analysis of nicotine regulation. Psychol Bull 93:302–327

Meiser J, Weindl D, Hiller K (2013) Complexity of dopamine metabolism. Cell Commun Signal: CCS 11:34

Merens W, Willem Van der Does AJ, Spinhoven P (2007) The effects of serotonin manipulations on emotional information processing and mood. J Affect Disord 103:43–62

Messina ES, Tyndale RF, Sellers EM (1997) A major role for CYP2A6 in nicotine C-oxidation by human liver microsomes. J Pharmacol Exp Ther 282:1608–1614

Miksys S, Lerman C, Shields PG, Mash DC, Tyndale RF (2003) Smoking, alcoholism and genetic polymorphisms alter CYP2B6 levels in human brain. Neuropharmacology 45:122–132

Mitchell SH (2004) Measuring impulsivity and modeling its association with cigarette smoking. Behav Cogn Neurosci Rev 3:261–275

Mizuno S, Ito H, Hamajima N, Tamakoshi A, Hirose K, Tajima K (2004) Association between smoking habits and tryptophan hydroxylase gene C218A polymorphism among the Japanese population. J Epidemiol 14:94–99

Mooney ME, Li ZZ, Murphy SE, Pentel PR, Le C, Hatsukami DK (2008) Stability of the nicotine metabolite ratio in ad libitum and reducing smokers. Cancer Epidemiol Biomark Prev 17:1396–1400

Mori A, Maruo Y, Iwai M, Sato H, Takeuchi Y (2005) UDP-glucuronosyltransferase 1A4 polymorphisms in a Japanese population and kinetics of clozapine glucuronidation. Drug Metab Dispos 33:672–675

Munafo M, Clark T, Johnstone E, Murphy M, Walton R (2004) The genetic basis for smoking behavior: a systematic review and meta-analysis. Nicotine Tob Res 6:583–597

Munafo MR, Johnstone EC, Guo B, Murphy MF, Aveyard P (2008) Association of COMT Val108/158Met genotype with smoking cessation. Pharmacogenet Genomics 18:121–128

Munafo MR, Johnstone EC, Murphy MF, Aveyard P (2009a) Lack of association of DRD2 rs1800497 (Taq1A) polymorphism with smoking cessation in a nicotine replacement therapy randomized trial. Nicotine Tob Res 11:404–407

Munafo MR, Timpson NJ, David SP, Ebrahim S, Lawlor DA (2009b) Association of the DRD2 gene Taq1A polymorphism and smoking behavior: a meta-analysis and new data. Nicotine Tob Res 11:64–76

Munafo MR, Johnstone EC, Walther D, Uhl GR, Murphy MF, Aveyard P (2011) CHRNA3 rs1051730 genotype and short-term smoking cessation. Nicotine Tob Res 13:982–988

Munafo MR, Timofeeva MN, Morris RW, Prieto-Merino D, Sattar N, Brennan P, Johnstone EC, Relton C, Johnson PC, Walther D, Whincup PH, Casas JP, Uhl GR, Vineis P, Padmanabhan S, Jefferis BJ, Amuzu A, Riboli E, Upton MN, Aveyard P, Ebrahim S, Hingorani AD, Watt G, Palmer TM, Timpson NJ, Group ES, Davey Smith G (2012) Association between genetic variants on chromosome 15q25 locus and objective measures of tobacco exposure. J Natl Cancer Inst 104:740–748

Mwenifumbo JC, Sellers EM, Tyndale RF (2007) Nicotine metabolism and CYP2A6 activity in a population of black African descent: impact of gender and light smoking. Drug Alcohol Depend 89:24–33

Mwenifumbo JC, Al Koudsi N, Ho MK, Zhou Q, Hoffmann EB, Sellers EM, Tyndale RF (2008) Novel and established CYP2A6 alleles impair in vivo nicotine metabolism in a population of Black African descent. Hum Mutat 29:679–688

Mwenifumbo JC, Zhou Q, Benowitz NL, Sellers EM, Tyndale RF (2010) New CYP2A6 gene deletion and conversion variants in a population of Black African descent. Pharmacogenomics 11:189–198

Nader MA, Morgan D, Gage HD, Nader SH, Calhoun TL, Buchheimer N, Ehrenkaufer R, Mach RH (2006) PET imaging of dopamine D2 receptors during chronic cocaine self-administration in monkeys. Nat Neurosci 9:1050–1056

Nakajima M, Yokoi T (2005) Interindividual variability in nicotine metabolism: C-oxidation and glucuronidation. Drug Metab Pharmacokinet 20:227–235

Nakajima M, Yamamoto T, Nunoya K, Yokoi T, Nagashima K, Inoue K, Funae Y, Shimada N, Kamataki T, Kuroiwa Y (1996) Characterization of CYP2A6 involved in 3′-hydroxylation of cotinine in human liver microsomes. J Pharmacol Exp Ther 277:1010–1015

Nakajima M, Fukami T, Yamanaka H, Higashi E, Sakai H, Yoshida R, Kwon JT, McLeod HL, Yokoi T (2006) Comprehensive evaluation of variability in nicotine metabolism and CYP2A6 polymorphic alleles in four ethnic populations. Clin Pharmacol Ther 80:282–297

Nakamura Y, Koyama K, Matsushima M (1998) VNTR (variable number of tandem repeat) sequences as transcriptional, translational, or functional regulators. J Hum Genet 43:149–152

Nedic G, Nikolac M, Borovecki F, Hajnsek S, Muck-Seler D, Pivac N (2010) Association study of a functional catechol-O-methyltransferase polymorphism and smoking in healthy Caucasian subjects. Neurosci Lett 473:216–219

Neville MJ, Johnstone EC, Walton RT (2004) Identification and characterization of ANKK1: a novel kinase gene closely linked to DRD2 on chromosome band 11q23.1. Hum Mutat 23:540–545

Nielsen DA, Jenkins GL, Stefanisko KM, Jefferson KK, Goldman D (1997) Sequence, splice site and population frequency distribution analyses of the polymorphic human tryptophan hydroxylase intron 7. Brain Res Mol Brain Res 45:145–148

Norregaard J, Tonnesen P, Petersen L (1993) Predictors and reasons for relapse in smoking cessation with nicotine and placebo patches. Prev Med 22:261–271

O'Loughlin J, Paradis G, Kim W, DiFranza J, Meshefedjian G, McMillan-Davey E, Wong S, Hanley J, Tyndale RF (2004) Genetically decreased CYP2A6 and the risk of tobacco dependence: a prospective study of novice smokers. Tob Control 13:422–428

Oak JN, Oldenhof J, Van Tol HH (2000) The dopamine D(4) receptor: one decade of research. Eur J Pharmacol 405:303–327

Omidvar M, Stolk L, Uitterlinden AG, Hofman A, Van Duijn CM, Tiemeier H (2009) The effect of catechol-O-methyltransferase Met/Val functional polymorphism on smoking cessation: retrospective and prospective analyses in a cohort study. Pharmacogenet Genomics 19:45–51

Patterson F, Schnoll RA, Wileyto EP, Pinto A, Epstein LH, Shields PG, Hawk LW, Tyndale RF, Benowitz N, Lerman C (2008) Toward personalized therapy for smoking cessation: a randomized placebo-controlled trial of bupropion. Clin Pharmacol Ther 84:320–325

Phelps EA (2006) Emotion and cognition: insights from studies of the human amygdala. Annu Rev Psychol 57:27–53

Pianezza ML, Sellers EM, Tyndale RF (1998) Nicotine metabolism defect reduces smoking. Nature 393:750

Picciotto MR, Zoli M, Rimondini R, Lena C, Marubio LM, Pich EM, Fuxe K, Changeux JP (1998) Acetylcholine receptors containing the beta2 subunit are involved in the reinforcing properties of nicotine. Nature 391:173–177

Pomerleau OF, Pomerleau CS, Namenek RJ (1998) Early experiences with tobacco among women smokers, ex-smokers, and never-smokers. Addiction 93:595–599

Porras G, Di Matteo V, Fracasso C, Lucas G, De Deurwaerdere P, Caccia S, Esposito E, Spampinato U (2002) 5-HT2A and 5-HT2C/2B receptor subtypes modulate dopamine release induced in vivo by amphetamine and morphine in both the rat nucleus accumbens and striatum. Neuropsychopharmacology 26:311–324

Quaak M, van Schayck CP, Postma DS, Wagena EJ, van Schooten FJ (2012) Genetic variants in the serotonin transporter influence the efficacy of bupropion and nortriptyline in smoking cessation. Addiction 107:178–187

Rao Y, Hoffmann E, Zia M, Bodin L, Zeman M, Sellers EM, Tyndale RF (2000) Duplications and defects in the CYP2A6 gene: identification, genotyping, and in vivo effects on smoking. Mol Pharmacol 58:747–755

Ribeiro EB, Bettiker RL, Bogdanov M, Wurtman RJ (1993) Effects of systemic nicotine on serotonin release in rat brain. Brain Res 621:311–318

Ring HZ, Valdes AM, Nishita DM, Prasad S, Jacob P 3rd, Tyndale RF, Swan GE, Benowitz NL (2007) Gene-gene interactions between CYP2B6 and CYP2A6 in nicotine metabolism. Pharmacogenet Genomics 17:1007–1015

Rubinstein ML, Benowitz NL, Auerback GM, Moscicki AB (2008) Rate of nicotine metabolism and withdrawal symptoms in adolescent light smokers. Pediatrics 122:e643–e647

Rubinstein ML, Shiffman S, Rait MA, Benowitz NL (2013) Race, gender, and nicotine metabolism in adolescent smokers. Nicotine Tob Res 15:1311–1315

Sabol SZ, Hu S, Hamer D (1998) A functional polymorphism in the monoamine oxidase A gene promoter. Hum Genet 103:273–279

Scherer G, Engl J, Urban M, Gilch G, Janket D, Riedel K (2007) Relationship between machine-derived smoke yields and biomarkers in cigarette smokers in Germany. Regul Toxicol Pharmacol 47:171–183

Schnoll RA, Patterson F, Wileyto EP, Tyndale RF, Benowitz N, Lerman C (2009) Nicotine metabolic rate predicts successful smoking cessation with transdermal nicotine: a validation study. Pharmacol Biochem Behav 92:6–11

Schnoll RA, George TP, Hawk L, Cinciripini P, Wileyto P, Tyndale RF (2014) The relationship between the nicotine metabolite ratio and three self-report measures of nicotine dependence across sex and race. Psychopharmacology

Schoedel KA, Hoffmann EB, Rao Y, Sellers EM, Tyndale RF (2004) Ethnic variation in CYP2A6 and association of genetically slow nicotine metabolism and smoking in adult Caucasians. Pharmacogenetics 14:615–626

Sekiguchi F, Ando D, Yamakita M, Yamagata Z (2012) An association between the serotonin transporter gene promoter polymorphism and smoking cessation among Japanese males. Asia Pac J Public Health 24:288–295

Shields PG, Lerman C, Audrain J, Bowman ED, Main D, Boyd NR, Caporaso NE (1998) Dopamine D4 receptors and the risk of cigarette smoking in African-Americans and Caucasians. Cancer Epidemiol Biomark Prev 7:453–458

Sieminska A, Buczkowski K, Jassem E, Tkacz E (2008) Lack of association between serotonin transporter gene polymorphism 5-HTTLPR and smoking among Polish population: a case-control study. BMC Med Genet 9:76

Sieminska A, Buczkowski K, Jassem E, Niedoszytko M, Tkacz E (2009) Influences of polymorphic variants of DRD2 and SLC6A3 genes, and their combinations on smoking in Polish population. BMC Med Genet 10:92

Silverman MA, Neale MC, Sullivan PF, Harris-Kerr C, Wormley B, Sadek H, Ma Y, Kendler KS, Straub RE (2000) Haplotypes of four novel single nucleotide polymorphisms in the nicotinic acetylcholine receptor beta2-subunit (CHRNB2) gene show no association with smoking initiation or nicotine dependence. Am J Med Genet 96:646–653

Sofuoglu M, Herman AI, Nadim H, Jatlow P (2012) Rapid nicotine clearance is associated with greater reward and heart rate increases from intravenous nicotine. Neuropsychopharmacology 37:1509–1516

St Helen G, Novalen M, Heitjan DF, Dempsey D, Jacob P 3rd, Aziziyeh A, Wing VC, George TP, Tyndale RF, Benowitz NL (2012) Reproducibility of the nicotine metabolite ratio in cigarette smokers. Cancer Epidemiol Biomark Prev 21:1105–1114

St Helen G, Jacob P 3rd, Benowitz NL (2013) Stability of the nicotine metabolite ratio in smokers of progressively reduced nicotine content cigarettes. Nicotine Tob Res 15:1939–1942

Stapleton JA, Sutherland G, O'Gara C (2007) Association between dopamine transporter genotypes and smoking cessation: a meta-analysis. Addict Biol 12:221–226

Stapleton JA, Sutherland G, O'Gara C, Spirling LI, Ball D (2011) Association between DRD2/ANKK1 Taq1A genotypes, depression and smoking cessation with nicotine replacement therapy. Pharmacogenet Genomics 21:447–453

Su T, Bao Z, Zhang QY, Smith TJ, Hong JY, Ding X (2000) Human cytochrome P450 CYP2A13: predominant expression in the respiratory tract and its high efficiency metabolic activation of a tobacco-specific carcinogen, 4-(methylnitrosamino)-1-(3-pyridyl)-1-butanone. Cancer Res 60:5074–5079

Sullivan PF, Jiang Y, Neale MC, Kendler KS, Straub RE (2001) Association of the tryptophan hydroxylase gene with smoking initiation but not progression to nicotine dependence. Am J Med Genet 105:479–484

Suriyaprom K, Tungtrongchitr R, Harnroongroj T (2013) Impact of COMT Val 108/158 Met and DRD2 Taq1B gene polymorphisms on vulnerability to cigarette smoking of Thai males. J Mol Neurosci 49:544–549

Taioli E, Wynder EL (1991) Effect of the age at which smoking begins on frequency of smoking in adulthood. N Engl J Med 325:968–969

Takami K, Saito H, Okuda M, Takano M, Inui KI (1998) Distinct characteristics of transcellular transport between nicotine and tetraethylammonium in LLC-PK1 cells. J Pharmacol Exp Ther 286:676–680

Tang DW, Hello B, Mroziewicz M, Fellows LK, Tyndale RF, Dagher A (2012) Genetic variation in CYP2A6 predicts neural reactivity to smoking cues as measured using fMRI. NeuroImage 60:2136–2143

Tapper AR, McKinney SL, Nashmi R, Schwarz J, Deshpande P, Labarca C, Whiteaker P, Marks MJ, Collins AC, Lester HA (2004) Nicotine activation of alpha4* receptors: sufficient for reward, tolerance, and sensitization. Science 306:1029–1032

Thorgeirsson TE, Geller F, Sulem P, Rafnar T, Wiste A, Magnusson KP, Manolescu A, Thorleifsson G, Stefansson H, Ingason A, Stacey SN, Bergthorsson JT, Thorlacius S, Gudmundsson J, Jonsson T, Jakobsdottir M, Saemundsdottir J, Olafsdottir O, Gudmundsson LJ, Bjornsdottir G, Kristjansson K, Skuladottir H, Isaksson HJ, Gudbjartsson T, Jones GT, Mueller T, Gottsater A, Flex A, Aben KK, de Vegt F, Mulders PF, Isla D, Vidal MJ, Asin L, Saez B, Murillo L, Blondal T, Kolbeinsson H, Stefansson JG, Hansdottir I, Runarsdottir V, Pola R, Lindblad B, van Rij AM, Dieplinger B, Haltmayer M, Mayordomo JI, Kiemeney LA, Matthiasson SE, Oskarsson H, Tyrfingsson T, Gudbjartsson DF, Gulcher JR, Jonsson S, Thorsteinsdottir U, Kong A, Stefansson K (2008) A variant associated with nicotine dependence, lung cancer and peripheral arterial disease. Nature 452:638–642

Tobacco and Genetics Consortium (2010) Genome-wide meta-analyses identify multiple loci associated with smoking behavior. Nat Genet 42:441–447

Tochigi M, Suzuki K, Kato C, Otowa T, Hibino H, Umekage T, Kato N, Sasaki T (2007) Association study of monoamine oxidase and catechol-O-methyltransferase genes with smoking behavior. Pharmacogenet Genomics 17:867–872

Trifilieff P, Feng B, Urizar E, Winiger V, Ward RD, Taylor KM, Martinez D, Moore H, Balsam PD, Simpson EH, Javitch JA (2013) Increasing dopamine D2 receptor expression in the adult nucleus accumbens enhances motivation. Mol Psychiatry 18:1025–1033

True WR, Heath AC, Scherrer JF, Waterman B, Goldberg J, Lin N, Eisen SA, Lyons MJ, Tsuang MT (1997) Genetic and environmental contributions to smoking. Addiction 92:1277–1287

Trummer O, Koppel H, Wascher TC, Grunbacher G, Gutjahr M, Stanger O, Ramschak-Schwarzer S, Boehm BO, Winkelmann BR, Marz W, Renner W (2006) The serotonin transporter gene polymorphism is not associated with smoking behavior. Pharmacogenomics J 6:397–400

Tsai SJ, Hong CJ, Yu YW, Chen TJ (2004) Association study of catechol-O-methyltransferase gene and dopamine D4 receptor gene polymorphisms and personality traits in healthy young chinese females. Neuropsychobiology 50:153–156

Tyndale RF, Sellers EM (2002) Genetic variation in CYP2A6-mediated nicotine metabolism alters smoking behavior. Ther Drug Monit 24:163–171

Urakami Y, Akazawa M, Saito H, Okuda M, Inui K (2002) cDNA cloning, functional characterization, and tissue distribution of an alternatively spliced variant of organic cation transporter hOCT2 predominantly expressed in the human kidney. J Am Soc Nephrol 13:1703–1710

Ursin R (2002) Serotonin and sleep. Sleep Med Rev 6:55–69

van Dyck CH, Malison RT, Jacobsen LK, Seibyl JP, Staley JK, Laruelle M, Baldwin RM, Innis RB, Gelernter J (2005) Increased dopamine transporter availability associated with the 9-repeat allele of the SLC6A3 gene. J Nucl Med 46:745–751

Vandenbergh DJ, Zonderman AB, Wang J, Uhl GR, Costa PT Jr (1997) No association between novelty seeking and dopamine D4 receptor (D4DR) exon III seven repeat alleles in Baltimore Longitudinal Study of Aging participants. Mol Psychiatry 2:417–419

Vandenbergh DJ, Bennett CJ, Grant MD, Strasser AA, O'Connor R, Stauffer RL, Vogler GP, Kozlowski LT (2002) Smoking status and the human dopamine transporter variable number of tandem repeats (VNTR) polymorphism: failure to replicate and finding that never-smokers may be different. Nicotine Tob Res 4:333–340

Vandenbergh DJ, O'Connor RJ, Grant MD, Jefferson AL, Vogler GP, Strasser AA, Kozlowski LT (2007) Dopamine receptor genes (DRD2, DRD3 and DRD4) and gene-gene interactions associated with smoking-related behaviors. Addict Biol 12:106–116

VanNess SH, Owens MJ, Kilts CD (2005) The variable number of tandem repeats element in DAT1 regulates in vitro dopamine transporter density. BMC Genet 6:55

Vink JM, Willemsen G, Boomsma DI (2005) Heritability of smoking initiation and nicotine dependence. Behav Genet 35:397–406

Volkow ND, Wang GJ, Fowler JS, Tomasi D, Telang F (2011) Addiction: beyond dopamine reward circuitry. Proc Natl Acad Sci USA 108:15037–15042

Wada E, McKinnon D, Heinemann S, Patrick J, Swanson LW (1990) The distribution of mRNA encoded by a new member of the neuronal nicotinic acetylcholine receptor gene family (alpha 5) in the rat central nervous system. Brain Res 526:45–53

Wang H, Tan W, Hao B, Miao X, Zhou G, He F, Lin D (2003) Substantial reduction in risk of lung adenocarcinoma associated with genetic polymorphism in CYP2A13, the most active cytochrome P450 for the metabolic activation of tobacco-specific carcinogen NNK. Cancer Res 63:8057–8061

Warner C, Shoaib M (2005) How does bupropion work as a smoking cessation aid? Addict Biol 10:219–231

Wassenaar CA, Dong Q, Wei Q, Amos CI, Spitz MR, Tyndale RF (2011) Relationship between CYP2A6 and CHRNA5-CHRNA3-CHRNB4 variation and smoking behaviors and lung cancer risk. J Natl Cancer Inst 103:1342–1346

Wellman RJ, Savageau JA, Godiwala S, Savageau N, Friedman K, Hazelton J, Difranza JR (2006) A comparison of the hooked on nicotine checklist and the fagerstrom test for nicotine dependence in adult smokers. Nicotine Tob Res 8:575–580

Wiesbeck GA, Wodarz N, Weijers HG, Dursteler-MacFarland KM, Wurst FM, Walter M, Boening J (2006) A functional polymorphism in the promoter region of the monoamine oxidase A gene is associated with the cigarette smoking quantity in alcohol-dependent heavy smokers. Neuropsychobiology 53:181–185

Wyen C, Hendra H, Vogel M, Hoffmann C, Knechten H, Brockmeyer NH, Bogner JR, Rockstroh J, Esser S, Jaeger H, Harrer T, Mauss S, van Lunzen J, Skoetz N, Jetter A, Groneuer C, Fatkenheuer G, Khoo SH, Egan D, Back DJ, Owen A, German Competence Network for HA (2008) Impact of CYP2B6 983T>C polymorphism on non-nucleoside reverse transcriptase inhibitor plasma concentrations in HIV-infected patients. J Antimicrob Chemother 61:914–918

Xian H, Scherrer JF, Madden PA, Lyons MJ, Tsuang M, True WR, Eisen SA (2003) The heritability of failed smoking cessation and nicotine withdrawal in twins who smoked and attempted to quit. Nicotine Tob Res 5:245–254

Yamanaka H, Nakajima M, Nishimura K, Yoshida R, Fukami T, Katoh M, Yokoi T (2004) Metabolic profile of nicotine in subjects whose CYP2A6 gene is deleted. Eur J Pharm Sci 22:419–425

Yamazaki H, Inoue K, Hashimoto M, Shimada T (1999) Roles of CYP2A6 and CYP2B6 in nicotine C-oxidation by human liver microsomes. Arch Toxicol 73:65–70

Yoon H, Cho HY, Yoo HD, Kim SM, Lee YB (2013) Influences of organic cation transporter polymorphisms on the population pharmacokinetics of metformin in healthy subjects. AAPS J 15:571–580

Young SN, Leyton M (2002) The role of serotonin in human mood and social interaction. Insight from altered tryptophan levels. Pharmacol Biochem Behav 71:857–865

Yudkin P, Munafo M, Hey K, Roberts S, Welch S, Johnstone E, Murphy M, Griffiths S, Walton R (2004) Effectiveness of nicotine patches in relation to genotype in women versus men: randomised controlled trial. BMJ 328:989–990

Zevin S, Schaner ME, Giacomini KM (1998) Nicotine transport in a human choriocarcinoma cell line (JAR). J Pharm Sci 87:702–706

Zhou X, Nonnemaker J, Sherrill B, Gilsenan AW, Coste F, West R (2009) Attempts to quit smoking and relapse: factors associated with success or failure from the ATTEMPT cohort study. Addict Behav 34:365–373

Zhu AZ, Cox LS, Nollen N, Faseru B, Okuyemi KS, Ahluwalia JS, Benowitz NL, Tyndale RF (2012) CYP2B6 and bupropion's smoking-cessation pharmacology: the role of hydroxybupropion. Clin Pharmacol Ther 92:771–777

Zhu AZ, Binnington MJ, Renner CC, Lanier AP, Hatsukami DK, Stepanov I, Watson CH, Sosnoff CS, Benowitz NL, Tyndale RF (2013a) Alaska Native smokers and smokeless tobacco users with slower CYP2A6 activity have lower tobacco consumption, lower tobacco-specific nitrosamine exposure and lower tobacco-specific nitrosamine bioactivation. Carcinogenesis 34:93–101

Zhu AZ, Renner CC, Hatsukami DK, Swan GE, Lerman C, Benowitz NL, Tyndale RF (2013b) The ability of plasma cotinine to predict nicotine and carcinogen exposure is altered by differences in CYP2A6: the influence of genetics, race, and sex. Cancer Epidemiol Biomark Prev 22:708–718

Zhu AZ, Zhou Q, Cox LS, Ahluwalia JS, Benowitz NL, Tyndale RF (2013c) Variation in trans-3′-hydroxycotinine glucuronidation does not alter the nicotine metabolite ratio or nicotine intake. PLoS ONE 8:e70938

Zhu AZ, Zhou Q, Cox LS, David SP, Ahluwalia JS, Benowitz NL, Tyndale RF (2014) Association of CHRNA5-A3-B4 SNP rs2036527 with smoking cessation therapy response in African-American smokers. Clin Pharmacol Ther 96:256–265

Heterogeneity Across Brain Regions and Neurotransmitter Interactions with Nicotinic Effects on Memory Function

Edward D. Levin, Brandon J. Hall and Amir H. Rezvani

Abstract Nicotinic acetylcholine receptors have been shown in many studies to be critically involved in memory function. The precise roles these receptors play depend on the receptor subtype, their anatomic localization, their interactions with other parts of the neural systems underlying cognition and the particular domain of cognitive function. Nicotinic agonists can significantly improve learning, memory, and attention. Nicotinic receptors in the hippocampus are innervated by cholinergic projections from the medial septum and diagonal band. Local infusions of either α7 or α4β2 nicotinic antagonists into either the dorsal or ventral hippocampus produce amnestic effects in rats navigating about a radial arm maze. There is cholinergic innervation of nicotinic receptors in other components of the limbic system as well. In the basolateral amygdala and the anterior thalamus, similar amnestic effects of nicotinic α7 and α4β2 antagonists are seen. Interestingly, there are no additive amnestic effects observed in these limbic areas when α7 and α4β2 receptor antagonists are combined. The particular expression patterns of α7 and α4β2 nicotinic receptors in these limbic and cortical areas may explain this nonadditivity, but further research is needed to determine the specific cause of this phenomenon. Nicotinic receptor mechanisms in the limbic system play an important role in cognitive impairment for a variety of neurological disorders, including Alzheimer's disease and schizophrenia. Alzheimer's disease results in a dramatic decrease in hippocampal nicotinic receptor density, affecting α4β2 receptor expression most prominently. In schizophrenia, there are anomalies in α7 nicotinic receptor expression, which seem to be crucial for the cognitive impairment of the disorder. Chronic nicotine exposure, such as seen with tobacco use, results in an increase in nicotinic receptor density in the limbic system. This effect appears to be related to the desensitization of nicotinic receptors seen after agonist application. Open questions remain concerning the role of desensitization versus activation of nicotinic receptors in cognitive improvement.

E.D. Levin (✉) · B.J. Hall · A.H. Rezvani
Department of Psychiatry and Behavioral Sciences, Duke University Medical Center, Box #104790, Durham, NC 27710, USA
e-mail: edlevin@duke.edu

D.J.K. Balfour and M.R. Munafò (eds.), *The Neurobiology and Genetics of Nicotine and Tobacco*, Current Topics in Behavioral Neurosciences 23,
DOI 10.1007/978-3-319-13665-3_4

Keywords Nicotinic · Memory · Hippocampus · Amygdala · Thalamus · Serotonergic · Dopaminergic · Glutaminergic

Contents

1 Introduction 88
2 Activation and Desensitization of Nicotinic Receptors 89
3 Nicotinic Receptors in Diverse Brain Regions Involved in Memory 90
3.1 Limbic System 90
3.2 Frontal Cortex 91
3.3 Thalamus 92
3.4 Brainstem 93
4 Nicotinic Interactions with Other Neurotransmitter Receptor Systems and Memory 94
4.1 Acetylcholine 94
4.2 Dopamine 94
4.3 Serotonin 95
4.4 Glutamate and GABA 95
5 Importance of Complex Interactions for Nicotinic Therapeutics 96
5.1 Alzheimer's Disease and Related Cognitive Impairments of Aging 96
5.2 Schizophrenia 96
5.3 Attention Deficit Hyperactivity Disorder (ADHD) 97
6 Conclusions and Needs for Further Study 97
References 98

1 Introduction

The brain is a heterogeneous organ of communication with a rich diversity of neuronal and glial cell types and interconnections. These complex systems provide robust and widely adaptable behavioral function. The same type of physiological component, whether it is a cell, an enzyme, or a receptor, performs the same immediate physiological function in the brain. However, in the context of the larger integration of brain activity, this immediate biological function may produce very different behavioral outcomes. This is certainly true for nicotinic receptors and cognitive function. For example, the α4β2 nicotinic receptor subtype located in the hippocampus (see chapter entitled Nicotinic Receptors, Memory, and Hippocampus; this volume) plays a very different role in memory function than the same receptor subtype located in the mediodorsal thalamic nucleus (see below for details). It is therefore not surprising that systemic administration of even the most selective drugs targeting nicotinic systems would have complex effects; they act like endocrine hormones being carried throughout the brain and the rest of the body, generating neural activity everywhere they encounter their target receptor. It is important to understand the diversity of nicotinic receptor actions in different brain areas and the differential interactions of nicotinic cholinergic signaling with related neurotransmitter systems in the brain systems involved with cognitive function to properly appreciate the complex mechanisms of nicotinic involvement in cognition. This will aid in understand how these systems go awry with disease and cause cognitive dysfunction and how we can most expeditiously develop effective nicotinic therapeutics.

Nicotine and other nicotinic ligands have been shown to improve memory function and to effectively reverse memory dysfunction. However, the precise neural mechanisms for these effects remain to be elucidated. In the process of drug development for cognitive dysfunction, nicotinic receptor subtype involvement has received the principal attention, since this is how drugs can be specifically targeted. However, the extended effects of receptor activation, their diverse anatomic localization, and the complex interactions with other components of neural circuits they may have are also quite important. Over the past 25 years, we have investigated the complex anatomic and pharmacologic interactions of nicotinic systems and memory function. This chapter reviews this work.

2 Activation and Desensitization of Nicotinic Receptors

Many neuronal nicotinic receptors desensitize very rapidly after stimulation. Thus, nicotinic agonists also serve as net antagonists through their potential to desensitize nicotinic receptors. The relative roles of nicotinic receptor activation, desensitization, and blockade in cognitive function have more recently received research attention (Buccafusco et al. 2009; Picciotto et al. 2008). Discerning the relative effects of nicotinic receptor activation versus inactivation is complicated by the fact that nicotinic ligands, like all drugs causing cognitive improvement, have a non-monotonic dose-effect function, where low to moderate doses cause improvement and higher doses do not and can even impair cognition. With this inverted J-shaped dose-effect function, the reversal of nicotinic agonist-induced cognitive improvement with antagonist coadministration does not provide proof that it is the stimulatory effect of the agonist that underlies the cognitive improvement. Given that agonists also desensitize nicotinic receptors and that typically the maximally effective agonist dose is tested with antagonist coadministration, the antagonist itself may reverse the agonist-induced cognitive improvement by either blocking the stimulatory effect of the agonist or by adding to the receptor inactivating effect of desensitization to a degree that exceeds the maximally effective degree of inactivation.

We have found that acute administration of the $\alpha4\beta2$ nicotinic desensitizing agent sazetidine-A significantly attenuates the attentional impairments caused by administration of either the muscarinic cholinergic antagonist scopolamine or the NMDA glutamate antagonist dizocilpine (MK-801) (Rezvani et al. 2011). Chronic infusion of sazetidine-A also significantly attenuates dizocilpine-induced attentional impairment (Rezvani et al. 2012). While it is true that sazetidine-A does have a short-lasting partial agonist effect, the attentional improvement it causes appears likely to be related to its prolonged nicotinic $\alpha4\beta2$ receptor desensitizing effect, inasmuch as a very similar reversal of dizocilpine-induced attentional impairment is seen after administration of the α4β2 selective antagonist Dihydro-β-erythroidine (DHβE) (Levin et al. 2013). The relative significance of desensitization of the receptor versus pure agonist effect regarding memory function remains to be fully investigated. However, we have found that low doses of the nonselective nicotinic receptor antagonist mecamylamine given systemically can improve working memory performance in rats using the radial arm maze; evidence which provides support for the former mechanism (Levin et al. 1993).

3 Nicotinic Receptors in Diverse Brain Regions Involved in Memory

Nicotinic receptors in different brain areas play different roles in neurobehavioral function in general and cognition in particular (see Chaps. Nicotinic Receptors and Attention, Nicotinic Receptors, Memory, and Hippocampus, and Smoking Abstinence and Neurocognition: Implications for Cessation and Relapse; this volume). The infusion of nicotinic antagonists into specific brain regions in animals performing cognitive and behavioral tasks can be used to discern the specific roles that nicotinic receptors play in various aspects of cognitive function. The most thoroughly investigated cognitive domain is working memory.

3.1 Limbic System

The hippocampus is the brain structure that has been most extensively characterized to be critically involved in memory function (Aggleton and Pearce 2001). In particular, cholinergic innervation from the medial septum and diagonal band to the dentate gyrus of the hippocampus has been shown to be significantly important for memory function. Both nicotinic and muscarinic cholinergic receptors in the hippocampus are involved in memory function (Bymaster et al. 1993). Within the nicotinic family α7 and α4β2 nicotinic receptors in the hippocampus are critically involved in memory function. Local brain infusions of the α7 receptor selective antagonist methyllycaconitine (MLA) and the α4β2 antagonist DHβE into either the ventral or dorsal hippocampus significantly impair working memory of rats in the radial arm maze (Fig. 1) (Arthur and Levin 2002; Felix and Levin 1997; Levin et al. 2002). Interestingly, hippocampal infusions of a combination of MLA and DHβE did not produce an additive effect. This may be due to the specific cellular localization of α7 and α4β2 nicotinic receptors as well as the particular expression patterns in the local circuits of the hippocampus. Further research is needed to elucidate the specifics of this disparity. Systemic nicotine treatment effectively reverses the memory impairment caused by infusion of DHβE into the ventral hippocampus (Bancroft and Levin 2000). In contrast, the same systemic nicotine

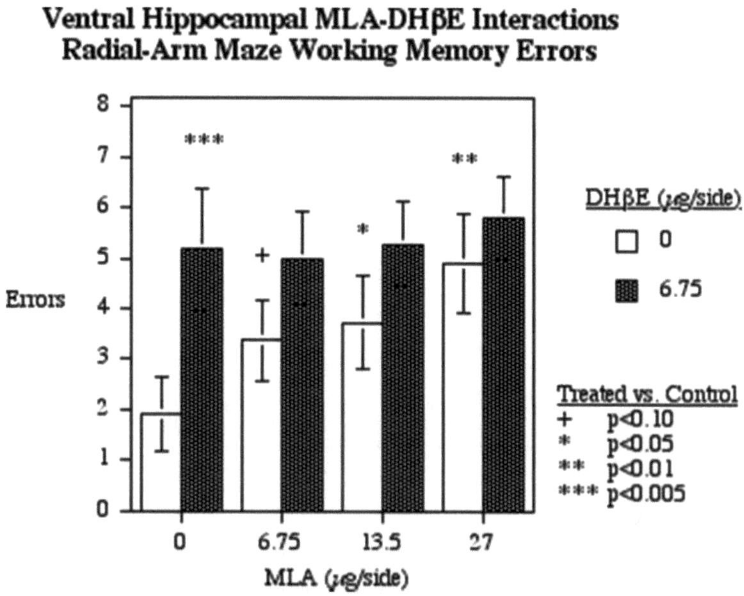

Fig. 1 Ventral hippocampal infusions of methyllycaconitine and dihydro-β-erythoidine to block α7 and α4β2 nicotinic receptors, effects on working memory errors in the 16-arm radial maze in rats (Levin et al. 2002)

dose has not been found to reverse working memory impairment caused by ventral hippocampal infusion of MLA, perhaps highlighting the particular importance of α7 receptors on working memory function in this brain region (Bettany and Levin 2001). Systemic clozapine administration, a treatment that by itself has amnestic effects, was shown to significantly reduce the memory impairment in rats caused by hippocampal infusions of DHβE (Pocivavsek et al. 2006).

The amygdala is another prominent component of the limbic system. It is classically thought of as being central to the processing of emotional function. Local infusions of MLA and DHβE into the basolateral amygdala (BLA) each caused significant impairments in working memory function in rats in the radial arm maze (Fig. 2). Interestingly, in the case of the BLA, the combination of the α4β2 and α7 antagonists diminished each other's effects (Addy et al. 2003).

3.2 Frontal Cortex

Frontal cortical involvement in memory has been the focus of abundant studies, which have demonstrated this region's involvement in maintaining accurate memory function (Steckler et al. 1998). Involvement of nicotinic receptors in the frontal cortex in memory function was tested using similar local infusion studies as used previously with nicotinic antagonists administered into the limbic system.

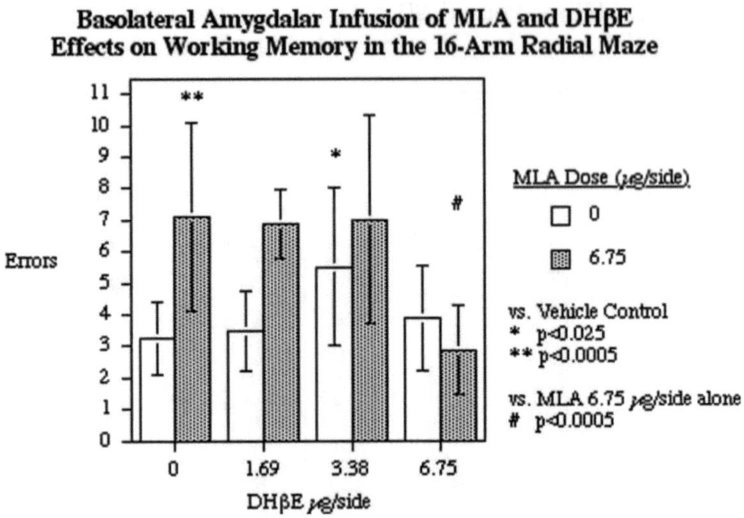

Fig. 2 Basolateral amygdalar infusions of methyllycaconitine and dihydro-β-erythoidine to block α7 and α4β2 nicotinic receptors, effects on working memory errors in the 16-arm radial maze in rats (Addy et al. 2003)

Local infusions of the α7 and α4β2 nicotinic antagonists MLA and DHβE at the same doses that were effective in impairing working memory with hippocampal or amygdalar infusion did not by themselves affect memory performance when infused into the medial frontal cortex (Levin et al. 2009). Local infusions of DHβE in the medial frontal cortex did significantly potentiate the amnestic effect of systemic clozapine (Levin et al. 2009). This is the opposite interaction as seen with the amnestic effect of DHβE that was significantly reduced by systemic administration of clozapine (Pocivavsek et al. 2006). The different roles that the frontal cortex and hippocampus play in the neural circuits underlying memory function likely account for the differential effects on memory of the same nicotinic receptor subtype blockade in these two areas.

3.3 Thalamus

The thalamus is a crossroads of connections with diverse parts of the neocortex, limbic system, and brainstem. Interestingly, acute local infusions of DHβE into the mediodorsal thalamic nucleus significantly improves working memory function as measured by performance in the radial arm maze (Fig. 3) (Cannady et al. 2009). This memory improvement was reversed by co-infusion of the α7 antagonist MLA, which when infused by itself into the mediodorsal thalamic nucleus was not found to impact memory function. Chronic infusions of DHβE into the mediodorsal thalamic nucleus, much like acute infusions, was found to significantly improve

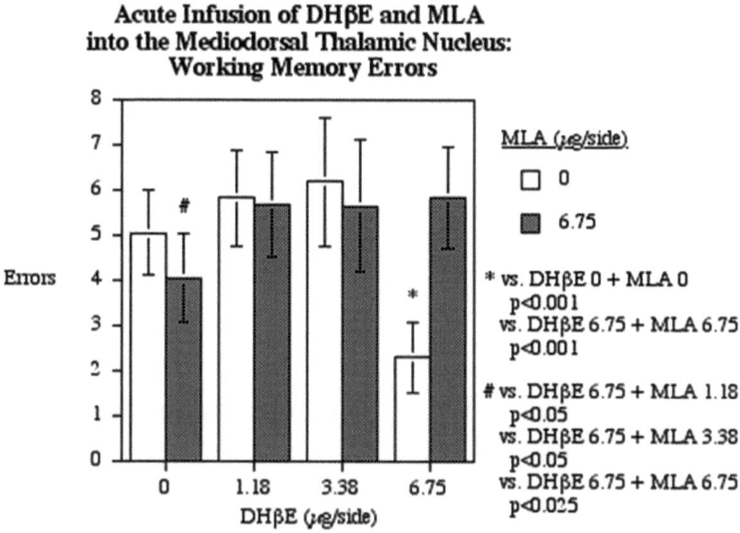

Fig. 3 Mediodorsal thalamic nucleus infusions of methyllycaconitine and dihydro-β-erythoidine to block α7 and α4β2 nicotinic receptors, effects on working memory errors in the 16-arm radial maze in rats (Cannady et al. 2009)

working memory function. This improvement was reversed by chronic systemic infusions of nicotine. It may be the case that the dose of nicotine infused was able to overcome the competitive antagonist DHβE or that nicotine-induced desensitization in combination with mediodorsal thalamic DHβE reduced receptor activity below the level optimal for memory function.

The habenula is an epithalamic structure, located on the mediodorsal surface of the thalamus, which serves to relay connections from the telencephalon to the brainstem (Sutherland 1980). The habenula contains among the highest density of nicotinic receptor concentration of any area in the brain (Clarke et al. 1985). In addition, the nicotinic receptors in the habenula are of a great variety of subtypes (Mulle et al. 1991). Chronic habenular infusions of mecamylamine, a nonselective nicotinic receptor channel blocker, significantly impair spatial memory in the radial arm maze (Sanders et al. 2010). This impairment was reversed by acute systemic administration of nicotine. Given that mecamylamine is a noncompetitive antagonist at nicotinic receptors, the reversal effect of acute nicotine in this case was likely the result of activity in a different brain area than the habenula.

3.4 Brainstem

From the brainstem arise the monoaminergic (dopamine, norepinephrine, and serotonin) neurons that project broadly to the telencephalon. Nicotinic interactions are particularly relevant for the dopaminergic nuclei, the substantia nigra, and ventral tegmental area, due to the fact that these areas are directly innervated by cholinergic projections from the pedunculopontine and dorsolateral tegmental nuclei with nicotinic receptors playing key roles in these connections. Infusion of the nonselective nicotinic antagonist mecamylamine into the ventral tegmental area (Levin et al. 1994) or substantia nigra (Levin et al. 1994) significantly impaired working memory of rats in the radial arm maze. The cholinergic projection to these areas, particularly those involving nicotinic receptors, may be key parts of the circuits underlying nicotinic dopaminergic interactions and cognitive function. However, nicotinic interactions with dopamine innervation of the hippocampus may be another key area for this interaction (see below).

4 Nicotinic Interactions with Other Neurotransmitter Receptor Systems and Memory

4.1 Acetylcholine

Nicotinic receptors are only one of the two main receptor subtypes in cholinergic receptor family. The other main type of cholinergic receptor is the muscarinic receptor. The prototypic nonselective muscarinic antagonist is scopolamine. Scopolamine is a classic amnestic drug. Low doses of scopolamine and mecamylamine have mutually augmenting effects impairing working memory function (Levin et al. 1989b). The $\alpha 4\beta 2$ nicotinic desensitizing agent sazetidine-A significantly reversed scopolamine-induced attentional impairment in rats (Rezvani et al. 2011). The effects of sazetidine-A on memory function remain to be tested.

4.2 Dopamine

D_1 and D_2 dopamine receptors appear to have important interactions with nicotinic effects on memory function (Levin and Rose 1995). Nicotinic interactions with D_1 receptors appear to be more relevant to reference memory function (Levin et al. 1996b). Chronic systemic nicotine infusions blocked the reference memory improvement seen in rats given the D_1 antagonist SCH-23390 and potentiated the memory impairment caused by the D_1 agonist dihydrexidine in the 16-arm radial maze. No such interactions were seen with working memory performance (Levin et al. 1996b).

D_2 interactions with nicotinic receptor systems appear to be more closely related to working memory. The D_2 agonist quinpirole given systemically reversed the working memory impairment in the radial arm maze caused by high dose application of the general nicotinic antagonist mecamylamine (Levin et al. 1989a). Also, the finding that the D_2 antagonists raclopride and haloperidol potentiated working memory impairment with a subthreshold dose of mecamylamine supports the interaction of the nicotinic receptor system with D_2 receptors (McGurk et al. 1989a, b). The hippocampus might be an important site for these D_2/nicotinic interactions regarding working memory. We found that local hippocampal infusions of the D_2 antagonist raclopride impaired working memory in the radial arm maze and that local hippocampal infusion of the D_2 agonist quinpirole improves memory (Wilkerson and Levin 1999). We have also found that local hippocampal infusion of quinpirole significantly attenuates the working memory impairment caused by hippocampal infusion of the $\alpha4\beta2$ nicotinic antagonist DHβE (unpublished data).

4.3 Serotonin

Serotonergic systems in the brain also interact with nicotinic systems involved in memory function. Nicotine-induced improvement in working memory is blocked by coadministration of the serotonin $5HT_2$ antagonist ketanserin (Levin et al. 2005). The same is true with attentional function. Ketanserin blocked the nicotine-induced reversal of dizocilpine-induced attentional impairment (Rezvani et al. 2005). This $5HT_2$ interaction with nicotinic effects may also explain why clozapine also blocks nicotine reversal of dizocilpine-induced attentional impairment given the substantial $5HT_2$ antagonist effects of clozapine (Rezvani et al. 2008).

4.4 Glutamate and GABA

The most widespread excitatory and inhibitory neurotransmitter systems glutamate and GABA have been shown to interact with nicotinic effects on memory function. Blockade of NMDA glutamate receptors with dizocilpine (MK-801) causes impairments in working and reference memory in the 16-arm radial maze (Levin et al. 1998; Timofeeva and Levin 2008). These impairments are largely reversed by coadministration of nicotine. Local infusions of dizocilpine into the ventral hippocampus at doses that by themselves do not cause memory impairment reversed the effect of systemic nicotine from improving memory function to impairing it (Levin et al. 2003). The same dose range of dizocilpine infused into the amygdala significantly impaired memory function, an effect that was reversed by systemic nicotine (May-Simera and Levin 2003). The noradrenergic alpha2 receptor antagonist idazoxan blocked the nicotine-induced reversal of dizocilpine-induced memory impairment in rats (Timofeeva and Levin 2008).

Interactions between GABA and nicotinic systems are important as well. Acute nicotine administration reversed the working memory impairment caused by higher dose (1 mg/kg) administration of baclofen, a GABA-B receptor agonist. Interestingly, a lower (0.25 mg/kg) dose administration of baclofen caused a significant memory improvement. This was not additive with the nicotine-induced improvement. In fact, the low-dose baclofen-induced improvement blocked the nicotine-induced improvement in memory (Levin et al. 2004).

5 Importance of Complex Interactions for Nicotinic Therapeutics

Drugs acting at nicotinic receptors including nicotine itself have been tested as treatments for cognitive impairments, including those seen in Alzheimer's disease, schizophrenia, and attention deficit hyperactivity disorder (ADHD). Because these drug treatments are given systemically they go to all parts of the brain, including areas in which nicotinic receptors may have opposing effects on cognitive function, including memory. In many types of cognitive impairment nicotinic receptors in various brain areas are impaired in numbers and function, indicating the central role nicotinic receptors play in cognition. Finally, because these syndromes involve a spectrum of impairments, and the people affected often have other ailments, other drugs are often taken concomitantly with the nicotinic treatments and we must beware of drug–drug interactions related to nicotinic interactions with other neurotransmitter systems.

5.1 Alzheimer's Disease and Related Cognitive Impairments of Aging

Nicotinic receptors in the brain have been found to be dramatically decreased in people with Alzheimer's disease (Court et al. 2001). This is particularly apparent in the hippocampus and frontal cortex (London et al. 1989; Perry et al. 1986). In contrast, nicotinic receptors in the thalamus are often found to be relatively unaffected. Nicotine treatment significantly improves cognitive function in people with mild-to-moderate Alzheimer's disease (White and Levin 1999). In addition, nicotine skin patches significantly improve cognitive performance in people with age-associated memory impairment (AAMI) (White and Levin 2004) or Mild Cognitive Impairment (MCI) (Newhouse et al. 2012).

5.2 Schizophrenia

Schizophrenia is characterized by substantial impairment in cognitive function related to abnormalities of $\alpha 7$ nicotinic receptors (Leonard et al. 1996; Martin et al. 2004). The great majority of people affected with schizophrenia smoke tobacco

(Ripoll et al. 2004). There is some indication that they may use tobacco at least in part as a form of self-medication. Nicotine skin patch treatment has been shown to improve attention and memory in people with schizophrenia (Levin et al. 1996c). People with schizophrenia also take antipsychotic medications, many of which have antagonist effects at dopamine D_2, serotonin $5HT_2$, and histamine H1 receptors among others (Schotte et al. 1993). These receptor actions are likely to have interactive effects with nicotinic treatment. Indeed, we have found that nicotine has important interactions with the antipsychotic drugs haloperidol and clozapine with regard to memory function. Haloperidol significantly impairs memory function in people with schizophrenia, an effect that was reversed by nicotine (Levin et al. 1996c). Many antipsychotic drugs, especially atypical antipsychotics, have effects of blocking serotonin $5HT_2$ receptors. This effect seems to block nicotine-induced cognitive improvement. As described above, the $5HT_2$ antagonist ketanserin when given to rats significantly blocks nicotine-induced working memory improvement (Levin et al. 2005). Thus, when testing cognitive improving effects of nicotinic drugs in people with schizophrenia, it is important to consider the interactions of antipsychotic drugs with nicotinic systems and how they may block nicotinic-induced cognitive improvement.

5.3 Attention Deficit Hyperactivity Disorder (ADHD)

ADHD is the most prevalently diagnosed cognitive impairment among children and adolescents, and ADHD residual type is being increasingly diagnosed in adults. Tobacco use among people affected with ADHD is around double the rate of the general population (Pomerleau et al. 1995). As with schizophrenia, there is evidence that people with ADHD may be self-medicating by smoking tobacco, albeit in a particularly hazardous way (Newhouse et al. 2004). Nicotine skin patch treatment has been shown to significantly improve cognitive function in people with ADHD (Levin et al. 1996a). Interactions of nicotinic treatment with stimulant therapy, which affects dopamine and norepinephrine systems, needs to be more completely evaluated before new nicotinic treatments for ADHD can be advanced effectively.

6 Conclusions and Needs for Further Study

Clearly, nicotinic receptors, even the same receptor subtypes, in different parts of the brain play different roles in memory function. Since the brain is a heterogeneous organ of communication, it would be unusual if anatomically disparate receptor populations would play the same roles in any particular behavioral function, much less a complex one like memory. In addition, another consequence of the complex circuitry of the brain is that nicotinic acetylcholine systems interact with numerous other transmitter systems with regard to memory function. What does it say about development of nicotinic therapeutics other than it is quite complex? For one thing, the differential involvement of anatomically distinct nicotinic receptors of the same subtype means that even drugs with great receptor subtype selectivity will have a complex of effects and side effects, particularly when they are administered systemically and get distributed to all parts of the brain. Also, since these drug treatments are given to people who have neural dysfunctions, the anatomic heterogeneity of the dysfunctions are likely to play important roles in the therapeutic effects and side effects of nicotinic treatments. For example, people with deficiencies of nicotinic receptors in the frontal cortex may have different responsivity to nicotinic treatments than people with deficiencies of nicotinic receptors in the hippocampus or the thalamus.

Nicotinic interaction with the variety of other neural systems presents both cautions and opportunities for therapeutics. Cautions include the problems with polypharmacy that is likely with some of the therapeutic targets. For example, with schizophrenia it is posited that nicotinic treatment might provide improvement for cognitive dysfunction. There is no clear hypothesis currently that nicotinic treatments would combat psychosis. Therefore, nicotinic treatments for cognitive improvement would be given together with antipsychotic medication. There are a variety of different antipsychotic medications in use with many different arrays of neural effects. Some have effects on dopaminergic, serotonergic, noradrenergic, histaminergic, and other receptors. Some of these actions such as blockade of serotonin $5HT_2$ receptors appear to counteract therapeutic effects of nicotinic drug action on cognitive function including memory and attention. Other types of polypharmacy could include nicotinic interactions with medications to treat hypertension such as adrenergic β-blockers, which may themselves cause cognitive impairment. In clinical studies of nicotinic drug effects, it is important to consider what other drugs are being taken concurrently.

Acknowledgments Funding for these studies was from: NIDA, NIMH, NSF, Alzheimer's Association, Abbott Pharmaceuticals, AstraZeneca Pharmaceuticals, Targacept Pharmaceuticals, and an unrestricted grant from Philip Morris-USA.

References

Addy NA, Nakijama A, Levin ED (2003) Nicotinic mechanisms of memory: effects of acute local DHβE and MLA infusions in the basolateral amygdala. Cogn Brain Res 16:51–57

Aggleton JP, Pearce JM (2001) Neural systems underlying episodic memory: insights from animal research. Philos Trans R Soc Lond B Biol Sci 356:1467–1482

Arthur D, Levin ED (2002) Chronic inhibition of alpha4beta2 nicotinic receptors in the ventral hippocampus of rats: Impacts on memory and nicotine response. Psychopharmacology 160:140–145

Bancroft A, Levin ED (2000) Ventral hippocampal α4β2 nicotinic receptors and chronic nicotine effects on memory. Neuropharmacology 39:2770–2778

Bettany JH, Levin ED (2001) Ventral hippocampal α7 nicotinic receptor blockade and chronic nicotine effects on memory performance in the radial-arm maze. Pharmacol Biochem Behav 70:467–474

Buccafusco JJ, Brach JW, Terry AV (2009) Desensitization of nicotinic acetylcholine receptors as a strategy for drug development. J Pharmacol Exp Ther 328:364–370

Bymaster FP, Heath I, Hendrix JC, Shannon HE (1993) Comparative behavioral and neurochemical activities of cholinergic antagonists in rats. J Pharmacol Exp Ther 267:16–24

Cannady R, Weir R, Wee B, Gotschlich E, Kolia N, Lau E, Brotherton J, Levin ED (2009) Nicotinic antagonist effects in the mediodorsal thalamic nucleus: regional heterogeneity of nicotinic receptor involvement in cognitive function. Biochem Pharmacol 78:788–794

Clarke PB, Schwartz RD, Paul SM, Pert CB, Pert A (1985) Nicotinic binding in rat brain: autoradiographic comparison of [3H]acetylcholine, [3H]nicotine, and [125I]-alpha-bungarotoxin. J Neurosci 5:1307–1315

Court J, Martin-Ruiz C, Piggott M, Sperden D, Griffiths M, Perry E (2001) Nicotinic receptor abnormalities in Alzheimer's disease. Biol Psychiatry 49:175–184

Felix R, Levin ED (1997) Nicotinic antagonist administration into the ventral hippocampus and spatial working memory in rats. Neuroscience 81:1009–1017

Leonard S, Adams C, Breese CR, Adler LE, Bickford P, Byerley W, Coon H, Griffith JM, Miller C, Myles-Worsley M, Nagamoto HT, Rollins Y, Stevens KE, Waldo M, Freedman R (1996) Nicotinic receptor function in schizophrenia. Schizophrenia Bull 22:431–445

Levin ED, Bettegowda C, Weaver T, Christopher NC (1998) Nicotine-dizocilpine interactions and working and reference memory performance of rats in the radial-arm maze. Pharmacol Biochem Behav 61:335–340

Levin ED, Bradley A, Addy N, Sigurani N (2002) Hippocampal α7 and α4β2 nicotinic receptors and working memory. Neuroscience 109:757–765

Levin ED, Briggs SJ, Christopher NC, Auman JT (1994) Working memory performance and cholinergic effects in the ventral tegmental area and substantia nigra. Brain Res 657:165–170

Levin ED, Briggs SJ, Christopher NC, Rose JE (1993) Chronic nicotinic stimulation and blockade effects on working memory. Behav Pharmacol 4:179–182

Levin ED, Cauley M, Rezvani AH (2013) Improvement of attentional function with antagonism of nicotinic receptors in female rats. Eur J Pharmacol 702:269–274

Levin ED, Conners CK, Sparrow E, Hinton SC, Erhardt D, Meck WH, Rose JE, March J (1996a) Nicotine effects on adults with attention-deficit/hyperactivity disorder. Psychopharmacology 123:55–63

Levin ED, Kim P, Meray R (1996b) Chronic nicotine working and reference memory effects in the 16-arm radial maze: interactions with D1 agonist and antagonist drugs. Psychopharmacology 127:25–30

Levin ED, Icenogle L, Farzad A (2005) Ketanserin attenuates nicotine-induced working memory improvement in rats. Pharmacol Biochem Behav 82:289–292

Levin ED, McGurk SR, Rose JE, Butcher LL (1989a) Reversal of a mecamylamine-induced cognitive deficit with the D_2 agonist, LY 171555. Pharmacol Biochem Behav 33:919–922

Levin ED, McGurk SR, South D, Butcher LL (1989b) Effects of combined muscarinic and nicotinic blockade on choice accuracy in the radial-arm maze. Behav Neural Biol 51:270–277

Levin ED, Perkins A, Brotherton T, Qazi M, Berez C, Montalvo-Ortiz J, Davis K, Williams P, Christopher NC (2009) Chronic underactivity of medial frontal cortical β2-containing nicotinic receptors increases clozapine-induced working memory impairment in female rats. Prog Neuropsychopharmacol Biol Psychiatry 33:296–302

Levin ED, Rose JE (1995) Acute and chronic nicotinic interactions with dopamine systems and working memory performance. Ann N Y Acad Sci 757:245–252

Levin ED, Sledge D, Baruah A, Addy NA (2003) Ventral hippocampal NMDA blockade and nicotinic effects on memory function. Brain Res Bull 61:489–495

Levin ED, Weber E, Icenogle L (2004) Baclofen interactions with nicotine in rats: effects on memory. Pharmacol Biochem Behav 79:343–348

Levin ED, Wilson W, Rose JE, McEvoy J (1996c) Nicotine-haloperidol interactions and cognitive performance in schizophrenics. Neuropsychopharmacology 15:429–436

London ED, Ball MJ, Waller SB (1989) Nicotinic binding sites in cerebral cortex and hippocampus in Alzheimer's disease. Neurochem Res 14:745–750

Martin LF, Kem WR, Freedman R (2004) Alpha-7 nicotinic receptor agonists: potential new candidates for the treatment of schizophrenia. Psychopharmacology 174:54–64

May-Simera H, Levin ED (2003) NMDA systems in the amygdala and piriform cortex and nicotinic effects on memory function. Cogn Brain Res 17:475–483

McGurk SR, Levin ED, Butcher LL (1989a) Nicotinic-dopaminergic relationships and radial-arm maze performance in rats. Behav Neural Biol 52:78–86

McGurk SR, Levin ED, Butcher LL (1989b) Radial-arm maze performance in rats is impaired by a combination of nicotinic-cholinergic and D2 dopaminergic antagonist drugs. Psychopharmacology 99:371–373

Mulle C, Vidal C, Benoit P, Changeux JP (1991) Existence of different subtypes of nicotinic acetylcholine receptors in the rat habenulo-interpeduncular system. J Neurosci 11:2588–2597

Newhouse P, Singh A, Potter A (2004) Nicotine and nicotinic receptor involvement in neuropsychiatric disorders. Curr Top Med Chem 4:267–282

Newhouse PA, Kellar K, Aisen P, White H, Wesnes K, Coderre E, Pfaff A, Wilkins H, Howard D, Levin ED (2012) Transdermal nicotine treatment of mild cognitive impairment: a six-month double-blind pilot clinical trial. Neurology 78:91–101

Perry EK, Perry RH, Smith CJ, Purohit D, Bonham J, Dick DJ, Candy JM, Edwardson JA, Fairbairn A (1986) Cholinergic receptors in cognitive disorders. Can J Neurol Sci 13:521–527

Picciotto MR, Addy NA, Mineur YS, Brunzell DH (2008) It is not "either/or": activation and desensitization of nicotinic acetylcholine receptors both contribute to behaviors related to nicotine addiction and mood. Prog Neurobiol 84:329–342

Pocivavsek A, Icenogle L, Levin ED (2006) Hippocampal α7 and α4β2 nicotinic receptors and clozapine effects on memory. Psychopharmacology 188:596–604

Pomerleau OF, Downey KK, Stelson FW, Pomerleau CS (1995) Cigarette smoking in adult patients diagnosed with attention deficit hyperactivity disorder. J Subst Abuse 7:373–378

Rezvani AH, Caldwell DP, Levin ED (2005) Nicotinic-serotonergic drug interactions and attentional performance in rats. Psychopharmacology 179:521–528

Rezvani AH, Cauley M, Sexton H, Xiao X, Brown ML, Paige MA, McDowell BE, Kellar KL, Levin ED (2011) Sazetidine-A, a selective α4β2 nicotinic acetylcholine receptor desensitizing agent reverses dizocilpine and scopolamine-induced attentional impairments in rats. Psychopharmacology 215:621–630

Rezvani AH, Cauley M, Xiao Y, Kellar KJ, Levin ED (2012) Effects of chronic sazetidine-A, a selective β2* nicotinic receptor desensitizing agent on pharmacologically-induced impaired sustained attention in rats. Psychopharmacology 222:269–276

Rezvani AH, Kholdebarin E, Dawson E, Levin ED (2008) Nicotine and clozapine effects on attentional performance impaired by the NMDA antagonist dizocilpine in female rats. Int J Neuropsychopharmacol 11:63–70

Ripoll N, Bronnec M, Bourin M (2004) Nicotinic receptors and schizophrenia. Curr Med Res Opin 20:1057–1074

Sanders D, Simkiss D, Braddy D, Baccus S, Morton T, Cannady R, Weaver N, Rose JE, Levin ED (2010) Nicotinic receptors in the habenula: importance for memory. Neuroscience 166:386–390

Schotte A, Janssen PF, Megens AA, Leysen JE (1993) Occupancy of central neurotransmitter receptors by risperidone, clozapine and haloperidol, measured ex vivo by quantitative autoradiography. Brain Res 631:191–202

Steckler T, Drinkenburg WH, Sahgal A, Aggleton JP (1998) Recognition memory in rats–II. Neuroanatomical substrates. Prog Neurobiol 54:313–332

Sutherland RJ (1980) The dorsal diencephalic conduction system: a review of the anatomy and functions of the habenular complex. Neurosci Biobehav Rev 6:1–13

Timofeeva OA, Levin ED (2008) Idazoxan blocks the nicotine-induced reversal of the memory impairment caused by the NMDA glutamate receptor antagonist dizocilpine. Pharmacol Biochem Behav 90:372–381
White HK, Levin ED (1999) Four-week nicotine skin patch treatment effects on cognitive performance in Alzheimer's disease. Psychopharmacology 143:158–165
White HK, Levin ED (2004) Chronic transdermal nicotine patch treatment effects on cognitive performance in age-associated memory impairment. Psychopharmacology 171:465–471
Wilkerson A, Levin ED (1999) Ventral hippocampal dopamine D1 and D2 systems and spatial working memory in rats. Neuroscience 89:743–749

Nicotinic Receptors and Attention

Britta Hahn

Abstract Facilitation of different attentional functions by nicotinic acetylcholine receptor (nAChR) agonists may be of therapeutic potential in disease conditions such as Alzheimer's disease or schizophrenia. For this reason, the neuronal mechanisms underlying these effects have been the focus of research in humans and in preclinical models. Attention-enhancing effects of the nonselective nAChR agonist nicotine can be observed in human nonsmokers and in laboratory animals, suggesting that benefits go beyond a reversal of withdrawal deficits in smokers. The ultimate aim is to develop compounds acting with greater selectivity than nicotine at a subset of nAChRs, with an effects profile narrowly matching the targeted cognitive deficits and minimizing unwanted effects. To date, compounds tested clinically target the nAChR subtypes most abundant in the brain. To help pinpoint more selectively expressed subtypes critical for attention, studies have aimed at identifying the secondary neurotransmitter systems whose stimulation mediates the attention-enhancing properties of nicotine. Evidence indicates that noradrenaline and glutamate, but not dopamine release, are critical mediators. Thus, attention-enhancing nAChR agents could spare the system central to nicotine dependence. Neuroimaging studies suggest that nAChR agonists act on a variety of brain systems by enhancing activation, reducing activation, and enhancing deactivation by attention tasks. This supports the notion that effects on different attentional functions may be mediated by distinct central mechanisms, consistent with the fact that nAChRs interact with a multitude of brain sites and neurotransmitter systems. The challenge will be to achieve the optimal tone at the right subset of nAChR subtypes to modulate specific attentional functions, employing not just direct agonist properties, but also positive allosteric modulation and low-dose antagonism.

Keywords Nicotinic acetylcholine receptor · Attention · Nicotine · Subtypes · Animal models · Neuroimaging

B. Hahn (✉)
Maryland Psychiatric Research Center, University of Maryland School of Medicine, Baltimore, MD, USA
e-mail: bhahn@mprc.umaryland.edu

D.J.K. Balfour and M.R. Munafò (eds.), *The Neurobiology and Genetics of Nicotine and Tobacco*, Current Topics in Behavioral Neurosciences 23,
DOI 10.1007/978-3-319-13665-3_5

Contents

1 Introduction 104
2 Withdrawal Confounds 105
3 Dissociating Attentional Effects from Other Cognitive Functions 106
4 Types of Attention Most Affected 107
5 Rodent Models of Attention 109
6 Chronic Exposure to nAChR Agonists 112
7 Nicotinic Receptor Subtypes Mediating Effects on Attention 113
7.1 Preclinical Studies 114
7.2 Human Drug Development and Clinical Trials 118
8 Secondary Neurotransmitter Systems Mediating Effects of nAChR Agonists on Attention 119
9 Large-Scale Systems Mediating the Effects of nAChR Agonists on Attention 120
9.1 Human EEG 120
9.2 Human Neuroimaging 121
10 Conclusions 124
References 125

1 Introduction

Over the last half century, much evidence has built up suggesting that nicotine has beneficial effects on the performance of cognitive tasks. Such effects include enhancement of motor functions, sensory processing, alertness, attention, and memory (Heishman et al. 1994, 2010). In the human literature, effects on attentional functions have been reported with particular consistency (Stolerman et al. 1995; Newhouse et al. 2004; Heishman et al. 2010), and facilitation of higher cognitive operations may in part be secondary to enhancement of more basal functions in the attention domain. The characterization of the attention-enhancing effects of nicotine and other nicotinic acetylcholine receptor (nAChR) agonists, both at the behavioral and neuronal levels, has been the subject of much research. This interest has been fueled mainly by the clinical potential of these effects for the treatment of disorders marked by cognitive dysfunction, in particular those with demonstrated nAChR hypofunction such as Mild Cognitive Impairment, Alzheimer's disease, schizophrenia, and Attention-Deficit/Hyperactivity Disorder (ADHD). Attentional impairments are an early and reliable symptom of Alzheimer's disease, for which cholinergic deficits are a neuropathological hallmark (Kendziorra et al. 2011; Perry et al. 2000; Terriere et al. 2008; Nordberg et al. 1992, 1995; O'Brien et al. 2007; Pimlott et al. 2004). Early studies indicated that acutely administered nicotine reduces attentional deficits of Alzheimer's disease patients (White and Levin 1999; Sahakian et al. 1989; Jones et al. 1992), leading to suggestions that the illness may benefit from nAChR agonist treatment (Levin and Rezvani 2002; Singh et al. 2004). Similar lines of argument exist for the potential benefit of nicotinics to treat attentional deficits in schizophrenia (Martin et al. 2004; Adams and Stevens 2007; Freedman et al. 1997; Leonard et al. 2002; Petrovsky et al. 2010) and ADHD

(Levin et al. 1996; Lerman et al. 2001). Drug development efforts have already been invested into more selective nAChR agonists (Haydar and Dunlop 2010; Radek et al. 2010; Wallace et al. 2011a; Hurst et al. 2013) for the treatment of these disorders. Understanding more about the neuropharmacological substrates mediating these effects has therefore wide-ranging clinical relevance and may help develop more efficacious compounds.

The aim of the following chapter is to review research studies on the attentional functions that are modulated by nAChR agonists, on neuropharmacological mechanisms, and on neuroanatomical and neural systems mediators. Specific questions that will be considered are nAChR subtypes and secondary neurotransmitter systems of particular relevance to the effects of nAChR agonists on attentional functions, the role of nAChR activation versus desensitization, EEG markers, and mediating brain regions and networks. Both human and animal studies will be considered, where possible in direct comparison.

2 Withdrawal Confounds

For a clinical application of nAChR agonists, a critical question is whether the effects reflect absolute improvement relative to an unimpaired baseline, or merely the reversal of deficits induced by tobacco deprivation. The majority of early studies on the performance effects of nicotine or smoking were conducted in smokers tobacco-deprived for more than 4 h (Heishman et al. 1994)—enough time for some deprivation-induced impairment to develop (Hatsukami et al. 1989; Snyder et al. 1989). On those aspects of performance most consistently enhanced by nicotine, the effects were reported with larger probability in studies assessing deprived smokers than nonsmokers (Heishman et al. 1994), supporting the view that there are little benefits beyond reversing withdrawal. However, more recent studies designed to address the issue with more stringent methodology repeatedly showed performance enhancement with nicotine in healthy, unimpaired nonsmokers (Heishman et al. 2010).

Contributing to the frequent difficulty demonstrating attentional and other performance-enhancing effects of nicotine in nonsmokers are disruptive side effects seen with nicotine upon initial exposure. Studies administering nicotine skin patches, gum or nasal spray to nonsmokers report negative physiological effects such as dizziness, nausea, increased heart rate and blood pressure, impaired hand-steadiness, lower scores in self-report measures of alertness and vigor, and aversive effects as reflected by higher scores in scales of head rush, jitteriness, tension, confusion, sedation, and dysphoria (Heishman et al. 1993; Perkins et al. 1994; Heishman and Henningfield 2000; Griesar et al. 2002). These effects were sometimes accompanied by nicotine-induced impairment in cognitive tasks (Heishman et al. 1990; Heishman and Henningfield 2000). Thus, the possibility that disruptive effects of nicotine in drug-naïve subjects can mask cognitive-enhancing effects offers an explanation for the greater difficulty in demonstrating such effects in nonsmokers. Accordingly, studies that have been successful in demonstrating performance enhancement with nicotine in nonsmokers are characterized by the use of small doses, or by having pre-administered additional nicotine doses (Foulds et al. 1996) to promote acute tolerance to any adverse and potentially disruptive side effects. Indeed, it has been demonstrated that acute tolerance develops to such aversive effects after four administrations of nicotine nasal spray within 2 h. Simultaneously with the development of tolerance to these effects, improvements in short-term recognition memory by nicotine became stronger (Perkins et al. 1994). In the same study, responses of smokers on measures of aversive effects were generally profoundly smaller than in nonsmokers, indicating persistent tolerance. Consistent with this notion, tolerance was observed to some aversive effects of nicotine in nonsmokers over 8 days of repeated nicotine gum consumption (Heishman and Henningfield 2000).

3 Dissociating Attentional Effects from Other Cognitive Functions

It is generally difficult to dissociate different aspects of the performance-enhancing effects of nicotine. To give an example at the most basic level, it is difficult to separate facilitation of sensory or motor functions from improvements in attention, learning and memory, or higher cognitive control functions. Nicotine consistently increases finger-tapping rate, and sensory processing speed as assessed by the critical flicker fusion task (Heishman et al. 1994). Both of these effects could account for faster reaction times in any cognitive paradigm. Furthermore, basal psychophysical gains and the processing of less noisy sensory signals is likely to result in stronger and more accurate cortical representations being fed forward, enabling more efficient engagement of higher functions through reduced competition for processing resources (Mahncke et al. 2006; Adcock et al. 2009). There is also evidence from EEG recordings that nicotine enhances processes of early sensory integration (reviewed below under 9.1), which may facilitate attention in a bottom-up manner. In a similar manner, facilitation of basal attentional functions such as alertness and processing speed may facilitate higher attentional functions such as attentional selection or control processes of attention, as well as functions beyond the attention domain.

There has been much discussion as to whether beneficial effects of nicotine in tasks of learning and memory may be secondary to effects on attentional functions, with attention being a necessary gateway for memory performance. Supporting this notion, improvements in recall were seen when subjects smoked prior to the presentation of a word list, but not when they smoked in the retention interval immediately following list presentation (Peeke and Peeke 1984). Thus, improvements may have been due to increased attention during encoding rather than enhanced retention. Other studies did report improved recall with post-acquisition cigarette smoking (Mangan and Golding 1983; Colrain et al. 1992; Warburton et al. 1992a), but this was shown to depend on the opportunity for stimulus rehearsal after presentation and may thus reflect enhanced attention to the encoded material during the retention period (Rusted and Warburton 1992). In another study, nicotine-induced improvements occurred mainly toward the end of long word lists, when attention toward the presented material may have been taxed most (Warburton et al. 1992b). Furthermore, improvements were seen mainly in those aspects of the input material that subjects were explicitly instructed to remember (Andersson and Hockey 1977), i.e., in those aspects to which attention was directed. Such findings suggest that effects of nicotine on attentional functions have a major influence on human performance in paradigms of learning and memory, although evidence from preclinical studies also suggests primary mnemonic effects (Kenney and Gould 2008).

4 Types of Attention Most Affected

Despite the above interpretative limitations, performance-enhancing effects of nicotine in paradigms of attention are generally seen with greater reliability than effects in paradigms targeting mnemonic functions. Thus, over decades, the majority of human studies testing attention reported improvement, while studies of learning and memory yielded more mixed results (Stolerman et al. 1995; Newhouse et al. 2004). Among studies of attention, the consistency with which performance benefits are observed appears to differ depending on the type of paradigm employed. Nicotine reliably improves performance in tasks of vigilance and simple stimulus detection (Koelega 1993; Mancuso et al. 1999, 2001). A 1993 review of the effects of psychostimulants in vigilance-demanding tasks found that the majority of studies with nicotine reported improvement (Koelega 1993). The most positive findings were made with the Rapid Visual Information Processing (RVIP) task, in which subjects continuously monitor a series of rapidly presented numbers and respond to target sequences of three consecutive odd or even digits. Smoking or nicotine can improve detectability and response latency and prevent performance decrement over time where it occurs, effects not restricted to deprived smokers (Wesnes and Warburton 1984; Revell 1988; Warburton and Arnall 1994; Warburton and Mancuso 1998; Foulds et al. 1996; Juliano et al. 2011; Gilbert et al. 2005).

The sensitivity of the RVIP task to the performance-enhancing effects of nicotine may be due to a number of features characteristic to this task. Resource demands are created by processing signals presented at a high event rate (100/min), and by the need to integrate information across trials, which adds a working memory component. These intense processing requirements have to be upheld continuously without even the shortest break, thus taxing sustained attention through mechanisms of fatigue or overload (Parasuraman et al. 1987). The increased detectability and reduced reaction times reported with nicotine could be a result of faster information processing speed by nicotine, as supported by the finding that nicotine selectively increased target hits that were emitted rapidly (Foulds et al. 1996). Reports that nicotine is able to alleviate performance decrements over time in this paradigm (Wesnes and Warburton 1984) may suggest additional effects on alertness and sustained attention, although faster processing speed could also facilitate the maintenance of performance over time by reducing low-level resource demands and freeing up processing resources for higher operations. However, improvements with nicotine were also found with Mackworth's Continuous Clock Test (Mackworth 1950), in which subjects have to observe the seconds pointer of a clock and detect brief pauses in its movements that occur very infrequently. Nicotine administered via tablets significantly reduced the decline in stimulus sensitivity (correct detections in relation to false alarms) over 80-min test sessions while not affecting performance early in the session (Wesnes et al. 1983). Effects on processing speed cannot account for these effects.

The results from studies not employing classic vigilance or sustained attention paradigms are consistent with the notion that effects of nicotine can most easily be explained by enhanced alertness and intensity aspects of attention. Tasks that require the maintenance of stimulus detection or other simple operations tend to report improvement (Bates et al. 1995; Levin et al. 1998; Spilich et al. 1992; Williams 1980; Parrott and Craig 1992; Mancuso et al. 2001). In contrast, there is less consistent evidence for improvement by nicotine in paradigms designed to tax selectivity aspects of attention (Mancuso et al. 1999; Parrott and Craig 1992; Foulds et al. 1996; Poltavski and Petros 2006). Similarly, rodent studies support enhancement with nicotine and other nicotinic agonists in low-level stimulus detection tasks (Grilly et al. 2000; Hahn et al. 2002b, 2003; Grottick et al. 2003), while improvements have been somewhat more difficult to show when discriminatory processes are required (Turchi et al. 1995, 1996; Bushnell et al. 1997):

Although comparatively less robust, several findings do suggest improvements specific to processes of selective attention, in that the performance-enhancing effects of nicotine were relatively greater in mitigating the effects of distractors. For example, nicotine or cigarette smoking reduced the Stroop effect, that is, performance costs of naming the ink color of an incongruent color word, in about half the investigations studying such effects (Wesnes and Warburton 1983; Provost and Woodward 1991; Hasenfratz and Battig 1992; Parrott and Craig 1992; Foulds et al. 1996; Poltavski and Petros 2006; Domier et al. 2007). Nicotine also reduced the Garner effect, that is, performance costs due to changes in the irrelevant stimulus dimension per se (Waters 1998). Furthermore, introducing sensory distractor stimuli can help reveal performance-enhancing effects of nicotine or other nicotinic agonists in humans (Grobe et al. 1998), rodents (Hahn and Stolerman 2002; Hahn et al. 2002b; Howe et al. 2010) and nonhuman primates (Prendergast et al. 1998). Reduced interference from irrelevant stimuli may reflect enhanced attentional filtering, or facilitation of attentional control processes. Indirect evidence for enhanced filtering can be deduced from the finding that nicotine impaired incidental memory of material that subjects had not been attempting to remember, and to which attention had presumably not been directed, while improving recall of attended material (Andersson and Hockey 1977). This finding suggests that nicotine increased attentional resources allocated to the attended material and enhanced the filtering of the unattended material.

There is some limited evidence suggesting that nicotine may affect control components of attention, that is, the coordination of attentional resource allocation. In Posner-type tasks of covert orienting, reaction time to a peripheral target stimulus is measured after this stimulus had been validly or invalidly cued or uncued. In invalidly cued trials, attention has to be disengaged from one and reallocated to another location upon presentation of the target stimulus. The reaction time cost of such invalid cueing is termed the validity effect. Nicotine or smoking has been reported to reduce the validity effect in humans, rhesus monkeys, and rats (Witte et al. 1997; Murphy and Klein 1998; Phillips et al. 2000; Stewart et al. 2001), and thus appears to enhance stimulus detection at previously unattended locations, perhaps reflecting facilitation of spatial reorienting of attention. This may reflect control processes of attention, but also a broadened attentional focus (Bentley et al. 2004), or improved sensory orienting.

Additional evidence for effects on attentional control aspects comes from a rat version of the Intra-/Extra-dimensional Set Shifting Task. In this task, rats choose one of two bowls which differ in surface texture and scent. Either the texture or the scent indicates which of the bowls is baited. An intradimensional reversal would be a reversal either in the scent or in the texture marking the baited bowl. For an intradimensional shift, new scents or textures would be introduced. An extra-dimensional shift would be a reversal of the relevant stimulus feature; that is, rats now have to pay attention to scent rather than texture or vice versa. Nicotine, given acutely or after repeated preexposure, facilitated both intra- and extradimensional set shifting performance, suggesting beneficial effects on control aspects of attentional resource allocation (Allison and Shoaib 2013). However, mediation through lower level, e.g., alerting effects again cannot entirely be excluded.

In summary, there is a large amount of evidence that nicotine-induced improvements in attention are robust in paradigms requiring simple operations, in which performance-limiting factors are likely to be processing speed, alertness, and staying on task. Higher order processes such as selective attention may also be enhanced, and there is growing evidence for benefits of nAChR agonists on certain control aspects of attentional resource allocation. However, it is always difficult to fully exclude action on lower level gateway functions.

5 Rodent Models of Attention

In the rodent literature, the two paradigms of attention that have been predominantly employed are based on creating stimulus detection demands:

(1) McGaughy and Sarter (1995) developed a task in which rats discriminate between signal and nonsignal trials by operating the appropriate one of two levers. During signal trials, a stimulus light of variable duration is presented; during nonsignal trials no stimulus occurs. Discrimination performance would be reflected

by a signal detectability index incorporating correct detections (%Hits) in relation to false alarms, although studies employing this paradigm increasingly limit themselves to reporting %Hits. Signal detection decreases with shorter stimuli, with a higher trial event rate, and with an increase in background noise implemented by flashing the houselight of the operant chamber. Some of these manipulations caused a performance decrement over time, suggesting they increased the task's sustained attention demands. In recent work, this task has been termed SAT (Sustained Attention Task) when employed without the flashing houselight, and distractor SAT (dSAT) when employed with this distractor (Howe et al. 2010).

(2) The 5-Choice Serial Reaction Time Task (5-CSRTT), in contrast, does not require discriminatory processes (if x, then y) but relies on the simple detection of light stimuli presented randomly in one of five horizontally arranged apertures. A nose-poke into an illuminated aperture is rewarded. Failures of attention manifest themselves as localization errors (i.e., a response into a wrong hole). In this scenario, the animal likely perceived the target light onset but was not sufficiently engaged in scanning the array of potential locations to correctly localize it. Correct localizations expressed as a percentage of all hole-pokes (correct and incorrect) emitted at the time of stimulus presentation yields response accuracy. This measure is independent of the animal's basal response rate. Failures of attention would also result in omission errors, i.e. a failure to respond to the light stimulus altogether. In this scenario, the animal likely was not engaged in performing the task but was "off task". It is easily seen how nonspecific (e.g., sedative or motor-depressant) drug effects could increase omission errors, or how psychomotor stimulant effects could decrease omission errors when noncontingent responses coincide with target stimuli. Nonspecific increases in response rate are typically observed with moderate doses of nicotine under interval schedules of reinforcement in the absence of any signal detection requirements (Spealman et al. 1981; Goldberg et al. 1989; Morrison and Stephenson 1973); thus, it is essential to disambiguate such effects from effects on attention. Response latency (i.e., how quickly after stimulus onset the animal responds) is similarly sensitive to nonspecific stimulant or depressant drug effects. It is therefore difficult to interpret drug-induced decreases in omission errors and response latency as attentional in nature unless they were accompanied by increases in response accuracy.

Early studies with either of these two paradigms had little success in demonstrating nicotine-induced improvements in attention. In the first investigations utilizing the SAT, the acute administration of nicotine (0.09–0.88 mg/kg i.p.) and three other nicotinic agonists (lobeline, ABT-418, A-82695) did not facilitate performance (Turchi et al. 1995). Since the nicotinic antagonist mecamylamine impaired signal detectability, the authors suggested that nAChRs are optimally activated in intact animals, thus preventing further benefits of acute agonists. In a later study, a large dose of nicotine (0.75 mg/kg s.c.) impaired signal detection in the first 100-trial ($\sim$20 min) block, while a small dose (0.083 mg/kg) slightly improved it in the second block (Bushnell et al. 1997). Similarly, the nAChR agonist ABT-418 enhanced performance in sham-lesioned rats at low doses (0.04 and 0.13 mg/kg) but not with a larger dose (0.39 mg/kg) (McGaughy et al. 1999), providing evidence for an inverted U-shaped dose–response function for attentional enhancement by nAChR agonists, as suggested previously for effects on other cognitive measures (Levin 1992). This was later replicated in the 5-Choice Serial Reaction Time Task (Hahn et al. 2002b, 2003). Thus, reasons for the early difficulty in showing improvement may include the use of large doses. Furthermore, ceiling effects may have limited further improvement in some cases. In one study (McGaughy et al. 1999), ABT-418 tended to cause larger improvement with longer stimulus durations, and the baseline hit-rate at the longest stimulus duration (500 ms) exceeded 90 % in the first unsuccessful study (Turchi et al. 1995). Introducing changes to task parameters so as to degrade baseline performance appears to help uncover improvements (Howe et al. 2010). In summary, performance benefits by nAChR agonists in the SAT were seen with select doses, under select task conditions.

In the first study testing nicotine in the 5-CSRTT, small doses of nicotine (0.06 and 0.1 mg/kg) reversed deficits in performance accuracy in basal forebrain lesioned rats, but no sign of improvement occurred in unlesioned animals (Muir et al. 1995). Response accuracy in the absence of drug was below 70 % in control animals; thus, ceiling effects were not the reason for this lack of improvement. In later studies improvements in response accuracy were observed occasionally, with only one dose of nicotine (0.1 mg/kg) under specific task conditions (Mirza and Stolerman 1998), only upon first exposure to the drug (Blondel et al. 1999), or only after chronic administration over several weeks (Grottick and Higgins 2000). More reliable improvements in task performance with nicotine were limited to reduced response latency and omission errors, i.e., rate- and speed-dependent measures.

At around the same time, clear nicotine-induced improvements in choice accuracy were reported in a simple two-choice stimulus detection task, in which the position of a light stimulus indicated the reinforced lever (Grilly et al. 2000; Grilly 2000). The authors compared their results to those obtained with the SAT and suggested that the requirement to report the location of a light, rather than its presence versus absence, may be critical for observing such effects. Findings that nicotine caused reductions in the validity effect of covert orienting (Phillips et al. 2000; Stewart et al. 2001) may be consistent with this suggestion, perhaps indicating facilitation of locating stimuli at previously unattended locations. Indeed, enhanced accuracy in the 5-CSRTT was seen more reliably with nicotine after a few changes were made to the task with the purpose of focusing task demands on simple stimulus localization (Hahn and Stolerman 2002, 2005; Hahn et al. 2002a, b, 2003; Bizarro and Stolerman 2003; Semenova et al. 2007; Quarta et al. 2007). In the original version, performance depended on several additional behavioral contingencies. For example, each trial had to be initiated by pushing the flap of the food hopper. Thus, in trials following incorrect responses or omission errors, trial initiation was an additional link in a chain of responses. Furthermore, noncontingent responses in the absence of a target light were punished by 5-s timeouts, making response inhibition an essential element of baseline task performance. Removing these contingencies reduced the number of different task demands, perhaps "clearing the way" for effects of nicotine on simple stimulus detection.

6 Chronic Exposure to nAChR Agonists

Any therapeutic use of nAChR agonists as cognitive enhancers in chronic disease states is dependent on minimal or no tolerance to the sought-after effects with repeated administration. The phenomenon of nAChR desensitization upon agonist exposure (Gentry and Lukas 2002), studied particularly well with nicotine itself, has led to the proposal that acute tolerance occurs to these beneficial effects of nicotine, thus annihilating its therapeutic potential (Harris et al. 2004; Martin et al. 2004). However, the picture appears to be more complex. Desensitization dynamics differ between nAChR subtypes, and in some cases desensitization is substantially less pronounced at physiological concentrations than at concentrations typically employed in vitro. The differential expression of nAChR subtypes across different brain systems can result in a complex interplay between effects of nAChR activation and desensitization, and shift the balance between circuits (Mansvelder et al. 2002; Wooltorton et al. 2003). Both nAChR activation and desensitization appear to contribute to the behavioral profile of nAChR agonists (Picciotto et al. 2008). What ultimately matters is how these interactions manifest themselves on the behavioral effect of interest.

The clearest evidence against acute tolerance to the alerting or attention-enhancing effects of nicotine comes from the fact that these effects are observed with slow-absorption delivery systems, such as transdermal patches, hours after they were applied (Barr et al. 2008; Griesar et al. 2002; Mancuso et al. 1999; Levin et al. 1998). These reports include studies in nonsmokers. On a longer time scale, nicotine given via transdermal patches to Alzheimer's disease patients reduced omission errors in the continuous performance task over the entire 4-week course of treatment (White and Levin 1999), providing evidence against chronic tolerance.

Rodent studies provided further evidence that the attention-enhancing effects of nicotine persist with chronic exposure. When infused by osmotic minipumps over 28 days, nicotine increased correct rejections (with a weaker effect also on hits) in the SAT relative to a control group without any sign of tolerance (Rezvani et al. 2005). In the 5-CSRTT, the attention-enhancing effects were larger after daily injections of 0.2 mg/kg for 20 days or longer than when nicotine was given to drug-naïve rats (Grottick and Higgins 2000). In a different attention task requiring rats to detect brief spatially invariant but temporally irregular visual stimuli, chronic injections of nicotine (0.1 mg/kg three times per day) improved performance by reducing inappropriate responding in the intertrial interval, with no sign of tolerance even after 4 weeks of administration. Tolerance developed only to the *increase* in omission errors observed upon the first exposure to nicotine (Nelsen and Goldstein 1972, 1973). This biphasic pattern was replicated with a different experimental design in the 5-CSRTT (Hahn and Stolerman 2002; Stolerman et al. 2009). In naïve rats, nicotine (0.2 mg/kg) profoundly increased omission errors. When testing the effects weekly thereafter, against a background of daily post-session injections of either vehicle or a large dose of nicotine (0.4 mg/kg), these rate-disruptive effects had disappeared and nicotine enhanced attentional performance. These beneficial effects did not differ significantly between post-session treatment groups, or weeks. Even after 2 months of daily exposure, nicotine robustly enhanced performance, providing evidence that, following rapid tolerance to initial disruptive effects, nicotine-induced attentional enhancement is stable across lengthy periods of chronic exposure.

The rapid tolerance to the response-suppressant effects of nicotine in the 5-CSRTT is reminiscent of the locomotor depressant action of nicotine in drug-naïve rats, which develops tolerance after only one exposure to a 0.4 mg/kg dose (Clarke and Kumar 1983). Even the acute time course matched between these two effects (Hahn et al. 2002b; Clarke and Kumar 1983). A range of studies in animals and humans also suggest that the aversive effects of nicotine tolerate out rapidly (Stolerman 1999). It was suggested above, in the context of why performance-enhancing effects seem more difficult to demonstrate in nonsmokers, that initial disruptive side effects of nicotine may mask its cognitive-enhancing action. The 5-CSRTT results are consistent with this hypothesis.

Despite the repeated finding that the attention-enhancing effects of nicotine persist with chronic administration, chronic exposure may cause baseline shifts that diminish the net benefits compared with a pre-drug baseline. Indeed, nicotine withdrawal appears to cause deficits in the very functions that it enhances acutely, including attention (Hatsukami et al. 1989; Snyder et al. 1989). Preclinical studies, which are able to quantify such baseline shifts with carefully controlled experimental control groups, did not identify such impairment in measures of attention (Hahn and Stolerman 2002; Rezvani et al. 2005). In humans, however, baseline decrements with chronic nicotine administration remain a valid concern, even if net benefits are not completely canceled out.

7 Nicotinic Receptor Subtypes Mediating Effects on Attention

Nicotinic receptors in the mammalian central nervous system are pentameric cation channels composed of α2–7 and β2–4 subunits. Different combinations of these subunits form nAChR subtypes with distinct pharmacological and physiological properties and distinct neuroanatomical distributions (Gotti et al. 2009; Clementi et al. 2000). Most nAChRs are α/β heteromers. Combinations of α2, α3, or α4 with β2 or β4 form functional nAChRs, whereas α5 and β3 assemble only, and α6 preferentially, with other functional α/β combinations (Chavez-Noriega et al. 1997; Lukas et al. 1999; Boorman et al. 2000; Millar and Gotti 2009). α7 mostly combines as homooligomers (Chen and Patrick 1997). α4β2* and α7 nAChR subtypes are the two most abundantly expressed subtypes in the brain, leaving few systems unaffected (Gotti et al. 2007; Seguela et al. 1993). The ability to selectively target nAChR subtype(s) that mediate the effects of nAChR agonists on attention would allow isolating desired effects with potential clinical benefit from undesired, such as dependence-related, effects.

7.1 Preclinical Studies

To determine which subtypes of the nAChR mediate the attention-enhancing effects of nicotine, as the prototypical nonselective nAChR agonist, several preclinical studies aimed at reproducing these effects with agonists or blocking them with antagonists that display varying binding profiles across nAChR subtypes.

In the SAT, a low dose of the selective α4β2* nAChR agonist S38232 increased %Hits in blocks of trials following visual distractor presentation, during which baseline performance was still impaired (Howe et al. 2010). Furthermore, in rats characterized by poor attentional performance and low task-induced cortical acetylcholine release, the relatively selective partial α4β2* agonist ABT-089 also improved detectability in the SAT (Paolone et al. 2013). Nicotine, however, was without beneficial effects under all conditions in either study, suggesting improvement in this paradigm predominantly with agonists displaying a preference for the α4β2* subtype.

Despite the initial absence of clearly interpretable attentional effects of nicotine in the 5-Choice Serial Reaction Time Task, early studies proceeded to test compounds with preferential action at specific nAChR subtypes. Dihydro-β-erythroidine (DHβE) is a competitive nAChRs antagonist that displays high affinity at α4β2, α4β4, α3β2 and α2β2 receptors but not at α3β4 and α7 receptors (Harvey and Luetje 1996; Harvey et al. 1996; Chavez-Noriega et al. 1997). Methyllycaconitine (MLA) is a competitive antagonist at the homomeric α7 nAChR (Macallan et al. 1988; Decker et al. 1995; Davies et al. 1999) but also blocks α6* receptors (Salminen et al. 2005). In single-dose interaction studies employing the original version of the 5-CSRTT, DHβE but not MLA antagonized reductions in response latency and increases in premature responding by nicotine (Blondel et al. 2000; Grottick and Higgins 2000). Because other response indices were not affected by nicotine, these experiments were inconclusive with respect to the nAChRs subtype involved in its attention-enhancing effects. Using the simplified version of the 5-CSRTT, DHβE did not antagonize the attention-enhancing effects of nicotine reflected by measures of accuracy and omission errors, suggesting that its reduction of effects on latency and anticipatory responses in the above studies reflected antagonism of the stimulant effects of nicotine (Hahn et al. 2011). This would be consistent with the observation that DHβE reverses the locomotor stimulant effects of nicotine (Grottick et al. 2000; Stolerman et al. 1997). However, a study in aged rats employing a 1-h version of the 5-CSRTT reported that nicotine reduced the decrement in accuracy, omission errors and response latency with time on task, and DHβE reduced all of these effects (Grottick et al. 2003). MLA had no effect in this study, but it dose-dependently reversed the nicotine-induced reduction in omission errors in the simplified 5-CSRTT (Hahn et al. 2011). This indicates greater involvement of α7 nAChRs in the performance effects of nicotine than first suggested, consistent with studies in α7 knockout mice suggesting α7 nAChR involvement in 5-CSRTT performance (Young et al. 2004; Hoyle et al. 2006). MLA also modulated response accuracy in unanticipated ways, to be discussed below. A general conclusion from the above results is that the nAChR subtypes involved in the performance-enhancing effects of nicotine appear to depend on the precise function assessed.

Several nAChR agonists have been examined in the original version of the 5-CSRTT. None of the tested compounds was found to enhance response accuracy. As mentioned, this usually included nicotine. Thus, there was no positive control to ensure that the tests were performed under conditions in which reliable increases in attentional performance could occur. In the simplified 5-CSRTT, the beneficial effects of nicotine on response accuracy, omission errors, and response latency were mimicked by epibatidine (Hahn et al. 2003), which has potent agonist activity at all nAChR subtypes examined, and whose behavioral effect profile generally resembles that of nicotine (Sullivan et al. 1994; Damaj et al. 1994; Gerzanich et al. 1995). Attentional effects of nicotine in aged rats were also mimicked by SIB 1765F which displays a preference for the α4β2* nAChR subtype, but not by AR-R17779, a selective agonist at the homomeric α7 nAChR (Grottick et al. 2003). Lack of effect of AR-R17779 was replicated in the simplified 5-CSRTT (Hahn et al. 2003), although this drug may have poorer CNS penetration than first thought (Grottick et al. 2000), which may explain the discrepancy between this finding and the effects of MLA described above (Hahn et al. 2011). Indeed, the partial α7 nAChR agonist and 5-HT3 antagonist RG3487 (MEM3454), at a large dose, enhanced %Hits in the SAT (Rezvani et al. 2009) and reversed PCP-induced impairment in attentional set shifting (Wallace et al. 2011b). The findings with SIB 1765F suggest α4β2* subtype involvement in the attention-enhancing effects of nicotine, consistent with results obtained in the SAT (Howe et al. 2010; Paolone et al. 2013). In contrast, SIB-1553A, a nAChR agonist with reputed selectivity at β4* subtypes, had no effect on any performance measure even under testing conditions that showed significant effects of nicotine on response accuracy (Grottick et al. 2001).

In the rat Intra-/Extra-dimensional Set Shifting Task describe above, the α7 nAChR agonist RG3487 (Wallace et al. 2011b) and the α7 nAChR positive allosteric modulator PNU-120596 (McLean et al. 2012) reversed PCP-induced impairment in extra-dimensional shifts. Similarly, MK801-induced attentional set shifting deficits in a maze-based paradigm were reversed by α7 nAChR agonists or partial agonists SSR-180711, PNU-282987 and GTS-21 (Jones et al. 2014). The benefits of nicotine on extra-dimensional set shifting in unimpaired animals (Allison and Shoaib 2013) may thus be mediated by α7 nAChRs; however, experiments employing α4β2*-selective compounds have to date not been reported.

Involvement of the α4β2* nAChR subtype in the attention-enhancing effects of nicotine in the SAT and 5-CSRTT may raise doubts about the possibility of dissociating these from dependence-related effects of nicotine. Behavioral assays suggest α4β2* involvement in effects mediated by accumbal dopamine release, including nicotine reinforcement (Tuesta et al. 2011). There is, however, a surprising absence of self-administration studies of α4β2*-selective or -favoring nAChR agonists. Thus, while selective α4β2* agonists appear to have clear benefits on attentional performance, it is not clear that α4β2* agonism would be sufficient to confer abuse potential when no other nAChR subtypes are targeted. Furthermore, other, still to be identified nAChR subtypes, expressed with greater regional selectivity, may also mediate the attention-enhancing effects of nAChR agonists and enable their dissociation from abuse-related brain mechanisms.

That such dissociation is possible in principle was shown with two nAChR agonists that appear to have no or reduced dopamine-releasing effects in the nucleus accumbens. ABT-418 enhanced mnemonic functions in monkeys with a similar potency to nicotine and was at least as efficacious in reducing performance impairments by visual distractors, indicating enhancement in attentional functions (Buccafusco et al. 1995b; Prendergast et al. 1998). This contrasts with a three- to fourfold lower potency than nicotine in most other assays (Decker et al. 1994; Damaj et al. 1995). In vitro evidence indicated that its potency to release dopamine was even further reduced (Arneric et al. 1994; Tani et al. 1997), which would be consistent with its failure to increase locomotor activity in nicotine-sensitized rats (Stolerman et al. unpublished observation), a dopamine-dependent effect. The second compound is isoarecolone, which is similarly devoid of locomotor stimulant action and elicits a smaller maximal release of dopamine from striatal and cortical synaptosomes than nicotine (Whiteaker et al. 1995). Furthermore, unlike nicotine, isoarecolone did not increase concentrations of dopamine in the nucleus accumbens of freely moving rats (Mirza et al. 1996), and it was not self-administered by rats that reliably self-administered nicotine (Shoaib 2006). In contrast, isoarecolone was as effective as nicotine in enhancing delayed matching to sample performance of monkeys (Buccafusco et al. 1995a). Importantly, both ABT-418 and isoarecolone induced dose-related enhancement in response accuracy in the simplified 5-CSRTT (Hahn et al. 2003). For ABT-418, the effect was limited to the first 10 min of testing, in line with its short plasma half-life in rats (Decker et al. 1994). The effects of both compounds on omission errors were reduced relative to nicotine, and ABT-418 had reduced and isoarecolone no effects on response latency. Consistent with isoarecolone's complete lack of effect on this measure, nicotine-induced reductions in response latency were shown to be dopamine-dependent (Hahn et al. 2002a). Increases in response accuracy by nicotine, in contrast, seemed insensitive to dopamine blockade (Hahn et al. 2002a), giving further support that the selectivity of effects of ABT-418 and isoarecolone in the simplified 5-CSRTT reflected their reduced action at, or sparing of, the dopamine system. While it is presently unclear which nAChR subtype-selectivity confers this neurochemical and behavioral selectivity, these results are very promising from a clinical perspective and should encourage the search for nAChR subtypes whose selective targeting enhances attentional functions but does not confer any abuse potential.

7.1.1 Paradoxical Effects of Low-Dose nAChR Antagonists

A growing number of studies report the paradoxical result that low doses of nAChR antagonists can mimic some performance-enhancing effects of agonists. Initial findings were in the memory domain: low doses of the nonselective nAChR antagonist mecamylamine improved working memory in rats and monkeys (Terry et al. 1999) and learning in rats with a U-shaped dose-response relationship (Levin and Caldwell 2006). In adults with ADHD, low-dose mecamylamine improved recognition memory (Potter et al. 2009). Later studies with low doses of more selective nAChR antagonists found indication for such effects also in the attention domain, suggesting that the performance-enhancing effects in mnemonic paradigms may have been mediated by effects on attention. Small doses of methyllycaconitine improved accuracy in the simplified 5-CSRTT (Hahn et al. 2011), and dihydro-beta-erythroidine improved %Hits in the SAT in an inversely dose-related manner (Levin et al. 2013). However, the maximally effective dose of DHβE in the SAT was without effect in the 5-CSRTT, where DHβE even produced small performance impairments. Vice versa, MLA at a dose effective in the 5-CSRTT was without effect in the SAT when tested in unimpaired animals. Thus, more work remains to be done to verify the robustness of the phenomenon and characterize the nature of the involvement of different nAChR subtypes.

The mechanism mediating performance-improvement by low-dose nAChR antagonism has to remain speculative at this point in time. The most intuitive explanation is based on the fact that nAChRs tend to desensitize following agonist exposure, and that many behavioral effects of nAChR agonists may in fact be mediated by nAChR inactivation rather than channel opening (Picciotto et al. 2008; Buccafusco et al. 2009). The above effects of nAChR antagonists may thus mimic effects of agonist-induced receptor desensitization. However, findings that some performance benefits of nAChR agonists are reversed by nAChR antagonists (Hahn et al. 2011; Grottick and Higgins 2000; Grottick et al. 2003; Blondel et al. 2000) limit the generality of this explanation and suggest that there may be an optimal nAChR tone that differs between different aspects of performance. Depending on nAChR localization on excitatory versus inhibitory afferents, the net effect of both activation and desensitization can be in opposite directions (Giniatullin et al. 2005). The modulation of excitatory and inhibitory neurons by nAChR subtypes displaying different desensitization and recovery kinetics (Dani et al. 2000; Giniatullin et al. 2005) depends on brain structure, system, and pre- versus postsynaptic location (Gotti et al. 2009; Alkondon and Albuquerque 2004). Thus, changes in the tone of any nAChR subtype combination can be expected to uniquely modulate an intricate system balance.

7.2 *Human Drug Development and Clinical Trials*

To date, the development of novel nAChR agonists for the treatment of cognitive deficits has been focused on compounds selective for either the α4β2* or the α7 nAChR subtype (Wallace et al. 2011a; Gundisch and Eibl 2011), the two most abundantly expressed subtypes in the brain. The targeting of the most widely expressed nAChR subtypes may reflect the fact that the system(s) critical for mediating attention-enhancing effects of nAChR agonists are still largely unknown.

Several α4β2*-selective nAChR agonists have advanced to clinical trial-stage for Alzheimer's disease, schizophrenia, and ADHD (Haydar and Dunlop 2010). ABT-089 (pozanicline) reversed scopolamine-induced deficits in attention in healthy adults and is being further evaluated for the treatment of ADHD, but development for Alzheimer's disease was terminated due to lack of efficacy. AZD-3480 (ispronicline) is being developed for the treatment of Alzheimer's disease and ADHD, but no longer for schizophrenia after a failed Phase-II trial (Velligan et al. 2012).

The clinical development of α7-selective nAChR agonists was originally dominated by the treatment of cognitive impairment associated with schizophrenia. This was based on findings that sensory gating deficits in people with schizophrenia and unaffected relatives, and the diagnosis of schizophrenia itself, are associated with polymorphisms in or near the α7 subunit gene and promoter region (Freedman et al. 1997; Leonard et al. 2002; Martin et al. 2004). The partial α7-selective agonist GTS-21 (DMXB-A) was indeed found to alleviate sensory gating deficits and produce small cognitive, in particular attentional, improvements in schizophrenia (Freedman et al. 2008; Olincy et al. 2006). However, the effects were too subtle to pass Phase-II trial stage. Subtle pro-cognitive, including attention-enhancing effects in this population were also found with the partial α7 agonists TC-5619 (Lieberman et al. 2013) and tropisetron (Shiina et al. 2010; Zhang et al. 2012) but not RG3487 (Umbricht et al. 2014). No positive outcome was reported from similar trials in Alzheimer's disease and ADHD (Haydar and Dunlop 2010). Several other full and partial α7 agonists have been evaluated for the treatment of schizophrenia and Alzheimer's disease. Some showed efficacy in Phase I and II trials, others did not (Haydar and Dunlop 2010).

Overall, performance benefits of subtype-selective compounds developed to date may have turned out to be more subtle than anticipated. Remarkably, benefits have been demonstrated with both α4β2*-selective agonists sparing α7 nAChRs, and with α7-selective agonists sparing α4β2* nAChRs, suggesting that enhanced efficacy may be expected with less selective compounds acting on both nAChR subtypes. Indeed, none of the novel compounds has undergone direct comparison with the prototypical nonselective nAChR agonist nicotine.

A clear benefit of the ability to evoke the cognitive benefits of nicotine with more selective nAChR agonists is the reduced potential for unwanted side effects. Some of the tested compounds appeared to meet this criterion, particularly with regard to peripheral side effects. However, novel compounds targeting the α4β2* or α7 nAChR leave few central systems unaffected (Gotti et al. 2007; Seguela et al. 1993) and do not spare the mesolimbic DA system (Mansvelder et al. 2006b) implicated in dependence-related effects of nAChR agonists (Di Chiara 2000; Rose and Corrigall 1997). To the author's knowledge, compounds selective for other subtypes of the nAChR have not been developed clinically. The development of positive allosteric modulator, which do not activate the receptor on their own but bind to a second, modulatory site and facilitate agonist-induced responses (Williams et al. 2011), may aid the effort to successfully isolate nAChR subtypes and enhance nicotinic tone while being more sparing of native circuit dynamics. In any case, a more selective targeting of the systems mediating the attention-enhancing effects of nAChR ligands requires knowledge of the neuroanatomical structures and secondary neurotransmitter systems responsible for mediating these specific aspects of their broad behavioral profile.

8 Secondary Neurotransmitter Systems Mediating Effects of nAChR Agonists on Attention

Nicotine potentiates the cortical and subcortical release of acetylcholine, dopamine, noradrenaline (NA), 5-HT, histamine, GABA, glutamate and glycine via action at pre- and postsynaptic nAChRs (Lopez et al. 2001; Role and Berg 1996; MacDermott et al. 1999; Wonnacott et al. 2006; Rollema et al. 2009). Despite overlap, these systems differ in nAChR subtype expression, enabling a degree of selective modulation by subtype-selective ligands. Evidence regarding the target system(s) central to the attention-enhancing effects of nicotine is still sparse, but some progress has been made. Such knowledge would encourage detailed analysis of the nAChR subtypes modulating these systems, and likely channel drug development towards more selectively expressed subtypes, allowing for a narrower, more targeted effects profile.

A series of systemic interaction studies employing the simplified 5-CSRTT tested the potential of dopaminergic, noradrenergic, glutamatergic, and serotonergic antagonists to reverse the attention-enhancing effects of nicotine. To differentiate any potential performance effects of the antagonists alone from their modulation of nicotine effects, these studies focused on antagonist doses that did not affect performance when given by themselves. Neither the D1-type dopamine receptor antagonist SCH23390 nor the D2-type antagonist raclopride reduced the effects of nicotine on response accuracy at doses that are known to readily block effects of nicotine dependent on subcortical dopamine release (Hahn et al. 2002a). Effects of nicotine on omission errors were also unaffected. However, raclopride dose-dependently reversed reductions in response latency by nicotine, consistent with the role of dopamine in the stimulant effects of nAChR agonists. This concords with the above-mentioned finding that a nAChR agonist devoid of subcortical DA-releasing properties (Mirza et al. 1996) had no effects on response latency but a preserved ability to enhance response accuracy (Hahn et al. 2003). Together, these two lines of evidence provide consistent evidence that subcortical DA release is not involved in the attention-enhancing effects of nAChR agonists, suggesting that these effects do not depend on the mechanism central to the psychomotor stimulant and dependence-related properties of nicotine.

By contrast, the attentional effects of nicotine in the 5-CSRTT were reversed by the β-adrenoceptor antagonist propranolol at doses that had no effects given alone. Specifically, propranolol antagonized the effects of nicotine on accuracy and omission errors (Hahn and Stolerman 2005), indicating a role of noradrenaline release acting via β-adrenoceptors in the attention-enhancing effects of nAChR agonists. The α_1, α_{2B}, and α_{2C} antagonist prazosin did not modulate effects of nicotine on attention. Similarly, the attention-enhancing effects of nicotine were antagonized by the NMDA glutamate receptor antagonist CPP but not by the mGlu5 antagonist MPEP (Quarta et al. 2007). The serotonergic antagonists WAY-100635 (5-HT1A) and SB-242084 (5-HT2C) modulated effects on rate- and speed-dependent 5-CSRTT measures in opposite ways, but not nicotine's effects on response accuracy (Quarta et al. 2012). However, the 5-HT2A antagonist ketanserin attenuated the increase in %Hits and %Correct Rejections by chronic nicotine in the SAT (Rezvani et al. 2005). Thus, evidence to date suggests that nAChRs on or afferent to noradrenergic and glutamatergic, but not dopaminergic projections are critical mediators, with a potential role also for parts of the 5-HT system.

9 Large-Scale Systems Mediating the Effects of nAChR Agonists on Attention

The effects of nAChR agonists on neural network activity reflect the summation of changes in both excitatory and inhibitory neurotransmission, pre- and postsynaptic mechanisms, and receptor activation and desensitization. Large-scale system effects can be measured as changes in network oscillations, event-related potentials (ERPs), and the relative energy requirement of brain regions and networks.

9.1 Human EEG

Synchronous neural network activity is reflected in specific EEG frequency bands, which have been linked to different behavioral states. Effects of nicotine on these electrocortical oscillations are generally marked by acceleration, with the power of lower frequency bands decreasing and the power of higher frequencies increasing (Mansvelder et al. 2006a). Such shifts may be reflective of heightened arousal and behavioral activation. The precise nature of this reported acceleration differs between studies. Nicotine or cigarette smoking has been reported to decrease delta power (Pritchard et al. 1995; Knott et al. 1999; Michel et al. 1988), increase (Michel et al. 1988) or decrease (Knott et al. 1999; Pritchard et al. 1995) theta power, increase alpha power (Knott et al. 1999; Pritchard et al. 1995; Michel et al. 1988), and shift activity within the alpha band toward higher frequencies (Lindgren et al. 1998). Increased power within the beta band is rarely reported (Pritchard et al. 1995; Knott et al. 1999). Inconsistencies between studies may be explained by findings suggesting that the effects of nicotine on EEG frequency bands, as well as inverse changes with cigarette withdrawal, depend on personality traits and behavioral state (Cinciripini 1986; Gilbert et al. 2004). The nAChR subtypes mediating electrocortical acceleration are largely unknown. However, the α4β2*-selective nAChR agonist AZD3480 has been reported to increase the alpha peak frequency (Dunbar et al. 2007), suggesting that α4β2* nAChRs are involved in the above effects.

In addition to spectral EEG indices of general behavioral activation, effects on event-related potential markers of specific attentional and preattentional processes have been investigated. As mentioned, several studies, especially in the context of cognitive impairment associated with schizophrenia, found benefits of nAChR agonists on preattentive processes reflective of sensory gating (Martin and Freedman 2007; Leiser et al. 2009). In particular, nicotine enhanced P50 inhibition in subsamples of healthy subjects (Knott et al. 2010a, b), and α7-selective nAChR agonists were found to enhance P50 gating in people with schizophrenia (Olincy et al. 2006; Shiina et al. 2010; Zhang et al. 2012). Non-α7 nAChR involvement in sensory gating is suggested by preclinical studies (Rollema et al. 2009; Schreiber et al. 2002; Radek et al. 2006; Wildeboer and Stevens 2008). The mismatch negativity (MMN), which occurs after an infrequent change in a repetitive stimulus sequence, is another preattentive ERP of sensory integration. Nicotine enhanced both visual and auditory MMN amplitude (Fisher et al. 2010; Martin et al. 2009), an effect that is also seen with the α4β2* nAChR agonist AZD3480 (Dunbar et al. 2007). In the realm of higher information processing functions potentially influenced by attention, nicotine has been reported to increase the amplitude of the P300 potential (Knott et al. 1999, 2011) thought to reflect processes of stimulus evaluation and categorization, although there is also a negative report (Lindgren et al. 1998). Another ERP finding that may reflect attentional modulation is the potentiation of the contingent negative variation, which was observed with cigarette smoking in deprived smokers (Hasenfratz et al. 1989).

9.2 Human Neuroimaging

In the last decade, a multitude of functional Magnetic Resonance Imaging (fMRI) studies, measuring changes in local blood-oxygen-level dependent (BOLD) signal as the index of neuronal activation, have investigated the neuroanatomical basis of attentional enhancement by nicotine and some other nAChR agonists. By quantifying BOLD signal changes in response to attention tasks in the presence and absence of nicotine, the aim was to identify the neuronal mediators of attention which nicotine modulates, and which might therefore also mediate its performance benefits. The results from these studies are heterogeneous, ranging from (a) increased regional activation to (b) decreased activation, and (c) increased deactivation by the task challenge.

(a) Increased task-induced activation of brain regions subserving attentional functions is the most intuitive mechanism of nicotine-induced attentional enhancement. An early study using the RVIP task described above found that a large dose of transdermal nicotine increased task-induced activation in parietal and occipital cortex, thalamus, and caudate in minimally deprived smokers (Lawrence et al. 2002). The parietal and thalamic changes were specific to blocks of RVIP versus a low-demand control task. The same dose of transdermal nicotine was shown to result in wide-spread increases in activation by both the RVIP and the control task in smokers with and without a diagnosis of schizophrenia (Hong et al. 2011). In the N-back working memory paradigm, in which subjects respond to stimuli that are the same as N trials ago, nicotine given to non-smokers (Kumari et al. 2003) and varenicline (acting mostly on α4β2* nAChRs as a partial agonist) given to deprived smokers (Loughead et al. 2010) increased the activation of regions responsive to working memory load, especially in the high-load conditions (Loughead et al. 2010). Similar results were obtained with nicotine in smokers with schizophrenia (Jacobsen et al. 2004). What the above studies have in common is the use of relatively resource-demanding task paradigms (RVIP and N-back), although effects of nicotine on the BOLD signal were not always selective for the high-load conditions. Furthermore, all of the above studies employed a block design; that is, they compared activation changes in ~30-s periods of task performance with that in periods of relative inactivity, or rest.

(b) Studies reporting reduced task-induced activation by nicotine, in contrast, tended to employ event-related designs, measuring activation changes of a more phasic nature, triggered by trial events. In visuospatial orienting paradigms, reduced frontoparietal, cingulate, and thalamic activation was reported by invalidly cued and uncued trials with low-dose nicotine gum in non-smokers (Thiel et al. 2005; Vossel et al. 2008; Giessing et al. 2006) and with large-dose transdermal nicotine in minimally deprived smokers (Hahn et al. 2007), and by validly cued trials with transdermal nicotine in smokers (Hahn et al. 2007; Rose et al. 2010). One of these experiments (Rose et al. 2010) also yielded evidence for *in*creased regional activation with nicotine when cues indicated the correct response hand, initiating processes of motor preparation. Thus, decreased activation by nicotine may be specific to attentional or early information processing functions. In a paradigm differentiating processes of simple stimulus detection, selective attention and divided attention, transdermal nicotine reduced activation in frontal, temporal, thalamic, and visual regions across all task conditions, although performance benefits were seen only in the stimulus detection and selective attention tasks (Hahn et al. 2009). This may suggest that the neuronal mechanisms reflected in the BOLD effects benefited specific attentional processes that played a minor role in the divided attention task. Furthermore, frontal eye field activation by an anti-saccade task was reduced by subcutaneous nicotine (Ettinger et al. 2009), and hippocampal activation induced by a smooth pursuit eye movement task was reduced by both nicotine and the partial α7 nAChR agonist DMXB-A in people with schizophrenia, who typically display hippocampal hyperactivity during such tasks (Tregellas et al. 2005, 2010). On an interindividual and trial-by-trial basis, decreases with nicotine in regional activation induced by oddball trials were accompanied by greater response speed benefits (Warbrick et al. 2011).

When studying smokers, it is difficult to distinguish, clearly and with certainty, performance-enhancing effects of nicotine from detrimental effects of nicotine withdrawal in the placebo condition. This applies equally to experiments aimed at investigating the effects of a nicotine challenge and to experiments designed to study the effects of nicotine withdrawal, given that in the latter scenario the non-withdrawn state is characterized by behaviorally relevant nicotine blood levels. Thus, the finding that overnight abstinence from cigarette smoking was associated with greater dorsolateral prefrontal cortex activation by a low-demand working memory task (Xu et al. 2005) is compatible with reduced task-induced activation by nicotine.

A common explanation of reduced activity accompanying equal or improved performance is that of enhanced functional efficiency. In other words, executing a cognitive operation requires less neuronal engagement and energy. The finding of reduced task-induced neuronal activation with another stimulant, methylphenidate, was interpreted in the same manner (Volkow et al. 2008). Enhanced functional efficiency may reflect greater automaticity, such that a cognitive operation is performed in a less effortful and resource-demanding manner. The cellular basis of this potential mechanism is unclear. However, reductions in activation do not appear to reflect non-specific drug effects on cerebral blood flow or the coupling between neuronal responses and brain hemodynamics. First, the observed reductions in activation do not represent absolute changes, but modulation of task-induced BOLD responses. Second, nicotine did not alter regional activation by visuomotor challenges (Jacobsen et al. 2002; Hahn et al. 2007, 2009), providing evidence against non-specific effects on coupling. Third, reduced activations were sometimes observed alongside increased activations in other task conditions or other brain regions (Giessing et al. 2006; Hahn et al. 2007; Rose et al. 2010). Interestingly, in one study (Hahn et al. 2007), the one instance of enhanced activation, observed in middle frontal gyrus, was negatively associated with reaction time benefits.

(c) Cognitive task challenges generally result not only in the activation of brain regions involved in mediating the functions engaged by the task, but also in the deactivation of a consistent set of brain regions. These regions were termed the default network of resting brain function and are thought to mediate task-independent thought processes related to mind-wandering (Gusnard and Raichle 2001). Employing a visuospatial cueing paradigm, an event-related functional Magnetic Resonance Imaging study found that effects of transdermal nicotine on cue-induced activity were concentrated on areas of the default network (Hahn et al. 2007). All effects consisted of inducing or enhancing cue-induced deactivation in these regions. Importantly, these effects displayed strong correlations with the performance-enhancing effects of nicotine, even when controlling for interindividual variability in nicotine plasma concentrations (Hahn et al. 2007), suggesting that nicotine's facilitation of default network deactivation plays an important role for its performance-enhancing properties. The finding of enhanced default network deactivation has since been replicated with transdermal nicotine in minimally deprived smokers during a selective attention/divided attention task (Hahn et al. 2009), with subcutaneous nicotine in smokers and nonsmokers during a pro-saccade task (Ettinger et al. 2009), and with DMXBA in nonsmokers with schizophrenia during smooth pursuit eye movements (Tregellas et al. 2011). The inverse pattern of reduced default network deactivation was found in tobacco withdrawal (Ashare et al. 2013), although the opposite has also been reported (Sweet et al. 2010). Several fMRI studies suggest that suppression of default network activity generally benefits the performance of external processing tasks (e.g., Polli et al. 2005; Weissman et al. 2006; Kelly et al. 2008; Buckner et al. 2008). Thus, a potentiation of this downregulation in response to task challenges can be expected to aid performance. The concept that nAChR agonists facilitate a shift from an internal to an external information processing mode by enhancing default network deactivation has a lot of face validity. However, not all fMRI studies reporting performance enhancement with nicotine found this effect; thus, it may be one of several mechanisms via which nAChR agonists improve attention.

10 Conclusions

Just as the facilitatory effects of nAChR agonists on attentional functions appear to be heterogeneous, likely including enhancement of early sensory integration, alertness, and vigilance, as well as processes of selective attention and attentional control, so do the brain mechanisms mediating these effects. This may not be surprising given the wide distribution of nAChRs throughout the brain and the fact that they interact with a multitude of brain sites and all major neurotransmitter systems. Thus, it appears likely that effects on different attentional functions may be mediated by distinct central mechanisms. Functional neuroimaging studies of the attentional effects of nicotine strongly suggest that this may be so, with nicotine modulating the brain sites engaged by the diverse processing tasks.

Inconsistencies with regard to which nAChR subtypes appear to mediate the attention-enhancing effects of nAChR agonists in preclinical models may be related to slight differences in the attentional demands of the task paradigm employed. The complex task-dependent interplay of different nAChR subtypes is exemplified by the finding that α7 nAChR blockade can sharpen nicotine-evoked cholinergic transients and enhance the attention-enhancing effects of nicotine, but only under task conditions that tax selective attention (Howe et al. 2010). Furthermore, while there is clear evidence for a role of dopamine in the psychomotor stimulant effects of nAChR agonists, effects on stimulus detection, per se, appear to depend on the modulation of noradrenergic and glutamatergic, but not dopaminergic neurotransmission. Depending on the task challenge, both types of effects may result in improved performance.

While the therapeutic potential of the proattentional effects of nAChR agonists is evident, much more needs to be known before these effects can be fully exploited clinically. The heterogeneity of nAChRs may allow targeting specific brain mechanisms and cognitive functions, thus remedying disease-specific deficits. A combination of nAChR subtypes may have to be targeted, with a different optimal receptor tone at each subtype, to harvest the full benefits. Subtype-selective positive allosteric nAChR modulators may enable even more fine-grained pharmacological manipulations. However, without detailed knowledge about the systems to be targeted and the precise nature of the modulation that would be beneficial to specific functions, the basis for the development of such tools is lacking. Thus, research to date may have only scratched the surface of the attentional effects potentially achievable with nicotinic compounds.

References

Adams CE, Stevens KE (2007) Evidence for a role of nicotinic acetylcholine receptors in schizophrenia. Front Biosci 12:4755–4772

Adcock RA, Dale C, Fisher M, Aldebot S, Genevsky A, Simpson GV et al (2009) When top-down meets bottom-up: auditory training enhances verbal memory in schizophrenia. Schizophr Bull 35:1132–1141

Alkondon M, Albuquerque EX (2004) The nicotinic acetylcholine receptor subtypes and their function in the hippocampus and cerebral cortex. Prog Brain Res 145:109–120

Allison C, Shoaib M (2013) Nicotine improves performance in an attentional set shifting task in rats. Neuropharmacology 64:314–320

Andersson K, Hockey GR (1977) Effects of cigarette smoking on incidental memory. Psychopharmacology 52:223–226

Arneric SP, Sullivan JP, Briggs CA, Donnelly-Roberts D, Anderson DJ, Raszkiewicz JL et al (1994) (S)-3-methyl-5-(1-methyl-2-pyrrolidinyl) isoxazole (ABT 418): a novel cholinergic ligand with cognition-enhancing and anxiolytic activities: I. In vitro characterization. J Pharmacol Exp Ther 270:310–318

Ashare RL, Valdez JN, Ruparel K, Albelda B, Hopson RD, Keefe JR et al (2013) Association of abstinence-induced alterations in working memory function and COMT genotype in smokers. Psychopharmacology 230:653–662

Barr RS, Culhane MA, Jubelt LE, Mufti RS, Dyer MA, Weiss AP et al (2008) The effects of transdermal nicotine on cognition in nonsmokers with schizophrenia and nonpsychiatric controls. Neuropsychopharmacology 33:480–490

Bates T, Mangan G, Stough C, Corballis P (1995) Smoking, processing speed and attention in a choice reaction time task. Psychopharmacology 120:209–212

Bentley P, Husain M, Dolan RJ (2004) Effects of cholinergic enhancement on visual stimulation, spatial attention, and spatial working memory. Neuron 41:969–982

Bizarro L, Stolerman IP (2003) Attentional effects of nicotine and amphetamine in rats at different levels of motivation. Psychopharmacology 170:271–277

Blondel A, Simon H, Sanger DJ, Moser P (1999) The effect of repeated nicotine administration on the performance of drug-naive rats in a five-choice serial reaction time task. Behav Pharmacol 10:665–673

Blondel A, Sanger DJ, Moser PC (2000) Characterisation of the effects of nicotine in the five-choice serial reaction time task in rats: antagonist studies. Psychopharmacology 149:293–305

Boorman JP, Groot-Kormelink PJ, Sivilotti LG (2000) Stoichiometry of human recombinant neuronal nicotinic receptors containing the b3 subunit expressed in Xenopus oocytes. J Physiol 529:565–577

Buccafusco JJ, Jackson WJ, Gattu M, Terry AV Jr (1995a) Isoarecolone-induced enhancement of delayed matching to sample performance in monkeys: role of nicotinic receptors. Neuroreport 6:1223–1227

Buccafusco JJ, Jackson WJ, Terry AV Jr, Marsh KC, Decker MW, Arneric SP (1995b) Improvement in performance of a delayed matching-to-sample task by monkeys following ABT-418: a novel cholinergic channel activator for memory enhancement. Psychopharmacology 120:256–266

Buccafusco JJ, Beach JW, Terry AV (2009) Desensitization of nicotinic acetylcholine receptors as a strategy for drug development. J Pharmacol Exp Ther 328:364–370

Buckner RL, Andrews-Hanna JR, Schacter DL (2008) The brain's default network: anatomy, function, and relevance to disease. Ann NY Acad Sci 1124:1–38

Bushnell PJ, Oshiro WM, Padnos BK (1997) Detection of visual signals by rats: effects of chlordiazepoxide and cholinergic and adrenergic drugs on sustained attention. Psychopharmacology 134:230–241

Chavez-Noriega LE, Crona JH, Washburn MS, Urrutia A, Elliott KJ, Johnson EC (1997) Pharmacological characterization of recombinant human neuronal nicotinic acetylcholine receptors h alpha 2 beta 2, h alpha 2 beta 4, h alpha 3 beta 2, h alpha 3 beta 4, h alpha 4 beta 2, h alpha 4 beta 4 and h alpha 7 expressed in Xenopus oocytes. J Pharmacol Exp Ther 280:346–356

Chen D, Patrick JW (1997) The alpha-bungarotoxin-binding nicotinic acetylcholine receptor from rat brain contains only the alpha7 subunit. J Biol Chem 272:24024–24029

Cinciripini PM (1986) The effects of smoking on electrocortical arousal in coronary prone (Type-A) and noncoronary prone (Type-B) subjects. Psychopharmacology 90:522–527

Clarke PB, Kumar R (1983) Characterization of the locomotor stimulant action of nicotine in tolerant rats. Br J Pharmacol 80:587–594

Clementi F, Fornasari D, Gotti C (2000) Neuronal nicotinic receptors, important new players in brain function. Eur J Pharmacol 393:3–10

Colrain IM, Mangan GL, Pellett OL, Bates TC (1992) Effects of post-learning smoking on memory consolidation. Psychopharmacology 108:448–451

Damaj MI, Creasy KR, Grove AD, Rosecrans JA, Martin BR (1994) Pharmacological effects of epibatidine optical enantiomers. Brain Res 664:34–40

Damaj MI, Creasy KR, Welch SP, Rosecrans JA, Aceto MD, Martin BR (1995) Comparative pharmacology of nicotine and ABT-418, a new nicotinic agonist. Psychopharmacology 120:483–490

Dani JA, Radcliffe KA, Pidoplichko VI (2000) Variations in desensitization of nicotinic acetylcholine receptors from hippocampus and midbrain dopamine areas. Eur J Pharmacol 393:31–38

Davies AR, Hardick DJ, Blagbrough IS, Potter BV, Wolstenholme AJ, Wonnacott S (1999) Characterisation of the binding of [3H]methyllycaconitine: a new radioligand for labelling alpha 7-type neuronal nicotinic acetylcholine receptors. Neuropharmacology 38:679–690

Decker MW, Brioni JD, Sullivan JP, Buckley MJ, Radek RJ, Raszkiewicz JL et al (1994) (S)-3-methyl-5-(1-methyl-2-pyrrolidinyl)isoxazole (ABT 418): a novel cholinergic ligand with cognition-enhancing and anxiolytic activities: II. In vivo characterization. J Pharmacol Exp Ther 270:319–328

Decker MW, Brioni JD, Bannon AW, Arneric SP (1995) Diversity of neuronal nicotinic acetylcholine receptors: lessons from behavior and implications for CNS therapeutics. Life Sci 56:545–570

Di Chiara G (2000) Role of dopamine in the behavioural actions of nicotine related to addiction. Eur J Pharmacol 393:295–314

Domier CP, Monterosso JR, Brody AL, Simon SL, Mendrek A, Olmstead R et al (2007) Effects of cigarette smoking and abstinence on stroop task performance. Psychopharmacology 195:1–9

Dunbar G, Boeijinga PH, Demazieres A, Cisterni C, Kuchibhatla R, Wesnes K et al (2007) Effects of TC-1734 (AZD3480), a selective neuronal nicotinic receptor agonist, on cognitive performance and the EEG of young healthy male volunteers. Psychopharmacology 191:919–929

Ettinger U, Williams SCR, Patel D, Michel TM, Nwaigwe A, Caceres A et al (2009) Effects of acute nicotine on brain function in healthy smokers and non-smokers: estimation of inter-individual response heterogeneity. Neuroimage 45:549–561

Fisher DJ, Scott TL, Shah DK, Prise S, Thompson M, Knott VJ (2010) Light up and see: enhancement of the visual mismatch negativity (vMMN) by nicotine. Brain Res 1313:162–171

Foulds J, Stapleton J, Swettenham J, Bell N, McSorley K, Russell MA (1996) Cognitive performance effects of subcutaneous nicotine in smokers and never-smokers. Psychopharmacology 127:31–38

Freedman R, Coon H, MylesWorsley M, OrrUrtreger A, Olincy A, Davis A et al (1997) Linkage of a neurophysiological deficit in schizophrenia to a chromosome 15 locus. Proc Natl Acad Sci USA 94:587–592

Freedman R, Olincy A, Buchanan RW, Harris JG, Gold JM, Johnson L et al (2008) Initial phase 2 trial of a nicotinic agonist in schizophrenia. Am J Psychiatry 165:1040–1047

Gentry CL, Lukas RJ (2002) Regulation of nicotinic acetylcholine receptor numbers and function by chronic nicotine exposure. Curr Drug Targets CNS Neurol Disord 1:359–385

Gerzanich V, Peng X, Wang F, Wells G, Anand R, Fletcher S et al (1995) Comparative pharmacology of epibatidine: a potent agonist for neuronal nicotinic acetylcholine receptors. Mol Pharmacol 48:774–782

Giessing C, Thiel CM, Rosler F, Fink GR (2006) The modulatory effects of nicotine on parietal cortex activity in a cued target detection task depend on cue reliability. Neuroscience 137:853–864

Gilbert DG, McClernon FJ, Rabinovich NE, Sugai C, Plath LC, Asgaard G et al (2004) Effects of quitting smoking on EEG activation and attention last for more than 31 days and are more severe with stress, dependence, DRD2 A1 allele, and depressive traits. Nicotine Tob Res 6:249–267

Gilbert DG, Izetelny A, Radtke R, Hammersley J, Rabinovich NE, Jameson TR et al (2005) Dopamine receptor (DRD2) genotype-dependent effects of nicotine on attention and distraction during rapid visual information processing. Nicotine Tob Res 7:361–379

Giniatullin R, Nistri A, Yakel JL (2005) Desensitization of nicotinic ACh receptors: shaping cholinergic signaling. Trends Neurosci 28:371–378

Goldberg SR, Risner ME, Stolerman IP, Reavill C, Garcha HS (1989) Nicotine and some related compounds: effects on schedule-controlled behaviour and discriminative properties in rats. Psychopharmacology 97:295–302

Gotti C, Moretti M, Gaimarri A, Zanardi A, Clementi F, Zoli M (2007) Heterogeneity and complexity of native brain nicotinic receptors. Biochem Pharmacol 74:1102–1111

Gotti C, Clementi F, Fornari A, Gaimarri A, Guiducci S, Manfredi I et al (2009) Structural and functional diversity of native brain neuronal nicotinic receptors. Biochem Pharmacol 78:703–711

Griesar WS, Zajdel DP, Oken BS (2002) Nicotine effects on alertness and spatial attention in non-smokers. Nicotine Tob Res 4:185–194

Grilly DM (2000) A verification of psychostimulant-induced improvement in sustained attention in rats: effects of d-amphetamine, nicotine, and pemoline. Exp Clin Psychopharmacol 8:14–21

Grilly DM, Simon BB, Levin ED (2000) Nicotine enhances stimulus detection performance of middle- and old-aged rats: a longitudinal study. Pharmacol Biochem Behav 65:665–670

Grobe JE, Perkins KA, Goettler-Good J, Wilson A (1998) Importance of environmental distractors in the effects of nicotine on short-term memory. Exp Clin Psychopharmacol 6:209–216

Grottick AJ, Higgins GA (2000) Effect of subtype selective nicotinic compounds on attention as assessed by the five-choice serial reaction time task. Behav Brain Res 117:197–208

Grottick AJ, Trube G, Corrigall WA, Huwyler J, Malherbe P, Wyler R et al (2000) Evidence that nicotinic alpha(7) receptors are not involved in the hyperlocomotor and rewarding effects of nicotine. J Pharmacol Exp Ther 294:1112–1119

Grottick AJ, Wyler R, Higgins GA (2001) A study of the nicotinic agonist SIB-1553A on locomotion, and attention as measured by the five-choice serial reaction time task. Pharmacol Biochem Behav 70:505–513

Grottick AJ, Haman M, Wyler R, Higgins GA (2003) Reversal of a vigilance decrement in the aged rat by subtype-selective nicotinic ligands. Neuropsychopharmacology 28:880–887

Gundisch D, Eibl C (2011) Nicotinic acetylcholine receptor ligands, a patent review (2006–2011). Expert Opin Ther Pat 21:1867–1896

Gusnard DA, Raichle ME (2001) Searching for a baseline: functional imaging and the resting human brain. Nat Rev Neurosci 2:685–694

Hahn B, Stolerman IP (2002) Nicotine-induced attentional enhancement in rats: effects of chronic exposure to nicotine. Neuropsychopharmacology 27:712–722

Hahn B, Stolerman IP (2005) Modulation of nicotine-induced attentional enhancement in rats by adrenoceptor antagonists. Psychopharmacology 177:438–447

Hahn B, Shoaib M, Stolerman IP (2002a) Effects of dopamine receptor antagonists on nicotine-induced attentional enhancement. Behav Pharmacol 13:621–632

Hahn B, Shoaib M, Stolerman IP (2002b) Nicotine-induced enhancement of attention in the five-choice serial reaction time task: the influence of task-demands. Psychopharmacology 162:129–137

Hahn B, Sharples CG, Wonnacott S, Shoaib M, Stolerman IP (2003) Attentional effects of nicotinic agonists in rats. Neuropharmacology 44:1054–1067

Hahn B, Ross TJ, Yang Y, Kim I, Huestis MA, Stein EA (2007) Nicotine enhances visuospatial attention by deactivating areas of the resting brain default network. J Neurosci 27:3477–3489

Hahn B, Ross TJ, Wolkenberg FA, Shakleya DM, Huestis MA, Stein EA (2009) Performance effects of nicotine during selective attention, divided attention, and simple stimulus detection: an fMRI study. Cereb Cortex 19:1990–2000

Hahn B, Shoaib M, Stolerman IP (2011) Selective nicotinic receptor antagonists: effects on attention and nicotine-induced attentional enhancement. Psychopharmacology 217:75–82

Harris JG, Kongs S, Allensworth D, Martin L, Tregellas J, Sullivan B et al (2004) Effects of nicotine on cognitive deficits in schizophrenia. Neuropsychopharmacology 29:1378–1385

Harvey SC, Luetje CW (1996) Determinants of competitive antagonist sensitivity on neuronal nicotinic receptor beta subunits. J Neurosci 16:3798–3806

Harvey SC, Maddox FN, Luetje CW (1996) Multiple determinants of dihydro-beta-erythroidine sensitivity on rat neuronal nicotinic receptor alpha subunits. J Neurochem 67:1953–1959

Hasenfratz M, Battig K (1992) Action profiles of smoking and caffeine: Stroop effect, EEG, and peripheral physiology. Pharmacol Biochem Behav 42:155–161

Hasenfratz M, Michel C, Nil R, Battig K (1989) Can smoking increase attention in rapid information-processing during noise—electrocortical, physiological and behavioral-effects. Psychopharmacology 98:75–80

Hatsukami D, Fletcher L, Morgan S, Keenan R, Amble P (1989) The effects of varying cigarette deprivation duration on cognitive and performance tasks. J Subst Abuse 1:407–416

Haydar SN, Dunlop J (2010) Neuronal nicotinic acetylcholine receptors—targets for the development of drugs to treat cognitive impairment associated with schizophrenia and Alzheimer's disease. Curr Top Med Chem 10:144–152

Heishman SJ, Henningfield JE (2000) Tolerance to repeated nicotine administration on performance, subjective, and physiological responses in nonsmokers. Psychopharmacology 152:321–333

Heishman SJ, Snyder FR, Henningfield JE (1990) Effect of repeated nicotine administration in nonsmokers. NIDA Res Monogr 105:314–315

Heishman SJ, Snyder FR, Henningfield JE (1993) Performance, subjective, and physiological effects of nicotine in non-smokers. Drug Alcohol Depend 34:11–18

Heishman SJ, Taylor RC, Henningfield JE (1994) Nicotine and smoking: a review of effects on human performance. Exp Clin Psychopharmacol 2:345–395

Heishman SJ, Kleykamp BA, Singleton EG (2010) Meta-analysis of the acute effects of nicotine and smoking on human performance. Psychopharmacology 210:453–469

Hong LE, Schroeder M, Ross TJ, Buchholz B, Salmeron BJ, Wonodi I et al (2011) Nicotine enhances but does not normalize visual sustained attention and the associated brain network in schizophrenia. Schizophr Bull 37:416–425
Howe WM, Ji J, Parikh V, Williams S, Mocaer E, Trocme-Thibierge C et al (2010) Enhancement of attentional performance by selective stimulation of alpha4beta2(*) nAChRs: underlying cholinergic mechanisms. Neuropsychopharmacology 35:1391–1401
Hoyle E, Genn RF, Fernandes C, Stolerman IP (2006) Impaired performance of alpha7 nicotinic receptor knockout mice in the five-choice serial reaction time task. Psychopharmacology 189:211–223
Hurst R, Rollema H, Bertrand D (2013) Nicotinic acetylcholine receptors: from basic science to therapeutics. Pharmacol Ther 137:22–54
Jacobsen LK, Gore JC, Skudlarski P, Lacadie CM, Jatlow P, Krystal JH (2002) Impact of intravenous nicotine on BOLD signal response to photic stimulation. Magn Reson Imaging 20:141–145
Jacobsen LK, D'Souza DC, Mencl WE, Pugh KR, Skudlarski P, Krystal JH (2004) Nicotine effects on brain function and functional connectivity in schizophrenia. Biol Psychiatry 55:850–858
Jones GM, Sahakian BJ, Levy R, Warburton DM, Gray JA (1992) Effects of acute subcutaneous nicotine on attention, information processing and short-term memory in Alzheimer's disease. Psychopharmacology 108:485–494
Jones KM, McDonald IM, Bourin C, Olson RE, Bristow LJ, Easton A (2014) Effect of alpha7 nicotinic acetylcholine receptor agonists on attentional set-shifting impairment in rats. Psychopharmacology 231:673–683
Juliano LM, Fucito LM, Harrell PT (2011) The Influence of nicotine dose and nicotine dose expectancy on the cognitive and subjective effects of cigarette smoking. Exp Clin Psychopharmacol 19:105–115
Kelly AM, Uddin LQ, Biswal BB, Castellanos FX, Milham MP (2008) Competition between functional brain networks mediates behavioral variability. Neuroimage 39:527–537
Kendziorra K, Wolf H, Meyer PM, Barthel H, Hesse S, Becker GA et al (2011) Decreased cerebral a4β2* nicotinic acetylcholine receptor availability in patients with mild cognitive impairment and Alzheimer's disease assessed with positron emission tomography. Eur J Nucl Med Mol Imaging 38:515–525
Kenney JW, Gould TJ (2008) Modulation of hippocampus-dependent learning and synaptic plasticity by nicotine. Mol Neurobiol 38:101–121
Knott V, Bosman M, Mahoney C, Ilivitsky V, Quirt K (1999) Transdermal nicotine: Single dose effects on mood, EEG, performance, and event-related potentials. Pharmacol Biochem Behav 63:253–261
Knott V, Millar A, Fisher D, Albert P (2010a) Effects of nicotine on the amplitude and gating of the auditory P50 and its influence by dopamine D2 receptor gene polymorphism. Neuroscience 166:145–156
Knott VJ, Fisher DJ, Millar AM (2010b) Differential effects of nicotine on P50 amplitude, its gating, and their neural sources in low and high suppressors. Neuroscience 170:816–826
Knott VJ, Millar AM, McIntosh JF, Shah DK, Fisher DJ, Blais CM et al (2011) Separate and combined effects of low dose ketamine and nicotine on behavioural and neural correlates of sustained attention. Biol Psychol 88:83–93
Koelega HS (1993) Stimulant drugs and vigilance performance: a review. Psychopharmacology 111:1–16
Kumari V, Gray JA, ffytche DH, Mitterschiffthaler MT, Das M, Zachariah E et al (2003) Cognitive effects of nicotine in humans: an fMRI study. Neuroimage 19:1002–1013
Lawrence NS, Ross TJ, Stein EA (2002) Cognitive mechanisms of nicotine on visual attention. Neuron 36:539–548
Leiser SC, Bowlby MR, Comery TA, Dunlop J (2009) A cog in cognition: how the alpha 7 nicotinic acetylcholine receptor is geared towards improving cognitive deficits. Pharmacol Ther 122:302–311

Leonard S, Gault J, Hopkins J, Logel J, Vianzon R, Short M et al (2002) Association of promoter variants in the alpha 7 nicotinic acetylcholine receptor subunit gene with an inhibitory deficit found in schizophrenia. Arch Gen Psychiatry 59:1085–1096

Lerman C, Audrain J, Tercyak K, Hawk LW Jr, Bush A, Crystal-Mansour S et al (2001) Attention-deficit hyperactivity disorder (ADHD) symptoms and smoking patterns among participants in a smoking-cessation program. Nicotine Tob Res 3:353–359

Levin ED (1992) Nicotinic systems and cognitive function. Psychopharmacology 108:417–431

Levin ED, Caldwell DP (2006) Low-dose mecamylamine improves learning of rats in the radial-arm maze repeated acquisition procedure. Neurobiol Learn Mem 86:117–122

Levin ED, Rezvani AH (2002) Nicotinic treatment for cognitive dysfunction. Curr Drug Targets CNS Neurol Disord 1:423–431

Levin ED, Conners CK, Sparrow E, Hinton SC, Erhardt D, Meck WH et al (1996) Nicotine effects on adults with attention-deficit/hyperactivity disorder. Psychopharmacology 123:55–63

Levin ED, Conners CK, Silva D, Hinton SC, Meck WH, March J et al (1998) Transdermal nicotine effects on attention. Psychopharmacology 140:135–141

Levin ED, Cauley M, Rezvani AH (2013) Improvement of attentional function with antagonism of nicotinic receptors in female rats. Eur J Pharmacol 702:269–274

Lieberman JA, Dunbar G, Segreti AC, Girgis RR, Seoane F, Beaver JS et al (2013) A randomized exploratory trial of an alpha-7 nicotinic receptor agonist (TC-5619) for cognitive enhancement in schizophrenia. Neuropsychopharmacology 38:968–975

Lindgren M, Stenberg G, Rosen I (1998) Effects of nicotine in a bimodal attention task. Neuropsychobiology 38:42–49

Lopez E, Arce C, Vicente S, Oset-Gasque MJ, Gonzalez MP (2001) Nicotinic receptors mediate the release of amino acid neurotransmitters in cultured cortical neurons. Cereb Cortex 11:158–163

Loughead J, Ray R, Wileyto EP, Ruparel K, Sanborn P, Siegel S et al (2010) Effects of the alpha 4 beta 2 partial agonist varenicline on brain activity and working memory in abstinent smokers. Biol Psychiatry 67:715–721

Lukas RJ, Changeux J-P, Le Novere N, Albuquerque EX, Balfour DJK, Berg DK et al (1999) International union of pharmacology. XX. Current status of the nomenclature for nicotinic acetylcholine receptors and their subunits. Pharmacol Rev 51:397–401

Macallan DR, Lunt GG, Wonnacott S, Swanson KL, Rapoport H, Albuquerque EX (1988) Methyllycaconitine and (+)-anatoxin-a differentiate between nicotinic receptors in vertebrate and invertebrate nervous systems. FEBS Lett 226:357–363

MacDermott AB, Role LW, Siegelbaum SA (1999) Presynaptic ionotropic receptors and the control of transmitter release. Annu Rev Neurosci 22:443–485

Mackworth NH (1950) Researches on the measurement of human performance. Medical Research Council, Special report series 268. His Majesty's stationary office, London, p 156

Mahncke HW, Bronstone A, Merzenich MM (2006) Brain plasticity and functional losses in the aged: scientific bases for a novel intervention. Reprogram Brain 157:81–109

Mancuso G, Warburton DM, Melen M, Sherwood N, Tirelli E (1999) Selective effects of nicotine on attentional processes. Psychopharmacology 146:199–204

Mancuso G, Lejeune M, Ansseau M (2001) Cigarette smoking and attention: processing speed or specific effects? Psychopharmacology 155:372–378

Mangan GL, Golding JF (1983) The effects of smoking on memory consolidation. J Psychol 115:65–77

Mansvelder HD, Keath JR, McGehee DS (2002) Synaptic mechanisms underlie nicotine-induced excitability of brain reward areas. Neuron 33:905–919

Mansvelder HD, van Aerde KI, Couey JJ, Brussaard AB (2006a) Nicotinic modulation of neuronal networks: from receptors to cognition. Psychopharmacology 184:292–305

Mansvelder HD, van Aerde KI, Couey JJ, Brussaard AB (2006b) Nicotinic modulation of neuronal networks: from receptors to cognition. Psychopharmacology 184:292–305

Martin LF, Freedman R (2007) Schizophrenia and the alpha 7 nicotinic acetylcholine receptor. Integr Neurobiol Schizophr 78:225–246

Martin LF, Kem WR, Freedman R (2004) Alpha-7 nicotinic receptor agonists: potential new candidates for the treatment of schizophrenia. Psychopharmacology 174:54–64

Martin LF, Davalos DB, Kisley MA (2009) Nicotine enhances automatic temporal processing as measured by the mismatch negativity waveform. Nicotine Tob Res 11:698–706

McGaughy J, Sarter M (1995) Behavioral vigilance in rats: task validation and effects of age, amphetamine, and benzodiazepine receptor ligands. Psychopharmacology 117:340–357

McGaughy J, Decker MW, Sarter M (1999) Enhancement of sustained attention performance by the nicotinic acetylcholine receptor agonist ABT-418 in intact but not basal forebrain-lesioned rats. Psychopharmacology 144:175–182

McLean SL, Idris NF, Grayson B, Gendle DF, Mackie C, Lesage AS et al (2012) PNU-120596, a positive allosteric modulator of alpha7 nicotinic acetylcholine receptors, reverses a sub-chronic phencyclidine-induced cognitive deficit in the attentional set-shifting task in female rats. J Psychopharmacol 26:1265–1270

Michel C, Hasenfratz M, Nil R, Battig K (1988) Cardiovascular, electrocortical, and behavioral-effects of nicotine chewing gum. Klin Wochenschr 66:72–79

Millar NS, Gotti C (2009) Diversity of vertebrate nicotinic acetylcholine receptors. Neuropharmacology 56:237–246

Mirza NR, Stolerman IP (1998) Nicotine enhances sustained attention in the rat under specific task conditions. Psychopharmacology 138:266–274

Mirza NR, Pei Q, Stolerman IP, Zetterstrom TS (1996) The nicotinic receptor agonists (−)-nicotine and isoarecolone differ in their effects on dopamine release in the nucleus accumbens. Eur J Pharmacol 295:207–210

Morrison CF, Stephenson JA (1973) Effects of stimulants on observed behaviour of rats on six operant schedules. Neuropharmacology 12:297–310

Muir JL, Everitt BJ, Robbins TW (1995) Reversal of visual attentional dysfunction following lesions of the cholinergic basal forebrain by physostigmine and nicotine but not by the 5-HT3 receptor antagonist, ondansetron. Psychopharmacology 118:82–92

Murphy FC, Klein RM (1998) The effects of nicotine on spatial and non-spatial expectancies in a covert orienting task. Neuropsychologia 36:1103–1114

Nelsen JM, Goldstein L (1972) Improvement of performance on an attention task with chronic nicotine treatment in rats. Psychopharmacology 26:347–360

Nelsen JM, Goldstein L (1973) Chronic nicotine treatment in rats. 1. Acquisition and performance of an attention task. Res Commun Chem Pathol Pharmacol 5:681–693

Newhouse PA, Potter A, Singh A (2004) Effects of nicotinic stimulation on cognitive performance. Curr Opin Pharmacol 4:36–46

Nordberg A, Alafuzoff I, Winblad B (1992) Nicotinic and muscarinic subtypes in the human brain: changes with aging and dementia. J Neurosci Res 31:103–111

Nordberg A, Lundqvist H, Hartvig P, Lilja A, Langstrom B (1995) Kinetic analysis of regional (S)(-)11C-nicotine binding in normal and Alzheimer brains—in vivo assessment using positron emission tomography. Alzheimer Dis Assoc Disord 9:21–27

O'Brien JT, Colloby SJ, Pakrasi S, Perry EK, Pimlott SL, Wyper DJ (2007) alpha4beta2 nicotinic receptor status in Alzheimer's disease using 123I-5IA-85380 single-photon-emission computed tomography. J Neurol Neurosurg Psychiatry 78:356–362

Olincy A, Harris JG, Johnson LL, Pender V, Kongs S, Allensworth D et al (2006) Proof-of-concept trial of an alpha7 nicotinic agonist in schizophrenia. Arch Gen Psychiatry 63:630–638

Paolone G, Angelakos CC, Meyer PJ, Robinson TE, Sarter M (2013) Cholinergic control over attention in rats prone to attribute incentive salience to reward cues. J Neurosci 33:8321–8335

Parasuraman R, Warm JS, Dember WN (1987) Vigilance: taxonomy and utility. In: Mark LS, Warm JS, Houston RL (eds) Ergonomics and human factors. Springer, New York

Parrott AC, Craig D (1992) Cigarette smoking and nicotine gum (0, 2 and 4 mg): effects upon four visual attention tasks. Neuropsychobiology 25:34–43

Peeke SC, Peeke HV (1984) Attention, memory, and cigarette smoking. Psychopharmacology 84:205–216

Perkins KA, Grobe JE, Fonte C, Goettler J, Caggiula AR, Reynolds WA et al (1994) Chronic and acute tolerance to subjective, behavioral and cardiovascular effects of nicotine in humans. J Pharmacol Exp Ther 270:628–638

Perry E, Martin-Ruiz C, Lee M, Griffiths M, Johnson M, Piggott M et al (2000) Nicotinic receptor subtypes in human brain ageing, Alzheimer and Lewy body diseases. Eur J Pharmacol 393:215–222

Petrovsky N, Quednow BB, Ettinger U, Schmechtig A, Mossner R, Collier DA et al (2010) Sensorimotor gating is associated with CHRNA3 polymorphisms in schizophrenia and healthy volunteers. Neuropsychopharmacology 35:1429–1439

Phillips JM, McAlonan K, Robb WG, Brown VJ (2000) Cholinergic neurotransmission influences covert orientation of visuospatial attention in the rat. Psychopharmacology 150:112–116

Picciotto MR, Addy NA, Mineur YS, Brunzell DH (2008) It is not "either/or": activation and desensitization of nicotinic acetylcholine receptors both contribute to behaviors related to nicotine addiction and mood. Prog Neurobiol 84:329–342

Pimlott SL, Piggott M, Owens J, Greally E, Court JA, Jaros E et al (2004) Nicotinic acetylcholine receptor distribution in Alzheimer's disease, dementia with Lewy bodies, Parkinson's disease, and vascular dementia: in vitro binding study using 5-[(125)I]-A-85380. Neuropsychopharmacology 29:108–116

Polli FE, Barton JJ, Cain MS, Thakkar KN, Rauch SL, Manoach DS (2005) Rostral and dorsal anterior cingulate cortex make dissociable contributions during antisaccade error commission. Proc Natl Acad Sci USA 102:15700–15705

Poltavski DV, Petros T (2006) Effects of transdermal nicotine on attention in adult non-smokers with and without attentional deficits. Physiol Behav 87:614–624

Potter AS, Ryan KK, Newhouse PA (2009) Effects of acute ultra-low dose mecamylamine on cognition in adult attention-deficit/hyperactivity disorder (ADHD). Hum Psychopharmacol 24:309–317

Prendergast MA, Jackson WJ, Terry AV Jr, Decker MW, Arneric SP, Buccafusco JJ (1998) Central nicotinic receptor agonists ABT-418, ABT-089, and (−)-nicotine reduce distractibility in adult monkeys. Psychopharmacology 136:50–58

Pritchard WS, Robinson JH, Debethizy JD, Davis RA, Stiles MF (1995) Caffeine and smoking—subjective, performance, and psychophysiological effects. Psychophysiology 32:19–27

Provost SC, Woodward R (1991) Effects of nicotine gum on repeated administration of the Stroop test. Psychopharmacology 104:536–540

Quarta D, Naylor CG, Morris HV, Patel S, Genn RF, Stolerman IP (2007) Different effects of ionotropic and metabotropic glutamate receptor antagonists on attention and the attentional properties of nicotine. Neuropharmacology 53:421–430

Quarta D, Naylor CG, Glennon JC, Stolerman IP (2012) Serotonin antagonists in the five-choice serial reaction time task and their interactions with nicotine. Behav Pharmacol 23:143–152

Radek RJ, Miner HM, Bratcher NA, Decker MW, Gopalakrishnan M, Bitner RS (2006) Alpha(4) beta(2) nicotinic receptor stimulation contributes to the effects of nicotine in the DBA/2 mouse model of sensory gating. Psychopharmacology 187:47–55

Radek RJ, Kohlhaas KL, Rueter LE, Mohler EG (2010) Treating the cognitive deficits of schizophrenia with alpha4beta2 neuronal nicotinic receptor agonists. Curr Pharm Des 16:309–322

Revell AD (1988) Smoking and performance–a puff-by-puff analysis. Psychopharmacology 96:563–565

Rezvani AH, Caldwell DP, Levin ED (2005) Nicotinic-serotonergic drug interactions and attentional performance in rats. Psychopharmacology 179:521–528

Rezvani AH, Kholdebarin E, Brucato FH, Callahan PM, Lowe DA, Levin ED (2009) Effect of R3487/MEM3454, a novel nicotinic alpha7 receptor partial agonist and 5-HT3 antagonist on sustained attention in rats. Prog Neuropsychopharmacol Biol Psychiatry 33:269–275

Role LW, Berg DK (1996) Nicotinic receptors in the development and modulation of CNS synapses. Neuron 16:1077–1085

Rollema H, Hajos M, Seymour PA, Kozak R, Majchrzak MJ, Guanowsky V et al (2009) Preclinical pharmacology of the alpha 4 beta 2 nAChR partial agonist varenicline related to effects on reward, mood and cognition. Biochem Pharmacol 78:813–824

Rose JE, Corrigall WA (1997) Nicotine self-administration in animals and humans: similarities and differences. Psychopharmacology 130:28–40

Rose EJ, Ross TJ, Kurup PK, Stein EA (2010) Nicotine modulation of information processing is not limited to input (attention) but extends to output (intention). Psychopharmacology 209:291–302

Rusted JM, Warburton DM (1992) Facilitation of memory by post-trial administration of nicotine: evidence for an attentional explanation. Psychopharmacology 108:452–455

Sahakian B, Jones G, Levy R, Gray J, Warburton D (1989) The effects of nicotine on attention, information processing, and short- term memory in patients with dementia of the Alzheimer type. Br J Psychiatry 154:797–800

Salminen O, Whiteaker P, Grady SR, Collins AC, McIntosh JM, Marks MJ (2005) The subunit composition and pharmacology of alpha-conotoxin MII-binding nicotinic acetylcholine receptors studied by a novel membrane-binding assay. Neuropharmacology 48:696–705

Schreiber R, Dalmus M, De Vry J (2002) Effects of alpha(4)/beta(2)- and alpha(7)-nicotine acetylcholine receptor agonists on prepulse inhibition of the acoustic startle response in rats and mice. Psychopharmacology 159:248–257

Seguela P, Wadiche J, Dineley-Miller K, Dani JA, Patrick JW (1993) Molecular cloning, functional properties, and distribution of rat brain alpha 7: a nicotinic cation channel highly permeable to calcium. J Neurosci 13:596–604

Semenova S, Stolerman IP, Markou A (2007) Chronic nicotine administration improves attention while nicotine withdrawal induces performance deficits in the 5-choice serial reaction time task in rats. Pharmacol Biochem Behav 87:360–368

Shiina A, Shirayama Y, Niitsu T, Hashimoto T, Yoshida T, Hasegawa T et al (2010) A randomised, double-blind, placebo-controlled trial of tropisetron in patients with schizophrenia. Ann Gen Psychiatry 9:27

Shoaib M (2006) Effects of isoarecolone, a nicotinic receptor agonist in rodent models of nicotine dependence. Psychopharmacology 188:252–257

Singh A, Potter A, Newhouse P (2004) Nicotinic acetylcholine receptor system and neuropsychiatric disorders. IDrugs. 7:1096–1103

Snyder FR, Davis FC, Henningfield JE (1989) The tobacco withdrawal syndrome: performance decrements assessed on a computerized test battery. Drug Alcohol Depend 23:259–266

Spealman RD, Goldberg SR, Gardner ML (1981) Behavioral effects of nicotine: schedule-controlled responding by squirrel monkeys. J Pharmacol Exp Ther 216:484–491

Spilich GJ, June L, Renner J (1992) Cigarette smoking and cognitive performance. Br J Addict 87:1313–1326

Stewart C, Burke S, Marrocco R (2001) Cholinergic modulation of covert attention in the rat. Psychopharmacology 155:210–218

Stolerman IP (1999) Inter-species consistency in the behavioural pharmacology of nicotine dependence. Behav Pharmacol 10:559–580

Stolerman IP, Mirza NR, Shoaib M (1995) Nicotine psychopharmacology: addiction, cognition and neuroadaptation. Med Res Rev 15:47–72

Stolerman IP, Chandler CJ, Garcha HS, Newton JM (1997) Selective antagonism of behavioural effects of nicotine by dihydro-beta-erythroidine in rats. Psychopharmacology 129:390–397

Stolerman IP, Naylor CG, Mesdaghinia A, Morris HV (2009) The duration of nicotine-induced attentional enhancement in the five-choice serial reaction time task: lack of long-lasting cognitive improvement. Behav Pharmacol 20:742–754

Sullivan JP, Decker MW, Brioni JD, Donnelly-Roberts D, Anderson DJ, Bannon AW et al (1994) (±)-Epibatidine elicits a diversity of in vitro and in vivo effects mediated by nicotinic acetylcholine receptors. J Pharmacol Exp Ther 271:624–631

Sweet LH, Mulligan RC, Finnerty CE, Jerskey BA, David SP, Cohen RA et al (2010) Effects of nicotine withdrawal on verbal working memory and associated brain response. Psychiatry Res-Neuroimaging 183:69–74

Tani Y, Saito K, Tsuneyoshi A, Imoto M, Ohno T (1997) Nicotinic acetylcholine receptor (nACh-R) agonist-induced changes in brain monoamine turnover in mice. Psychopharmacology 129:225–232

Terriere E, Sharman M, Donaghey C, Herrmann L, Lonie J, Strachan M et al (2008) Alpha4 beta2-nicotinic receptor binding with 5-IA in Alzheimer's disease: methods of scan analysis. Neurochem Res 33:643–651

Terry AV, Buccafusco JJ, Prendergast MA (1999) Dose-specific improvements in memory-related task performance by rats and aged monkeys administered the nicotinic-cholinergic antagonist mecamylamine. Drug Dev Res 47:127–136

Thiel CM, Zilles K, Fink GR (2005) Nicotine modulates reorienting of visuospatial attention and neural activity in human parietal cortex. Neuropsychopharmacology 30:810–820

Tregellas JR, Tanabe JL, Martin LF, Freedman R (2005) fMRI of response to nicotine during a smooth pursuit eye movement task in schizophrenia. Am J Psychiatry 162:391–393

Tregellas JR, Olincy A, Johnson L, Tanabe J, Shatti S, Martin LF et al (2010) Functional magnetic resonance imaging of effects of a nicotinic agonist in Schizophrenia. Neuropsychopharmacology 35:938–942

Tregellas JR, Tanabe J, Rojas DC, Shatti S, Olincy A, Johnson L et al (2011) Effects of an alpha 7-nicotinic agonist on default network activity in schizophrenia. Biol Psychiatry 69:7–11

Tuesta LM, Fowler CD, Kenny PJ (2011) Recent advances in understanding nicotinic receptor signaling mechanisms that regulate drug self-administration behavior. Biochem Pharmacol 82:984–995

Turchi J, Holley LA, Sarter M (1995) Effects of nicotinic acetylcholine receptor ligands on behavioral vigilance in rats. Psychopharmacology 118:195–205

Turchi J, Holley LA, Sarter M (1996) Effects of benzodiazepine receptor inverse agonists and nicotine on behavioral vigilance in senescent rats. J Gerontol Ser A-Biol Sci Med Sci 51: B225–B231

Umbricht D, Keefe RS, Murray S, Lowe DA, Porter R, Garibaldi G et al (2014) A randomized, placebo-controlled study investigating the nicotinic a7 Agonist, RG3487, for cognitive deficits in schizophrenia. Neuropsychopharmacology 39:1568–1577

Velligan D, Brenner R, Sicuro F, Walling D, Riesenberg R, Sfera A et al (2012) Assessment of the effects of AZD3480 on cognitive function in patients with schizophrenia. Schizophr Res 134:59–64

Volkow ND, Fowler JS, Wang GJ, Telang F, Logan J, Wong C et al (2008) Methylphenidate decreased the amount of glucose needed by the brain to perform a cognitive task. PLoS ONE 3: e2017

Vossel S, Thiel CM, Fink GR (2008) Behavioral and neural effects of nicotine on visuospatial attentional reorienting in non-smoking subjects. Neuropsychopharmacology 33:731–738

Wallace TL, Ballard TM, Pouzet B, Riedel WJ, Wettstein JG (2011a) Drug targets for cognitive enhancement in neuropsychiatric disorders. Pharmacol Biochem Behav 99:130–145

Wallace TL, Callahan PM, Tehim A, Bertrand D, Tombaugh G, Wang SJ et al (2011b) RG3487, a novel nicotinic alpha 7 receptor partial agonist, improves cognition and sensorimotor gating in rodents. J Pharmacol Exp Ther 336:242–253

Warbrick T, Mobascher A, Brinkmeyer J, Musso F, Stoecker T, Shah NJ et al (2011) Direction and magnitude of nicotine effects on the fMRI BOLD response are related to nicotine effects on behavioral performance. Psychopharmacology 215:333–344

Warburton DM, Arnall C (1994) Improvements in performance without nicotine withdrawal. Psychopharmacology 115:539–542

Warburton DM, Mancuso G (1998) Evaluation of the information processing and mood effects of a transdermal nicotine patch. Psychopharmacology 135:305–310

Warburton DM, Rusted JM, Fowler J (1992a) A comparison of the attentional and consolidation hypotheses for the facilitation of memory by nicotine. Psychopharmacology 108:443–447

Warburton DM, Rusted JM, Muller C (1992b) Patterns of facilitation of memory by nicotine. Behav Pharmacol 3:375–378

Waters AJ (1998) The effects of smoking on performance on the Garner speeded classification task. Hum Psychopharmacol Clin Exp 13:477–491

Weissman DH, Roberts KC, Visscher KM, Woldorff MG (2006) The neural bases of momentary lapses in attention. Nat Neurosci 9:971–978

Wesnes K, Warburton DM (1983) Smoking, nicotine and human performance. Pharmacol Ther 21:189–208

Wesnes K, Warburton DM (1984) Effects of scopolamine and nicotine on human rapid information processing performance. Psychopharmacology 82:147–150

Wesnes K, Warburton DM, Matz B (1983) Effects of nicotine on stimulus sensitivity and response bias in a visual vigilance task. Neuropsychobiology 9:41–44

White HK, Levin ED (1999) Four-week nicotine skin patch treatment effects on cognitive performance in Alzheimer's disease. Psychopharmacology 143:158–165

Whiteaker P, Garcha HS, Wonnacott S, Stolerman IP (1995) Locomotor activation and dopamine release produced by nicotine and isoarecolone in rats. Br J Pharmacol 116:2097–2105

Wildeboer KM, Stevens KE (2008) Stimulation of the alpha 4 beta 2 nicotinic receptor by 5-I A-85380 improves auditory gating in DBA/2 mice. Brain Res 1224:29–36

Williams G (1980) Effects of cigarette smoking on immediate memory and performance in different kinds of smoker. Br J Psychol 71:83–90

Williams DK, Wang JY, Papke RL (2011) Positive allosteric modulators as an approach to nicotinic acetylcholine receptor-targeted therapeutics: advantages and limitations. Biochem Pharmacol 82:915–930

Witte EA, Davidson MC, Marrocco RT (1997) Effects of altering brain cholinergic activity on covert orienting of attention: comparison of monkey and human performance. Psychopharmacology 132:324–334

Wonnacott S, Barik J, Dickinson J, Jones IW (2006) Nicotinic receptors modulate transmitter cross talk in the CNS. J Mol Neurosci 30:137–140

Wooltorton JR, Pidoplichko VI, Broide RS, Dani JA (2003) Differential desensitization and distribution of nicotinic acetylcholine receptor subtypes in midbrain dopamine areas. J Neurosci 23:3176–3185

Xu JS, Mendrek A, Cohen MS, Monterosso J, Rodriguez P, Simon SL et al (2005) Brain activity in cigarette smokers performing a working memory task: effect of smoking abstinence. Biol Psychiatry 58:143–150

Young JW, Finlayson K, Spratt C, Marston HM, Crawford N, Kelly JS et al (2004) Nicotine improves sustained attention in mice: evidence for involvement of the alpha7 nicotinic acetylcholine receptor. Neuropsychopharmacology 29:891–900

Zhang XY, Liu L, Liu SW, Hong XH, Chen DC, Xiu MH et al (2012) Short-term tropisetron treatment and cognitive and P50 auditory gating deficits in schizophrenia. Am J Psychiatry 169:974–981

Nicotinic Receptors, Memory, and Hippocampus

Munir Gunes Kutlu and Thomas J. Gould

Abstract Nicotinic acetylcholine receptors (nAChRs) modulate the neurobiological processes underlying hippocampal learning and memory. In addition, nicotine's ability to desensitize and upregulate certain nAChRs may alter hippocampus-dependent memory processes. Numerous studies have examined the effects of nicotine on hippocampus-dependent learning, as well as the roles of low- and high-affinity nAChRs in mediating nicotine's effects on hippocampus-dependent learning and memory. These studies suggested that while acute nicotine generally acts as a cognitive enhancer for hippocampus-dependent learning, withdrawal from chronic nicotine results in deficits in hippocampus-dependent memory. Furthermore, these studies demonstrated that low- and high-affinity nAChRs functionally differ in their involvement in nicotine's effects on hippocampus-dependent learning. In the present chapter, we reviewed studies using systemic or local injections of acute or chronic nicotine, nAChR subunit agonists or antagonists; genetically modified mice; and molecular biological techniques to characterize the effects of nicotine on hippocampus-dependent learning.

Keywords Acetylcholine · Learning · Memory · Hippocampus · Nicotine · Addiction

Contents

1 Introduction 138
2 The Role of nAChRs in Hippocampus-Dependent Learning and Memory: Evidence from Psychopharmacology Studies 140
  2.1 Systemic Effects of Nicotine and nAChR Agonists/Antagonists 140
  2.2 Brain Regions Involved in Nicotine's Effect on Learning and Memory: Evidence from Direct Infusion Studies 145

M.G. Kutlu · T.J. Gould (✉)
Temple University, 1701 N. 13th St, Weiss Hall, Philadelphia, PA 19122, USA
e-mail: tgould@temple.edu

D.J.K. Balfour and M.R. Munafò (eds.), *The Neurobiology and Genetics of Nicotine and Tobacco*, Current Topics in Behavioral Neurosciences 23,
DOI 10.1007/978-3-319-13665-3_6

3 The Role of nAChRs in Hippocampus-Dependent Learning and Memory: Evidence from Gene Knockout Studies 148
3.1 Contextual and Trace Fear Conditioning 148
3.2 Spatial Learning 149
4 Cell-Signaling Cascades Involved in Nicotine's Effect on Learning and Memory 149
5 Effects of Nicotine on Hippocampus-Dependent Learning and Memory During Adolescence 151
6 Effects of Nicotine on Hippocampus-Dependent Learning: Evidence from Human Studies 152
7 Conclusion 153
References 154

1 Introduction

The hippocampus is widely populated with nicotinic acetylcholine receptors (nAChRs), which are pentameric cholinergic receptors with ligand-gated ion channels that are expressed both pre- and postsynaptically (Le Novere et al. 2007; Rush et al. 2002; Cordero-Erausquin et al. 2000). Nicotinic receptors are located on diverse neuronal subtypes including acetylcholinergic, serotonergic, GABAergic, dopaminergic, norepinephrinergic, and glutamatergic neurons (Freund et al. 1988; Fu et al. 1998; Grady et al. 1992; Radcliffe et al. 1999; Rapier et al. 1990; Ribeiro et al. 1993; Rowell and Winkler 1984; Westfall et al. 1983). Those nAChRs located on GABAergic and glutamatergic neurons (Fabian-Fine et al. 2001; Radcliffe et al. 1999), can modulate excitatory and inhibitory circuitries. Also, pre- and postsynaptic nAChR localization can result in functional differences for the same nAChR subtypes. While activation of presynaptic nAChRs can lead to neurotransmitter release (Araujo et al. 1988; Clarke and Reuben 1996; Fabian-Fine et al. 2001; Gray et al. 1996; Radcliffe et al. 1999; Wilkie et al. 1996; Zarei et al. 1999), postsynaptic nAChR activation can result in depolarization that can lead to second messenger activation (Alkondon et al. 1996; Fabian-Fine et al. 2001; Zarei et al. 1999; see Kenney and Gould 2008a for a review). Therefore, nAChRs may be critically involved in synaptic plasticity and enhance memory by activating learning-associated cell-signaling cascades (Broide and Leslie 1999; Wonnacott 1997; Perry et al. 2002; Berg and Conroy 2002; Barrantes et al. 1994, 1995; Sorensen et al. 1998; Porter et al. 1999; see Kenney and Gould 2008a for a review). Importantly, activation of nAChRs can directly induce synaptic plasticity in the form of long-term potentiation (LTP; He et al. 2000; Matsuyama et al. 2000; Matsuyama and Matsumoto 2003), a process underlying long-term memory formation (Bliss and Collingridge 1993). LTP is thought to occur through the N-methyl-D-aspartic acid (NMDA) receptor-mediated calcium influx into the cell (Lynch et al. 1990; Nicoll and Malenka 1999). LTP induction requires simultaneous depolarization of the NMDA receptors at the pre- and postsynaptic terminals (Bliss and Collingridge 1993). Upon simultaneous activation, NMDA receptors allow calcium to flow into the cell and therefore, intracellular calcium levels increase, which induces changes in protein activation, mRNA synthesis, and protein translation (Akers et al. 1986; Klann et al. 1991; Frey et al. 1996; Nguyen et al. 1994; Frey et al. 1988; Krug et al. 1984; Otani et al. 1989). Similarly, nAChR activation can also induce calcium flow into the cell (Karadsheh et al. 2004; McKay et al. 2007) and therefore, NMDA receptors and nAChRs may modulate learning and memory through similar cell-signaling cascades. One of these cascades, the CREB-dependent gene transcription pathway, is thought to be triggered by initial calcium influx, which activates adenylyl cyclase to convert intracellular adenosine diphosphate (ADP) to cyclic adenosine monophosphate (cAMP; Poser and Storm 2001). Increased concentration of cAMP in the cell activates protein kinase A (PKA) that is critically involved in learning and memory (Abel and Nguyen 2008). In turn, PKA phosphorylates cAMP-response element binding protein (CREB), which subsequently leads to protein translation and therefore, storage of long-term memory (Impey et al. 1998a; Silva et al. 1998). In addition, studies have also suggested that PKA may activate the mitogen-activated protein kinase (MAPK) pathway, which contributes to the activation of CREB (Impey et al. 1998b). In support, pharmacological inhibition of both PKA (Bernabeu et al. 1997; Huang and Kandel 1994) and prototypical MAPK, extracellular-regulated kinase 1/2 (ERK1/2; Atkins et al. 1998; English and Sweatt 1997), was found to impair long-term memory formation. Nicotine, an agonist of nAChRs, can directly activate nAChRs and, therefore, trigger the above-mentioned pathways in neurons and modulate learning and memory.

In contrast to acute nicotine, chronic nicotine has been suggested to desensitize these receptors, which may contribute to upregulation of nAChRs (Hulihan-Giblin et al. 1990; Sharp and Beyer 1986; Marks et al. 1983; Schwartz and Kellar 1983; Marks 1999). Even though dissociation between tolerance and receptor upregulation has been suggested (Collins et al. 1990; McCallum et al. 2000), there is evidence indicating that both desensitization and upregulation may contribute to the tolerance for the somatic and locomotor effects of chronic nicotine (Marks et al. 1983, 1985; Robinson et al. 2006, 2007). Furthermore, others have also proposed a link between receptor upregulation associated with chronic nicotine exposure and withdrawal symptoms (Dani and Heinemann 1996; Gould et al. 2012; Wilkinson and Gould 2013).

Neuronal nAChRs are assembled as homomeric receptors, containing only α subunits (α7–α10), or heteromeric receptors, containing a combination of α (α2–α6) and β (β2–β4) subunits (McGhee 1999; Decker et al. 1995; Hogg et al. 2003; Jones et al. 1999; see Gould and Leach 2014 for a review; see also chapter entitled Structure of Neuronal Nicotinic Receptors; this volume). Among nAChRs, α7 and α4β2* nAChRs (* designates potential additional subunit) are the most common types in the central nervous system (Marks et al. 1986; Marks and Collins 1982; Orr-Urtreger et al. 1995, 1997; Perry et al. 2002; Wonnacott 1986) and they manifest different characteristics in terms of localization, affinity to nicotine binding, and behavioral outcome. For example, while α7 nAChRs characteristically show lower affinity to nicotine and desensitize rapidly, high-affinity α4β2* nAChRs desensitize slowly and the upregulation of these receptors is longer lasting (Olale et al. 1997; Marks et al. 1985). Low- and high-affinity nAChRs are also differentially located in the brain. Low-affinity α7 containing nAChRs are mainly located in the granule and pyramidal cells of the hippocampus as well as other limbic system regions such as amygdala and hypothalamus (Dominguez del Toro et al. 1994; Fabian-Fine et al. 2001; Seguela et al. 1993). On the other hand, high-affinity nAChRs are mostly expressed in the dentate gyrus and CA1 subregions of the hippocampus (Perry et al. 2002). These differences suggest that low- and high-affinity nAChRs may play a differential role in modulating hippocampal function and thus potentially, learning and memory also.

In the present chapter, we review the existing literature of studies using systemic and local pharmacological manipulations, gene knockout techniques and, molecular biology techniques to examine the effects of nicotine on hippocampus-dependent learning and memory; as well as the functional differences between low- and high-affinity nAChRs in learning and memory.

2 The Role of nAChRs in Hippocampus-Dependent Learning and Memory: Evidence from Psychopharmacology Studies

Typically, in pharmacological studies, pharmacological agents are administered systemically to affect a whole neurotransmitter system or infused locally to limit the effect of the agent to a specific brain region. To investigate the effects of the pharmacological agent of interest on learning and memory, various behavioral paradigms that test different types of memory functions (e.g., spatial, fear, or working memory) are employed. In the following sections, we will review the studies examining the effects of systemic and local injections of nicotinic agonists and antagonists on hippocampus-dependent learning and memory.

2.1 Systemic Effects of Nicotine and nAChR Agonists/ Antagonists

2.1.1 Fear Conditioning

A large number of studies have investigated the effects of systemic injections of nicotine on hippocampus-dependent fear learning (e.g., Davis et al. 2006; Gould 2003; Gould et al. 2004; Gould and Higgins 2003; Gould and Wehner 1999; Wehner et al. 2004). Results from multiple studies have consistently demonstrated that systemic administration of acute nicotine-enhanced hippocampus-dependent contextual fear conditioning (see Box 1) while having no effect on hippocampus-independent cued fear (e.g., Gould and Higgins 2003; Gould and Wehner 1999). However, there is at least one study in rats reporting a contextual fear-conditioning deficit as a result of acute nicotine administration (Szyndler et al. 2001). Though, unlike the studies that reported acute nicotine-induced enhancement of contextual fear conditioning using intraperitoneal injections in mice, Szyndler and colleagues (2001) administered nicotine subcutaneously in rats, which, along with species and dose differences, may contribute to the contradicting results. There are several possible mechanisms through which nicotine could alter fear conditioning. For instance, nicotine could promote contextual processing, enhance unconditioned stimulus (US) salience, or strengthens Context-US associations. To answer this question, Kenney and Gould (2008b) used a context pre-exposure paradigm in which the animals were given exposure to the context prior to receiving immediate shocks in the same contexts. The results demonstrated that acute nicotine facilitated contextual learning but not directly the Context-US association; that is, nicotine-enhanced context processing which led to stronger context-US association.

Systemic acute nicotine has also been shown to enhance hippocampus-dependent trace fear conditioning (Davis and Gould 2007; Gould et al. 2004). In trace conditioning, the hippocampus is thought to be responsible for maintaining the memory trace during the temporal gap between the cue and the US (McEchron et al. 1998). Therefore, it is possible that the acute nicotine acts as a cognitive enhancer as it strengthens the memory trace during temporal delay, a process that is not required for delayed conditioning.

Although, acute nicotine was found to enhance contextual fear conditioning, chronic administration of nicotine was shown to have no effect on the same type of learning (André et al. 2008; Davis et al. 2005; Portugal et al. 2012a; Portugal and Gould 2009; Raybuck and Gould 2009). It has been suggested that with chronic nicotine administration, the cholinergic system in the hippocampus adapts by desensitizing and upregulating nAChRs (Dani and Heinemann 1996; Gould et al. 2012; Marks et al. 1993; Wilkinson and Gould 2013). Even though chronic nicotine does not have an effect, multiple studies have consistently shown that withdrawal from nicotine resulted in hippocampus-dependent learning deficits while not affecting hippocampus-independent learning (Davis et al. 2005; Kenney et al. 2011; Portugal et al. 2012a). The neurobiological mechanisms of the withdrawal effects are unknown. However, it is possible that following cessation of nicotine, the upregulated nAChRs resensitize, which could result in a hypersensitive cholinergic system and contribute to the nicotine withdrawal-associated deficits in hippocampus-dependent learning and memory. In support, Wilkinson and Gould (2013) found that nicotine administration following withdrawal resulted in a greater enhancement of learning and memory. This suggests that upregulated but desensitized nAChRs may regain function when nicotine is no longer available and produce a greater activity when nicotine is reintroduced. Generally, withdrawal effects disappear over time (Hughes 2007). For example, Gould and colleagues (2012) found that nicotine withdrawal deficits in mice lasted about 4 days, which was correlated with the time course of the nAChR upregulation in the hippocampus. These results support a role of hippocampal nAChR upregulation in the nicotine withdrawal-associated deficits in learning and memory. Although nAChRs upregulation is involved in withdrawal, it has been suggested that the tolerance developed during chronic nicotine administration might have a different neurobiological basis. For example, Portugal and colleagues (2012a) showed that while inbred strains of mice developed tolerance to the cognitive-enhancing effects of nicotine, only some of these strains showed withdrawal deficit in contextual fear conditioning. Importantly, Gould and colleagues (2014a) showed that nAChR upregulation and withdrawal deficits in contextual fear conditioning occurred after tolerance effects developed. Also, in the same study, the onset of tolerance was found to be dependent on the chronic nicotine dose, which means desensitization of nAChR may be mediating the tolerance effects. Together, these results suggest while nAChR upregulation contributes to the withdrawal effects, tolerance development to the cognitive effects of nicotine may be independent from this mechanism.

Similar to the process of fear memory acquisition, extinction of fear memories is also considered as a new form of learning (Konorski 1967). During extinction, a stimulus previously paired with a US is presented in the absence of the US in order to eliminate the conditioned response (CR) to that stimulus. While in cued extinction the cue is repeatedly presented alone; in contextual extinction, the animal stays in the context for a period of time in the absence of the US. Studies have shown that extinction learning is context-specific (Bouton 2004; Bouton et al. 2006). In other words, the extinction memory is specific to the context in which the excitatory cue is extinguished and therefore, the CR reappears if the animal is given cue presentations outside the extinction context. This phenomenon is known as renewal (Bouton and Bolles 1979; Bouton and King 1983). There is evidence showing that the hippocampus also plays a significant role in extinction and context-specificity of extinction memories (Corcoran et al. 2005; Hobin et al. 2006; Ji and Maren 2007).

Given that nicotine modulates hippocampus-dependent learning and memory, several studies have investigated the effects of nicotine on extinction. For example, Elias and colleagues (2010) showed that acute nicotine-enhanced extinction of cued fear when administered only during extinction while nicotine administration during both acquisition and extinction impaired cued extinction. Moreover, the same study found that nicotine administered during extinction blocked renewal of the extinguished fear response whereas nicotine administered during both acquisition and extinction enhanced renewal. These results are important as they show nicotine modulates the hippocampal control over extinction learning and context-specificity of extinction. Recently, Kutlu and Gould (2014) investigated acute nicotine's effects on extinction of contextual fear when administered only during extinction. The results from this study showed that an acute dose of nicotine during extinction impaired contextual fear extinction by enhancing retrieval of the contextual fear memories. Even though these results seem to contradict previous results showing nicotine-enhanced cued extinction, Kutlu and Gould's (2014) further analysis of the results of Elias and colleagues (2010) revealed that Elias and colleagues (2010) also found impaired contextual extinction within the first extinction session. Overall, acute nicotine administered during extinction only enhances cued extinction while disrupting contextual extinction. On the other hand, if acute nicotine administered during both training and extinction, enhanced contextual memories may interfere with cued extinction.

In addition to the studies that examined the effect of acute nicotine on extinction, Tian and colleagues (2008) investigated the effects of prior chronic nicotine on both cued and contextual fear extinction. The result of this study showed that in rats, prior chronic nicotine administration impaired cued fear extinction but did not affect extinction of contextual fear conditioning. However, the same study also showed that chronic nicotine enhanced retention of contextual fear extinction. Importantly, in this study, acquisition and extinction sessions were given 2 weeks after the last nicotine injections. Thus, the results of Tian and colleagues (2008) may be attributed to the effects of nicotine withdrawal rather than chronic nicotine administration. Together with the studies investigating the effects of acute nicotine on extinction, these results suggest that acute and chronic nicotine may have differential effects on cued and contextual fear extinction. Nevertheless, more studies are required to reach a more substantial conclusion about the effects of chronic and withdrawal from chronic nicotine on extinction.

As discussed earlier, even though nAChRs interact with NMDA receptors during learning, nAChR activation is not sufficient for learning but necessary for the effects of nicotine on hippocampus-dependent learning. In support, multiple studies have shown that the systemic injections of the nonspecific nAChR antagonist, mecamylamine, had no effect on contextual fear conditioning (Feiro and Gould 2005; Gould and Higgins 2003; Gould and Wehner 1999). However, the NMDA receptor antagonist MK801 resulted in severe impairment of contextual fear conditioning (Gould et al. 2002) and systemic injections of mecamylamine disrupted contextual fear learning when paired with subthreshold doses of MK801 (Gould and Lewis 2005). This suggests that NMDA receptors and nAChRs interact during the formation of hippocampus-dependent fear memories. Therefore, it is possible that additional activation of nAChRs may lead to a greater calcium influx into the cell and consequently, alter the cell-signaling cascades involved in the learning and memory processes.

Having no effect on normal learning, mecamylamine has been shown to eliminate the nicotine-induced enhancement of contextual fear conditioning, which suggests that nAChRs are necessary for the nicotine's effects on learning and memory (Gould and Higgins 2003). Also, even though the hippocampus is widely populated both by low- and high-affinity nAChRs, there is evidence showing that nicotine's effects on hippocampus-dependent learning are mostly mediated through the high-affinity nAChRs. For example, Davis and Gould (2006; also Davis et al. 2007) found that a systemic dose of an antagonist of the high-affinity $\alpha4\beta2^*$ nAChRs, dihydro-beta-erythoidine (DhβE), blocked the enhancing effects of acute nicotine on contextual fear conditioning while having no effect on cued fear and baseline contextual fear conditioning. Interestingly, the same study showed that a selective antagonist of the $\alpha7$ nAChR, methyllycaconitine (MLA), had no effect on nicotine's effect on contextual fear conditioning. This study suggests a larger involvement of $\alpha4\beta2^*$ nAChRs in the enhancement of contextual fear conditioning by acute nicotine administration. In line with $\alpha4\beta2^*$ nAChRs modulation of learning, Karadsheh and colleagues (2004) found that nicotine can increase calcium influx through activation of high-affinity nAChRs. Thus, it is possible that nicotine's effect on the hippocampus alters cell-signaling cascades through high-affinity nAChR-mediated calcium influx.

2.1.2 Spatial Learning and Working Memory

The reports on the effects of acute nicotine on spatial memory are mixed. There is evidence suggesting enhanced spatial learning in Morris Water Maze (MWM, see Box 1) as a result of acute nicotine administration (Abdulla et al. 1996; Sharifzadeh et al. 2005). There are also studies showing no effect of nicotine on spatial memory (Attaway et al. 1999) and nicotine-induced spatial memory deficit (Bernal et al. 1999). However, it is difficult to reconcile these studies as they used different nicotine doses, injection regimens, species, and strains. Several studies have also examined the effects of chronic nicotine on spatial learning. While there is evidence for improved spatial working memory with chronic nicotine administration (Socci et al. 1995; Abdulla et al. 1996; Bernal et al. 1999), another study found that chronic nicotine administration for 10 days impaired acquisition and retention of spatial memory (Scerri et al. 2006). However, this discrepancy could be explained by the different chronic nicotine administration methods employed in these studies. For example, while Scerri and colleagues (2006) used a continual nicotine administration protocol, other studies (Socci et al. 1995; Abdulla et al. 1996; Bernal et al. 1999) that found enhancement of spatial learning used repeated injections of nicotine. Due to nicotine's short life in animals (Matta et al. 2007), it is possible that repeated injections do not mimic chronic tobacco use in humans. Therefore, repeated injections of nicotine may reflect the effects of acute or subchronic nicotine on spatial learning.

Working memory is a cognitive process that temporarily maintains information online, which can be a necessary component for spatial learning (Awh and Jonides 2001). Results from studies examining the effects of acute nicotine on working memory in the radial arm maze (see Box 1) are also mixed. While the majority of the studies found enhancement of spatial working memory (Levin et al. 1997, 1998, 2003, 2005a; Levin and Torry 1996; Addy and Levin 2002), other studies have reported no effect of acute nicotine administration (Kholdebarin et al. 2007; Arthur and Levin 2002). Moreover, Levin and Torry (1996) observed nicotine-induced enhancement in the radial arm maze performance only in aged but not in the young rats. Similar to these reports, the studies using chronic nicotine administration reported enhanced spatial working memory (Levin and Torry1996; Levin et al. 1990, 1993a, b, 1999) or no effect (Bancroft and Levin 2000; Bettany and Levin 2001) in radial arm maze. Interestingly, Levin and colleagues (1992) also reported enhanced spatial working memory even after chronic nicotine administration ceased, which suggests that the effect of chronic nicotine on spatial working memory persists after cessation of nicotine administration. Together, it appears that acute nicotine's effects on spatial working memory may be dependent on other factors such as age. In addition, chronic nicotine has the ability to enhance spatial working memory, which suggests that chronic nicotine does not produce tolerance for its effects on spatial working memory.

An interaction between nAChRs and NMDA receptors may also mediate spatial working memory. Several studies have investigated the relationship between NMDA receptor activity, nicotine, and working memory. For example, acute nicotine can reverse working memory impairment observed after administration of the NMDA receptor antagonists dizocilpine (Levin et al. 1998) and ketamine (Rushforth et al. 2011). These studies indicate that, as in contextual fear conditioning, nAChRs interact with NMDA receptor to modulate NMDA activity in spatial working memory. The effects of nicotine on spatial working memory seem to be working through low-affinity α7 nAChR activation. For instance, there is evidence demonstrating that dizocilpine increases α7 nAChR binding in the hippocampus and this effect is reversed by chronic nicotine administration (Levin et al. 2005b). Also, Castner and colleagues (2011) demonstrated that systemic injections of the α7 nAChR agonist, AZD0328, dose-dependently improved spatial working memory in primates. Therefore, these results suggest that in the absence of NMDA receptor activity, low-affinity α7 nAChRs in the hippocampus compensate for the lack of NMDA function to maintain normal spatial working memory. Moreover, with functioning NMDA receptors, α7 nAChR activity may have an additive and therefore, enhancing effect on spatial working memory.

2.2 Brain Regions Involved in Nicotine's Effect on Learning and Memory: Evidence from Direct Infusion Studies

2.2.1 Contextual and Trace Fear Conditioning

As discussed above, systemic injections of nicotine and nAChR agonists/antagonists suggest that nicotine affects hippocampus-dependent learning and memory through high-affinity nAChRs in the hippocampus. However, involvement of different subregions of hippocampus and the interaction between the hippocampus and other brain regions that are important for learning and memory cannot be specified through systemic manipulations. Therefore, a number of studies investigated the involvement of nAChRs in hippocampal learning using local injections of nicotine and nAChR agonists/antagonists into different subregions of the hippocampus and other related brain regions such as prefrontal cortex (PFC).

Parallel to the results from the studies using systemic administration of nicotine, Davis et al. (2007) found that acute nicotine infusion into the dorsal hippocampus dose-dependently enhanced contextual fear conditioning. This effect was reversed when the high-affinity nAChR antagonist DhβE was systemically administered but systemic injections of the preferential α7 nAChR antagonist MLA had no effect. Additionally, DhβE reversed the enhancing effect of systemic nicotine injections when infused into the dorsal hippocampus. These results add to the body of evidence suggesting that nicotine affects hippocampal processing through the high-affinity nAChRs. Moreover, direct infusion of nicotine into the dorsal hippocampus was found to reverse NMDA antagonist MK801-induced deficits in contextual fear learning and systemic injections of nicotine reversed the detrimental effects of direct infusions into the dorsal hippocampus of another NMDA receptor antagonist,

DL-2-Amino-5-phosphonovaleric acid (APV, Andre et al. 2011). This supports the conclusion that NMDA receptor functions can be compensated by nAChR activity in the dorsal hippocampal. Finally, Kenney et al. (2012a) showed that nicotine infusions into the ventral hippocampus impaired contextual fear conditioning and α7-specific nAChR antagonist MLA infusions into the ventral hippocampus reversed this effect. Together, these results suggest that dorsal and ventral subregions of the hippocampus are functionally in competition to produce a behavioral response, which is in line with other studies suggesting that dorsal hippocampus is more involved in contextual and spatial information processing while the ventral hippocampus may be more involved in anxiety (Fanselow and Dong 2010). Therefore, it is possible that when systemically administered, nicotine promotes dorsal hippocampus processing but when locally infused into the ventral hippocampus it facilitates anxiogenic processes and consequently impairs learning and memory. Due to the fact that nicotine desensitizes low-affinity nAChRs faster and has lower affinity for these receptors than their high-affinity counterparts (Alkondon et al. 2000), it is possible that systemic injections of nicotine activate high-affinity nAChRs with lower doses whereas activation of low-affinity nAChRs requires higher doses of nicotine. Therefore, the nicotine-induced enhancement of contextual and trace fear conditioning may be achieved with lower systemic doses while higher doses of nicotine activate low-affinity nAChRs, which could disrupt learning and explain the inverted U-shaped curve for the effects of nicotine found in previous studies (Gould and Higgins 2003).

Also in parallel with systemic injections' studies, Davis and Gould (2009) found that cessation of infusion of chronic nicotine into the dorsal hippocampus impaired contextual fear conditioning while the same infusions into the thalamus or cortex did not produce any effect. Moreover, studies showed that in systemic chronic nicotine-treated mice, the direct infusions of the high-affinity nAChR antagonist DHβE into the dorsal hippocampus precipitated withdrawal deficits in contextual (Davis and Gould 2009) and trace (Raybuck and Gould 2009) fear conditioning. In contrast, infusions of the α7 subtype nAChRs antagonist MLA into the dorsal hippocampus did not have any effect on trace fear conditioning (Raybuck and Gould 2009). These studies further support the suggestion that high-affinity α4β2* nAChRs in the hippocampus underlie cognitive symptoms of nicotine withdrawal.

Although both contextual and trace fear-conditioning tasks depend on hippocampal activity, the PFC is also necessary for trace fear conditioning. For example, similar to contextual fear conditioning, direct nicotine infusions into the dorsal hippocampus enhanced trace fear conditioning. In contrast, direct nicotine infusions into the ventral hippocampus impaired trace conditioning (Raybuck and Gould 2010). There is evidence, however, showing that unlike contextual fear conditioning, trace fear conditioning recruits the PFC (Knight et al. 2004; Runyan et al. 2004). Supporting this view, Raybuck and Gould (2010) found that nicotine infusion into the medial PFC enhanced trace but not contextual or cued fear conditioning. Furthermore, the same study showed that infusions of the high-affinity nAChR antagonist DHβE into the dorsal hippocampus impaired trace fear conditioning in nicotine naïve mice but did not have any effect on either contextual or cued fear conditioning. Raybuck and Gould (2010) also found that neither MLA nor infusions of DHβE into the ventral hippocampus had any effect on any type of fear conditioning. Finally, infusions of both DHβE and MLA into the medial PFC enhanced trace conditioning in nicotine naïve mice while having no effect on contextual fear conditioning. Because both nAChR agonists and antagonists infused into the ventromedial PFC have similar effect on trace fear conditioning, this effect of nicotine on trace fear conditioning may be related to PFC desensitization. Overall, these results add to the literature suggesting that unlike contextual fear conditioning, trace fear conditioning requires involvement of the PFC as well as the hippocampus. This result is not surprising as trace fear conditioning requires active maintenance of the cue representation during the temporal gap between the cue and the US while contextual fear conditioning does not. Nevertheless, high-affinity nAChRs are responsible for the enhancing effects of nicotine on both contextual and trace fear conditioning.

2.2.2 Spatial Learning and Working Memory

Direct infusion studies further clarify the role of low- and high-affinity nAChRs in spatial learning and working memory. For example, studies have shown that acute infusions of both DHβE and MLA into the ventral hippocampus impaired spatial working memory (Levin et al. 2002) while another study found that chronic infusions of the DHβE but not MLA into the ventral hippocampus disrupted spatial working memory (Pocivavsek et al. 2006). Also, several studies investigated the interaction between nAChR antagonism and nicotine-induced working memory deficits (Bancroft and Levin 2000; Bettany and Levin 2001). The results of these studies showed that the chronic nicotine-induced spatial working memory deficits were reversed by the local infusions of DhβE into the ventral hippocampus but not with MLA. Interestingly, these studies suggest that both α4β2 and α7 nAChR may be involved in spatial working memory but only α4β2 nAChRs are important for the effects of nicotine on this type of memory. Moreover, in contrast to the aforementioned results indicating that the dorsal hippocampus is crucial for nicotine's effects on contextual fear memories, these results suggest that the ventral hippocampus is dominant in the effects of nicotine on spatial learning and memory. However, the results from the studies using direct infusions of low- and high-affinity nAChR antagonists (Pocivavsek et al. 2006; Bancroft and Levin 2000; Bettany and Levin 2001) seem to be in contradiction with the reports from studies using systemic injections of nAChR agonists and antagonists suggesting that low-affinity nAChRs modulate spatial learning (Levin et al. 2005b; Castner et al. 2011). This may suggest that brain regions other than the hippocampus may play role in modulating spatial learning and memory through nAChRs.

3 The Role of nAChRs in Hippocampus-Dependent Learning and Memory: Evidence from Gene Knockout Studies

3.1 Contextual and Trace Fear Conditioning

As discussed above, nAChRs in the hippocampus and PFC play a modulatory role in both regular contextual and trace fear conditioning and nicotine differentially mediates these types of fear memories depending on brain subregion and learning task. Importantly, studies using genetically modified mice, also known as Knock-Out (KO) mice, which lack specific nAChR can further clarify the specific nAChR subunits involved in the effects of nicotine on hippocampus-dependent learning and memory as KO technique allows more specificity than pharmacological antagonists. For example, in line with the results from the studies using pharmacological manipulations, studies have demonstrated that α7 KO mice showed normal contextual and cued fear conditioning as well as acute nicotine-induced enhancement and withdrawal deficits in contextual fear learning (Davis and Gould 2007; Paylor et al. 1998; Wehner et al. 2004; Portugal et al. 2008). Moreover, Wehner and colleagues (2004) found that in the absence of the β3 and β4 nAChR subunits, mice showed normal contextual fear conditioning and enhancement of contextual fear by acute nicotine. Interestingly, some nAChR subunits show sex differences in terms of their roles in cued fear conditioning. For example, Semenova and colleagues (2012) found that male β4 KO mice showed deficits in cued fear conditioning while females did not. Recently, Lotfipour et al. (2013) showed that α2 nAChR KO mice showed normal contextual fear conditioning and no effect of nicotine on contextual or trace fear conditioning.

Also, in line with the results from the pharmacological studies cited earlier, KO mice studies suggest that the β2 nAChR subunit plays a critical role in the effects of nicotine on contextual fear conditioning. In support, β2 KO mice did not show either the acute nicotine-induced enhancement (Davis and Gould 2007) or nicotine withdrawal-induced impairment of contextual fear conditioning (Portugal et al. 2008) relative to wild-type control animals. Both studies saw no deficits in contextual fear conditioning in β2 KO mice treated with saline. Another study however, found slight contextual fear-conditioning deficits in the β2 KO mice (Wehner et al. 2004). Interestingly, Caldarone and colleagues (2000) found that age is a factor for how the β2 subunit contributes to contextual and cued fear conditioning. For example, while younger β2 KO mice showed normal contextual and cued conditioning, aged β2 KO mice showed deficits in both forms of learning. However, there is also evidence of a loss of hippocampal NMDA binding in aged rats and monkeys while no effect was observed in the levels of choline acetyltransferase (ChAT) activity in the hippocampus (Wenk et al. 1991). NMDA receptors and nAChRs mediate similar processes (Gould and Lewis 2005; Andre et al. 2011); this suggests that an aging-related decrease in NMDA receptor functioning may contribute to the learning deficit found in β2 KO mice because the ability of nAChRs to compensate for age-related changes in NMDA receptor function is lost in the β2 KO mice.

Using KO mice, multiple studies have shown that β2 containing nAChRs are also necessary for the effects of nicotine on trace fear conditioning. These studies showed that while α2, α7, and β2 nAChR subunit KO animals showed normal trace fear conditioning, β2 KO mice did not show the enhancement of trace conditioning by acute nicotine (Davis and Gould 2007; Lotfipour et al. 2013). In line with these results, Raybuck and Gould (2009) showed that systemic DHβE-precipitated withdrawal deficits were absent in β2 KO mice. Together, these results support the pharmacological studies showing that β2 containing nAChRs are necessary for nicotine's effect on hippocampus-dependent contextual and trace fear conditioning.

3.2 Spatial Learning

Unlike the relatively large literature of KO studies on hippocampus-dependent fear conditioning, there are only a few studies that investigated the role of different nAChR subunits in spatial learning using genetically modified mice. One of these studies showed that in the Barnes maze, a maze with multiple circular holes where only one hole opens to an escape box, β4 KO mice did not show impaired spatial learning (Semenova et al. 2012). However, these animals used spatial search strategies, which are the use of extra-maze cues rather than serial trial and error strategies, less than the wild-type controls. This suggests a small spatial learning impairment in the β4 KO mice. Furthermore, α7 KO mice did not differ from their wild-type litter mates in both Barnes maze and MWM (Azzopardi et al. 2013; Paylor et al. 1998). Overall, these results are in line with the pharmacological studies showing that high-affinity nAChRs may be involved in spatial memory (Levin et al. 2002; Pocivavsek et al. 2006). However, the lack of effect reported in α7 KO mice contradicts the studies that found improved spatial learning with systemic α7 nAChR agonist injections (Castner et al. 2011) and spatial learning deficits with direct infusions of α7 nAChR antagonist MLA into the ventral hippocampus (Levin et al. 2002).

4 Cell-Signaling Cascades Involved in Nicotine's Effect on Learning and Memory

Identifying cell-signaling cascades activated by nAChR binding is crucial to understand the ability of nicotine to modulate hippocampus-dependent learning. As discussed earlier, nAChRs and NMDA receptors interact in a way that enables nAChRs to facilitate and substitute NMDA receptor activity. This is only possible if the nAChRs can activate similar cell-signaling cascades involved in NMDA activity-induced plasticity. As stated earlier, calcium influx by way of NMDA receptor activation can activate downstream cell-signaling cascades that have been identified as critical for hippocampal plasticity and learning (Fanselow et al. 1994; Gould et al. 2002; Huerta et al. 2000; Morris et al. 1986; Nakazawa et al. 2002; Place et al. 2012; Tonegawa et al. 1996; Tsien et al. 1996; English and Sweatt 1997; Platenik et al. 2000; Poser and Storm 2001). Specifically, the calcium influx resulted by nAChR and NMDA receptor activation triggers PKA activation, which, in turn, directly or indirectly, through ERK1/2 phosphorylation (Impey et al. 1998b), activates CREB (Impey et al. 1998a; Silva et al. 1998). Several studies have linked PKA and ERK1/2 activation to hippocampus-dependent learning (Abel and Nguyen 2008; Abel et al. 1997; Atkins et al. 1998; Bernabeu et al. 1997). The effects of nicotine on hippocampus-dependent learning also involve the activation of PKA and ERK1/2. In support, there is evidence showing that subthreshold doses of PKA (Gould et al. 2014b) or ERK1/2 inhibitors (Raybuck et al. 2007) abolished the nicotine-induced enhancement of contextual fear conditioning. In other words, when a nicotine-induced increase in PKA or ERK1/2 activation is blocked, the acute effect of nicotine on contextual fear conditioning is eliminated. Also, Gould and colleagues (2014b) found that nicotine administration shifted the timing of PKA and ERK1/2 in the hippocampus, which suggests that nicotine modulates the consolidation pattern of the hippocampus-dependent memory. Interestingly, Gould and colleagues (2014b) also showed that nicotine increased learning-induced PKA and ERK1/2 activity in the dorsal hippocampus but not in the ventral hippocampus, which adds to the literature suggesting that the effects of nicotine on hippocampus-dependent learning are mediated by the dorsal hippocampus. Together, these results suggest that nicotine enhances hippocampus-dependent learning by facilitating and altering the activation patterns of the proteins that are involved in memory formation.

While nicotine alters the cell-signaling cascades involved in normal memory formation, it also activates unique pathways that would otherwise not be activated. For example, Kenney and colleagues (2010) found that nicotine administered at learning increased β2-containing nAChR-mediated c-Jun N-terminal kinase-1 (*Jnk1*) expression in the dorsal but not in the ventral hippocampus. *Jnk1* phosphorylates microtubule-associated proteins (Bjorkblom et al. 2005) and activates transcription factors, such as the JUN family, ATF-2, and ELK-1 (Bogoyevitch and Kobe 2006; Gupta et al. 1996), which are the processes that modulate and strengthen synaptic signaling (Li et al. 2007; Sananbenesi et al. 2002; Strekalova et al. 2003). Moreover, the same study suggested that the increased *Jnk1* activity was not observed when nicotine was administered in the absence of the contextual fear-conditioning training, or when mice were given the training in the absence of the nicotine administration. Kenney and colleagues (2010) also showed that pharmacological inhibition of *Jnk1* during consolidation of the contextual fear memories impaired nicotine enhancement of contextual fear conditioning. In line with this finding, Kenney and colleagues (2012b) showed that the increase in *Jnk1* expression in the hippocampus as a result of nicotine administration at learning was associated with CREB phosphorylation at the *Jnk1* promoter region. Interestingly,

Kenney and colleagues (2012b) also showed that both Jnk1 activation and CREB phosphorylation at the *Jnk1* promoter region were absent in the β2 lacking KO mice, which suggests that nicotine's specific effects on cell-signaling are associated with β2 containing high-affinity nAChRs, such as α4β2 nAChRs. Thus, learning in nicotine-treated animals resulted in CREB phosphorylation that, in turn, led to increased transcription of *Jnk1* in the dorsal hippocampus. Together with the results indicating a role of PKA and ERK1/2 activation in the nicotine-enhanced hippocampus-dependent learning, these results suggest that acute nicotine also requires activation of additional proteins in order to manifest its cognitive-enhancing characteristics.

5 Effects of Nicotine on Hippocampus-Dependent Learning and Memory During Adolescence

The role of nicotine in hippocampus-dependent learning has been studied mostly in adult animals. However, the adolescent brain and adult brain respond differentially to the effects of nicotine. For example, adolescent rats were found to be more sensitive to the rewarding effects of nicotine (Kota et al. 2009; Shram and Le 2010; Torres et al. 2008). Moreover, several studies in rats demonstrated that adolescents metabolize nicotine faster than adults (O'Dell et al. 2006; Trauth et al. 2000) and that nicotine causes longer lasting nAChR upregulation in adolescents than in adults (Abreu-Villaca et al. 2003; Trauth et al. 1999). There are a few studies examining the developmental effects of nicotine exposure on hippocampal learning. For example, Spaeth and colleagues (2010) found rats exposed to chronic nicotine during adolescence and trained in adulthood showed deficits in contextual fear memories. Using a more comprehensive approach, Portugal and colleagues (2012b) investigated the effects of acute, chronic, and withdrawal from chronic nicotine on contextual fear conditioning in preadolescent, adolescent, and adult mice. The results of this study demonstrated that acute nicotine-induced enhancement of contextual fear memories was stronger and the effect of 24 h withdrawal from nicotine was weaker in younger mice relative to adult mice. Moreover, animals that received chronic nicotine during preadolescence and adolescence showed deficits in contextual fear conditioning when trained and tested in adulthood relative to the animals that both received nicotine administration and behavioral training and testing during adulthood.

As explained above, previous studies have shown that CREB activity in adults increases after nicotine administration (Walters et al. 2005) and nicotine and learning interact to increase CREB phosphorylation at the *Jnk1* promoter region (Kenney et al. 2012b). Hence, it is possible that the differential effects of nicotine in young and adult mice are reflected in differences in CREB activity in the hippocampus. Portugal and colleagues (2012b) examined the gene transcription factor CREB in the hippocampus 24 h after cessation of chronic nicotine administration and showed that CREB was decreased in the chronic nicotine-treated young mice relative to the saline-treated young and nicotine or saline-treated adult mice. This result is consistent with previous reports showing age-dependent changes in the hippocampus as a result of chronic nicotine exposure in adolescent rats (Polesskaya et al. 2007). In sum, adolescence is a period of increased sensitive to the enhancing effects of nicotine while decreased susceptible to the withdrawal deficits in learning and memory, and yet adolescent nicotine exposure may alter hippocampus development resulting in adult cognitive deficits.

6 Effects of Nicotine on Hippocampus-Dependent Learning: Evidence from Human Studies

There are several parallels between animal and human studies investigating the effects of nicotine on nAChR regulation in the brain. As discussed above chronic nicotine administration has been associated with high-affinity β2-containing nAChR upregulation (Hulihan-Giblin et al. 1990; Sharp and Beyer 1986; Marks et al. 1983; Schwartz and Kellar 1983; Marks 1999). Similarly, postmortem studies showed a higher density of α4β2 nAChRs in the brains of chronic smokers in comparison to nonsmoker controls (Benwell et al. 1988; Breese et al. 1997). Also, there is evidence showing that the α4β2 nAChRs density did not differ between nonsmokers and former smokers who quit smoking at least 2 months prior (Breese et al. 1997). Furthermore, using positron emission tomography (PET) scanning in humans, Brody and colleagues (2013) found that α4β2 nAChRs density decreased in the PFC, brainstem, and cerebellum during smoking cessation treatment while participants who quit at the end of the treatment showed a greater reduction than nonquitters. Similarly, Staley and colleagues (2006) demonstrated that α4β2 nAChR density in the cerebellum and striatum was higher in the smokers than in the participants that recently (6.8 ± 1.9 days) quit smoking and the α4β2 nAChR density was correlated with the days of abstinence and craving behavior. Overall, these results reveal that similar to direct nicotine administration in rodents, cigarette smoking also upregulates high-affinity β2-containing nAChRs in the human brain and that this upregulation continues during abstinence with a return to baseline over time, which may suggest that the effects of acute, chronic, and withdrawal from chronic nicotine on human cognition may be similar to the effects reported in the animal studies.

Although it is possible to draw clear parallels between nAChR regulation between animal and human brain, there is only a few studies examining the effects of nicotine dependence on hippocampus-dependent learning and memory in humans. For example, while in rodent studies results suggest that a relationship exists between nAChR upregulation and withdrawal deficits in hippocampus-dependent learning (Gould et al. 2012; Wilkinson and Gould 2013; Gould et al. 2014a), in humans, the relationship between changes in α4β2 upregulation and changes in hippocampus-dependent learning has not been examined. However, numerous studies have demonstrated that abstinence is associated with changes in cognition (Jacobsen et al. 2005; Mendrek et al. 2006; Myers et al. 2008; Patterson et al. 2009, 2010). In addition, there is evidence suggesting that prenatal nicotine exposure in humans leads to deficits in visuospatial memory as measured using the Brief Visuospatial Memory Test-Revised task and reduction in task-associated temporal lobe activation (Jacobsen et al. 2006). Interestingly, there is also evidence demonstrating that spatial working memory, measured in a delayed response task, was impaired in smokers in comparison to nonsmoker controls (Park et al. 2000). Together, these results suggest that nicotine withdrawal may influence hippocampus-dependent learning and memory performance in humans but more studies are required to establish a link between the cognitive changes and nAChR upregulation in humans.

7 Conclusion

Nicotine, a direct agonist of nAChRs, alters processes that underlie hippocampus-dependent forms of learning and memory. Numerous studies have investigated the role of nAChR activation and nicotine administration on contextual and trace fear conditioning, spatial learning, and working memory. These studies suggested a cognitive-enhancing role for acutely administered nicotine in hippocampus-dependent learning and memory. However, with chronic administration, the hippocampus adapts to nicotine through receptor desensitization and therefore, the effects of nicotine are decreased. Although acute nicotine has pro-cognitive effects, cognitive deficits result when chronic nicotine administration is ceased. The neural correlates of the withdrawal effects are still unknown; however, studies suggest that the upregulation of nAChRs may be a contributing factor.

Effects of nicotine on hippocampus-dependent learning and memory are affected and modulated by various factors. Evidence from studies using systemic and local administration of pharmacological agents and genetically modified mice converge to suggest that high-affinity nAChRs play a critical role in the effects of acute, chronic, and withdrawal from chronic nicotine on many types of hippocampus-dependent learning. Importantly, multiple studies suggest that nAChRs facilitate hippocampus-dependent learning and memory by activating both similar cell-signaling cascades involved in NMDA receptor-mediated plasticity and additional pathways unique to the effects of nicotine. Finally, age appears to be a critical factor in determining the effects of nicotine. While nicotine acts as a stronger cognitive enhancer during adolescence, the animals in this age group are less affected by the impairing effects of withdrawal from chronic nicotine; however, chronic nicotine exposure during adolescence may alter adult cognitive function. These results demonstrate that adolescence is a critical developmental stage for nicotine addiction. Together, understanding the involvement of different nAChRs and cell-signaling cascades as well as the effects of nicotine exposure length and the age of onset of nicotine exposure on the effects of acute and chronic nicotine on learning and memory is crucial to develop better pharmacological and behavioral tools to modulate changes in cognition associated with altered-nAChR function.

Acknowledgments This work was funded with grant support from the National Institute on Drug Abuse (T.J.G., DA017949).

References

Abdulla FA, Bradbury E, Calaminici MR et al (1996) Relationship between up-regulation of nicotine binding sites in rat brain and delayed cognitive enhancement observed after chronic or acute nicotinic receptor stimulation. Psychopharmacology 124:323–331

Abel T, Nguyen PV (2008) Regulation of hippocampus-dependent memory by cyclic AMP-dependent protein kinase. Prog Brain Res 169:97–115

Abel T, Nguyen PV, Barad M, Deuel TA, Kandel ER, Bourtchouladze R (1997) Genetic demonstration of a role for PKA in the late phase of LTP and in hippocampus-based long-term memory. Cell 88:615–626

Abreu-Villaca Y, Seidler FJ, Qiao D, Tate CA, Cousins MM, Thillai I, Slotkin TA (2003) Short-term adolescent nicotine exposure has immediate and persistent effects on cholinergic systems: critical periods, patterns of exposure, dose thresholds. Neuropsychopharmacology 28:1935–1949

Addy N, Levin ED (2002) Nicotine interactions with haloperidol, clozapine and risperidone and working memory function in rats. Neuropsychopharmacology 27:534–541

Akers RF, Lovinger DM, Colley PA, Linden DJ, Routtenberg A (1986) Translocation of protein kinase C activity may mediate hippocampal long-term potentiation. Science 231:587–589

Alkondon M, Pereira EF, Albuquerque EX (1996) Mapping the location of functional nicotinic and gamma-aminobutyric acid A receptors on hippocampal neurons. J Pharmacol Exp Ther 279:1491–1506

Alkondon M, Braga MF, Pereira EF, Maelicke A, Albuquerque EX (2000) Alpha7 nicotinic acetylcholine receptors and modulation of gabaergic synaptic transmission in the hippocampus. Eur J Pharmacol 393:59–67

André JM, Gulick D, Portugal GS, Gould TJ (2008) Nicotine withdrawal disrupts both foreground and background contextual fear conditioning but not pre-pulse inhibition of the acoustic startle response in C57BL/6 mice. Behav Brain Res 190:174–181

Andre JM, Leach PT, Gould TJ (2011) Nicotine ameliorates NMDA receptor antagonist-induced deficits in contextual fear conditioning through highaffinity nicotinic acetylcholine receptors in the hippocampus. Neuropharmacology 60:617–625

Araujo DM, Lapchak PA, Collier B, Quirion R (1988) Characterization of N-[3H]methylcarbamylcholine binding sites and effect of Nmethylcarbamylcholine on acetylcholine release in rat brain. J Neurochem 51:292–299

Arthur D, Levin ED (2002) Chronic inhibition of alpha4beta2 nicotinic receptors in the ventral hippocampus of rats: impacts on memory and nicotine response. Psychopharmacology 160:140–145

Atkins CM, Selcher JC, Petraitis JJ, Trzaskos JM, Sweatt JD (1998) The MAPK cascade is required for mammalian associative learning. Nat Neurosci 1:602–609

Attaway CM, Compton DM, Turner MD (1999) The effects of nicotine on learning and memory: a neuropsychological assessment in young and senescent fischer 344 rats. Physiol Behav 67:421–431

Awh E, Jonides J (2001) Overlapping mechanisms of attention and spatial working memory. Trends Cogn Sci 5:119–126

Azzopardi E, Typlt M, Jenkins B, Schmid S (2013) Sensorimotor gating and spatial learning in a7-nicotinic receptor knockout mice. Genes Brain Behav 12:414–423

Bancroft A, Levin ED (2000) Ventral hippocampal alpha4beta2 nicotinic receptors and chronic nicotine effects on memory. Neuropharmacology 39:2770–2778

Barrantes GE, Westwick J, Wonnacott S (1994) Nicotinic acetylcholine receptors in primary cultures of hippocampal neurons: pharmacology and Ca++ permeability. Biochem Soc Trans 22:294S

Barrantes GE, Murphy CT, Westwick J, Wonnacott S (1995) Nicotine increases intracellular calcium in rat hippocampal neurons via voltage-gated calcium channels. Neurosci Lett 196:101–104

Benwell ME, Balfour DJK, Anderson JM (1988) Evidence that tobacco smoking increases the density of (-)-[3H]nicotine binding sites in human brain. J Neurochem 50:1243–1247

Berg DK, Conroy WG (2002) Nicotinic α7 receptors: synaptic options and downstream signaling in neurons. J Neurobiol 53:512–523

Bernabeu R, Bevilaqua L, Ardenghi P, Bromberg E, Schmitz P, Bianchin M et al (1997) Involvement of hippocampal cAMP/cAMP-dependent protein kinase signaling pathways in a late memory consolidation phase of aversively motivated learning in rats. Proc Natl Acad Sci USA 94:7041–7046

Bernal MC, Vicens P, Carrasco MC, Redolat R (1999) Effects of nicotine on spatial learning in C57BL mice. Behav Pharmacol 10:333–336

Bettany JH, Levin ED (2001) Ventral hippocampal alpha7 nicotinic receptor blockade and chronic nicotine effects on memory performance in the radial-arm maze. Pharmacol Biochem Behav 70:467–474

Bjorkblom B, Ostman N, Hongisto V et al (2005) Constitutively active cytoplasmic c-jun n-terminal kinase 1 is a dominant regulator of dendritic architecture: role of microtubule-associated protein 2 as an effector. J Neurosci 25:6350–6361

Bliss TV, Collingridge GL (1993) A synaptic model of memory: long-term potentiation in the hippocampus. Nature 361:31–39

Bogoyevitch MA, Kobe B (2006) Uses for JNK: the many and varied substrates of the c-jun n-terminal kinases. Microbiol Mol Biol Rev 70:1061–1095

Bouton ME (2004) Context and behavioral processes in extinction. Learn Mem 11:485–494

Bouton ME, Bolles RC (1979) Contextual control of the extinction of conditioned fear. Learn Motiv 10:445–466

Bouton ME, King DA (1983) Contextual control of the extinction of conditioned fear: tests for the associative value of the context. J Exp Psychol Anim Behav Process 9:248–265

Bouton ME, Westbrook RF, Corcoran KA, Maren S (2006) Contextual and temporal modulation of extinction: behavioral and biological mechanisms. Biol Psychiatry 60:352–360

Breese CR, Marks MJ, Logel J, Adams CE, Sullivan B, Collins AC et al (1997) Effect of smoking history on [3H]nicotine binding in human postmortem brain. J Pharmacol Exp Ther 282:7–13

Brody AL, Mukhin AG, Shulenberger S, Mamoun MS, Kozman M, Phuong J, Mandelkern MA (2013) Treatment for tobacco dependence: effect on brain nicotinic acetylcholine receptor density. Neuropsychopharmacology 38:1548–1556

Broide RS, Leslie FM (1999) The α7 nicotinic acetylcholine receptor in neuronal plasticity. Mol Neurobiol 20:1–16

Caldarone BJ, Duman CH, Picciotto MR (2000) Fear conditioning and latent inhibition in mice lacking the high affinity subclass of nicotinic acetylcholine receptors in the brain. Neuropharmacology 39:2779–2784

Castner SA, Smagin GN, Piser TM et al (2011) Immediate and sustained improvements in working memory after selective stimulation of α7 nicotinic acetylcholine receptors. Biol Psychiatry 69:12–18

Clarke PB, Reuben M (1996) Release of [3H]-noradrenaline from rat hippocampal synaptosomes by nicotine: Mediation by different nicotinic receptor subtypes from striatal [3H]-dopamine release. Br J Pharmacol 117:595–606

Collins AC, Romm E, Wehner JM (1990) Dissociation of the apparent relationship between nicotine tolerance and up-regulation of nicotinic receptors. Brain Res Bull 25:373–379

Corcoran KA, Desmond TJ, Frey KA, Maren S (2005) Hippocampal inactivation disrupts the acquisition and contextual encoding of fear extinction. J Neurosci 25:8978–8987

Cordero-Erausquin M, Marubio LM, Klink R, Changeux JP (2000) Nicotinic receptor function: new perspectives from knockout mice. Trends Pharmacol Sci 21:211–217

Dani JA, Heinemann S (1996) Molecular and cellular aspects of nicotine abuse. Neuron 16:905–908

Davis JA, Gould TJ (2006) The effects of DHBE and MLA on nicotine-induced enhancement of contextual fear conditioning in C57BL/6 mice. Psychopharmacology 184:345–352

Davis J, Gould T (2007) Beta2 subunit-containing nicotinic receptors mediate the enhancing effect of nicotine on trace cued fear conditioning in C57BL/6 mice. Psychopharmacology 190:343–352

Davis JA, Gould TJ (2009) Hippocampal nAChRs mediate nicotine withdrawal-related learning deficits. Eur Neuropsychopharmacol 19:551–561

Davis JA, James JR, Siegel SJ, Gould TJ (2005) Withdrawal from chronic nicotine administration impairs contextual fear conditioning in C57BL/6 mice. J Neurosci 25:8708–8713

Davis JA, Porter J, Gould TJ (2006) Nicotine enhances both foreground and background contextual fear conditioning. Neurosci Lett 394:202–205

Davis JA, Kenney JW, Gould TJ (2007) Hippocampal alpha4beta2 nicotinic acetylcholine receptor involvement in the enhancing effect of acute nicotine on contextual fear conditioning. J Neurosci 27:10870–10877

Decker MW, Brioni JD, Bannon AW, Arneric SP (1995) Diversity of neuronal nicotinic acetylcholine receptors: lessons from behavior and implications for CNS therapeutics. Life Sci 56:545–570

Dominguez del Toro E, Juiz JM, Peng X, Lindstrom J, Criado M (1994) Immunocytochemical localization of the alpha 7 subunit of the nicotinic acetylcholine receptor in the rat central nervous system. J Comp Neurol 349:325–342

Elias GA, Gulick D, Wilkinson DS, Gould TJ (2010) Nicotine and extinction of fearconditioning. Neuroscience 165:1063–1073

English JD, Sweatt JD (1997) A requirement for the mitogen-activated protein kinase cascade in hippocampal long term potentiation. J Biol Chem 272:19103–19106

Fabian-Fine R, Skehel P, Errington ML, Davies HA, Sher E, Stewart MG et al (2001) Ultrastructural distribution of the alpha7 nicotinic acetylcholine receptor subunit in rat hippocampus. J Neurosci 21:7993–8003

Fanselow MS, Dong HW (2010) Are the dorsal and ventral hippocampus functionally distinct structures? Neuron 65:7–19

Fanselow MS, Kim JJ, Yipp J, De Oca B (1994) Differential effects of the Nmethyl- D-aspartate antagonist DL-2-amino-5-phosphonovalerate on acquisition of fear of auditory and contextual cues. Behav Neurosci 108:235–240

Feiro O, Gould TJ (2005) The interactive effects of nicotinic and muscarinic cholinergic receptor inhibition on fear conditioning in young and aged C57BL/6 mice. Pharmacol Biochem Behav 80:251–262

Freund RK, Jungschaffer DA, Collins AC, Wehner JM (1988) Evidence for modulation of GABAergic neurotransmission by nicotine. Brain Res 453:215–220

Frey U, Krug M, Reymann KG, Matthies H (1988) Anisomycin, an inhibitor of protein synthesis, blocks late phases of LTP phenomena in the hippocampal CA1 region in vitro. Brain Res 452:57–65

Frey U, Frey S, Schollmeier F, Krug M (1996) Influence of actinomycin D, a RNA synthesis inhibitor, on long-term potentiation in rat hippocampal neurons in vivo and in vitro. J Physiol 490:703–711

Fu Y, Matta SG, James TJ, Sharp BM (1998) Nicotine-induced norepinephrine release in the rat amygdala and hippocampus is mediated through brainstem nicotinic cholinergic receptors. J Pharmacol Exp Ther 284:1188–1196

Gould TJ, McCarthy MM, Keith RA (2002) MK-801 disrupts acquisition of contextual fear conditioning but enhances memory consolidation of cued fear conditioning. Behav Pharmacol 13:287–294

Gould TJ (2003) Nicotine produces a within subject enhancement of contextual fear conditioning in C57BL/6 mice independent of sex. Integr Physiol Behav Sci 38:124–132

Gould TJ, Higgins JS (2003) Nicotine enhances contextual fear conditioning in C57BL/6J mice at 1 and 7 days post-training. Neurobiol Learn Mem 80:147–157

Gould TJ, Leach PT (2014) Cellular, molecular, and genetic substrates underlying the impact of nicotine on learning. Neurobiol Learn Mem 107:108–132

Gould TJ, Lewis MC (2005) Coantagonism of glutamate receptors and nicotinic acetylcholinergic receptors disrupts fear conditioning and latent inhibition of fear conditioning. Learn Mem 12:389–398

Gould TJ, Wehner JM (1999) Nicotine enhancement of contextual fear conditioning. Behav Brain Res 102:31–39

Gould TJ, McCarthy MM, Keith RA (2002) MK-801 disrupts acquisition of contextual fear conditioning but enhances memory consolidation of cued fear conditioning. Behav Pharmacol 13:287–294

Gould TJ, Feiro O, Moore D (2004) Nicotine enhancement of trace cued fear conditioning but not delay cued fear conditioning in C57BL/6J mice. Behav Brain Res 155:167–173

Gould TJ, Portugal GS, Andre JM, Tadman MP, Marks MJ, Kenney JW et al (2012) The duration of nicotine withdrawal-associated deficits in contextual fear conditioning parallels changes in hippocampal high affinity nicotinic acetylcholine receptor upregulation. Neuropharmacology 62:2118–2125

Gould TJ, Wilkinson DS, Yildirim E, Blendy JA, Adoff MD (2014a) Dissociation of tolerance and nicotine withdrawal-associated deficits in contextual fear. Brain Res 1559:1–10

Gould TJ, Wilkinson DS, Yildirim E, Poole RL, Leach PT, Simmons SJ (2014b) Nicotine shifts the temporal activation of hippocampal protein kinase a and extracellular signal-regulated kinase 1/2 to enhance long-term, but not short-term, hippocampus-dependent memory. Neurobiol Learn Mem 109:151–159

Grady S, Marks MJ, Wonnacott S, Collins AC (1992) Characterization of nicotinic receptor-mediated [H-3] dopamine release from synaptosomes prepared from mouse striatum. J Neurochem 59:848–856

Gray R, Rajan AS, Radcliffe KA, Yakehiro M, Dani JA (1996) Hippocampal synaptic transmission enhanced by low concentrations of nicotine. Nature 383:713–716

Gupta S, Barrett T, Whitmarsh AJ et al (1996) Selective interaction of JNK protein kinase isoforms with transcription factors. EMBO J 15:2760–2770

He J, Deng CY, Chen RZ, Zhu XN, Yu JP (2000) Long-term potentiation induced by nicotine in CA1 region of hippocampal slice is Ca(2+)-dependent. Acta Pharmacol Sin 21:429–432

Hobin JA, Ji J, Maren S (2006) Ventral hippocampal muscimol disrupts context-specific fear memory retrieval after extinction in rats. Hippocampus 16:174–182

Hogg RC, Raggenbass M, Bertrand D (2003) Nicotinic acetylcholine receptors: from structure to brain function. Reviews of physiology, biochemistry and pharmacology. Springer, Berlin, pp 1–46)

Huang YY, Kandel ER (1994) Recruitment of long-lasting and protein kinase A-dependent long-term potentiation in the CA1 region of hippocampus requires repeated tetanization. Learn Mem 1:74–82

Huerta PT, Sun LD, Wilson MA, Tonegawa S (2000) Formation of temporal memory requires NMDA receptors within CA1 pyramidal neurons. Neuron 25:473–480

Hughes JR (2007) Effects of abstinence from tobacco: valid symptoms and time course. Nicotine Tob Res 9:315–327

Hulihan-Giblin BA, Lumpkin MD, Kellar KJ (1990) Acute effects of nicotine on prolactin release in the rat: agonist and antagonist effects of a single injection of nicotine. J Pharmacol Exp Ther 252:15–20

Impey S, Smith DM, Obrietan K, Donahue R, Wade C, Storm DR (1998a) Stimulation of cAMP response element (CRE)-mediated transcription during contextual learning. Nat Neurosci 1:595–601
Impey S, Obrietan K, Wong ST, Poser S, Yano S, Wayman G et al (1998b) Cross talk between ERK and PKA is required for Ca2+ stimulation of CREB dependent transcription and ERK nuclear translocation. Neuron 21:869–883
Jacobsen LK, Krystal JH, Mencl WE, Westerveld M, Frost SJ, Pugh KR (2005) Effects of smoking and smoking abstinence on cognition in adolescent tobacco smokers. Biol Psychiatry 57:56–66
Jacobsen LK, Slotkin TA, Westerveld M, Menci WE, Pugh KR (2006) Visuospatial memory deficits emerging during nicotine withdrawal in adolescents with prenatal exposure to active maternal smoking. Neuropsychopharmacology 31:1550–1561
Ji J, Maren S (2007) Hippocampal involvement in contextual modulation of fear extinction. Hippocampus 17:749–758
Jones S, Sudweeks S, Yakel JL (1999) Nicotinic receptors in the brain: correlating physiology with function. Trends Neurosci 22:555–561
Karadsheh MS, Shah MS, Tang X, Macdonald RL, Stitzel JA (2004) Functional characterization of mouse alpha4beta2 nicotinic acetylcholine receptors stably expressed in HEK293T cells. J Neurochem 91:1138–1150
Kenney JW, Gould TJ (2008a) Modulation of hippocampus-dependent learning and synaptic plasticity by nicotine. Mol Neurobiol 38:101–121
Kenney JW, Gould TJ (2008b) Nicotine enhances context learning but not context-shock associative learning. Behav Neurosci 122:1158–1165
Kenney JW, Florian C, Portugal GS, Abel T, Gould TJ (2010) Involvement of hippocampal jun-N terminal kinase pathway in the enhancement of learning and memory by nicotine. Neuropsychopharmacology 35:483–492
Kenney JW, Adoff MD, Wilkinson DS, Gould TJ (2011) The effects of acute, chronic, and withdrawal from chronic nicotine on novel and spatial object recognition in male C57BL/6J mice. Psychopharmacology 217:353–365
Kenney JW, Raybuck JD, Gould TJ (2012a) Nicotinic receptors in the dorsal and ventral hippocampus differentially modulate contextual fear conditioning. Hippocampus 22:1681–1690
Kenney JW, Poole RL, Adoff MD, Logue SF, Gould TJ (2012b) Learning and nicotine interact to increase CREB phosphorylation at the jnk1 promoter in the hippocampus. PLoS One 7:e39939
Kholdebarin E, Caldwell DP, Blackwelder WP, Kao M, Christopher NC, Levin ED (2007) Interaction of nicotinic and histamine H3 systems in the radial-arm maze repeated acquisition task. Eur J Pharmacol 569:64–69
Klann E, Chen SJ, Sweatt JD (1991) Persistent protein kinase activation in the maintenance phase of long-term potentiation. J Biol Chem 266:24253–24256
Knight DC, Cheng DT, Smith CN, Stein EA, Helmstetter FJ (2004) Neural substrates mediating human delay and trace fear conditioning. J Neurosci 24:218–228
Konorski J (1967) Integrative activity of the brain. University of Chicago Press, Chicago
Kota D, Robinson SE, Imad Damaj M (2009) Enhanced nicotine reward in adulthood after exposure to nicotine during early adolescence in mice. Biochem Pharmacol 78:873–879
Krug M, Lossner B, Ott T (1984) Anisomycin blocks the late phase of long-term potentiation in the dentate gyrus of freely moving rats. Brain Res Bull 13:39–42
Kutlu MG, Gould TJ (2014) Acute nicotine delays extinction of contextual fear in mice. Behav Brain Res 263:133–137
Le Novere N, Grutter T, Changeux JP (2002) Models of the extracellular domain of the nicotinic receptors and of agonist-and Ca2+-binding sites. Proc Natl Acad Sci USA 99:3210–3215
Levin ED, Torry D (1996) Acute and chronic nicotine effects on working memory in aged rats. Psychopharmacology 123:88–97
Levin ED, Lee C, Rose JE et al (1990) Chronic nicotine and withdrawal effects on radial-arm maze performance in rats. Behav Neural Biol 53:269–276
Levin ED, Briggs SJ, Christopher NC, Rose JE (1992) Persistence of chronic nicotine-induced cognitive facilitation. Behav Neural Biol 58:152–158

Levin ED, Briggs SJ, Christopher NC, Rose JE (1993a) Chronic nicotinic stimulation and blockade effects on working memory. Behav Pharmacol 4:179–182
Levin ED, Christopher NC, Briggs SJ, Rose JE (1993b) Chronic nicotine reverses working memory deficits caused by lesions of the fimbria or medial basalocortical projection. Brain Res Cogn Brain Res. 1:137-43
Levin ED, Kaplan S, Boardman A (1997) Acute nicotine interactions with nicotinic and muscarinic antagonists: working and reference memory effects in the 16-arm radial maze. Behav Pharmacol 8:236–242
Levin ED, Bettegowda C, Weaver T, Christopher NC (1998) Nicotine-dizocilpine interactions and working and reference memory performance of rats in the radial-arm maze. Pharmacol Biochem Behav 61:335–340
Levin ED, Christopher NC, Weaver T, Moore J, Brucato F (1999) Ventral hippocampal ibotenic acid lesions block chronic nicotine-induced spatial working memory improvement in rats. Cogn Brain Res 7:405–410
Levin ED, Bradley A, Addy N, Sigurani N (2002) Hippocampal alpha 7 and alpha 4 beta 2 nicotinic receptors and working memory. Neuroscience 109:757–765
Levin ED, Sledge D, Baruah A, Addy NA (2003) Ventral hippocampal NMDA blockade and nicotinic effects on memory function. Brain Res Bull 61:489–495
Levin E, Icenogle L, Farzad A (2005a) Ketanserin attenuates nicotine-induced working memory improvement in rats. Pharmacol Biochem Behav 82:289–292
Levin ED, Tizabi Y, Rezvani AH, Caldwell DP, Petro A, Getachew B (2005b) Chronic nicotine and dizocilpine effects on regionally specific nicotinic and NMDA glutamate receptor binding. Brain Res 1041:132–142
Li XM, Li CC, Yu SS, Chen JT, Sabapathy K, Ruan DY (2007) JNK1 contributes to metabotropic glutamate receptor-dependent long-term depression and short-term synaptic plasticity in the mice area hippocampal CA1. Eur J Neurosci 25:391–396
Lotfipour S, Byun JS, Leach P et al (2013) Targeted deletion of the mouse alpha2 nicotinic acetylcholine receptor subunit gene (Chrna2) potentiates nicotine-modulated behaviors. J Neurosci 33:7728–7741
Lynch G, Kessler M, Arai A, Larson J (1990) The nature and causes of hippocampal long-term potentiation. Prog Brain Res 83:233–250
Marks MJ (1999) Desensitization and the regulation of neuronal nicotinic receptors. In: Arneric SP, Brioni JD (eds) Neuronal nicotinic receptors: pharmacology and therapeutic opportunities. Wiley-Liss, New York, pp 65–80
Marks MJ, Collins AC (1982) Characterization of nicotine binding in mouse brain and comparison with the binding of alpha-bungarotoxin and quinuclidinyl benzilate. Mol Pharmacol 22:554–564
Marks MJ, Burch JB, Collins AC (1983) Effects of chronic nicotine infusion on tolerance development and nicotinic receptors. J Pharmacol Exp Ther 226:817–825
Marks MJ, Stitzel JA, Collins AC (1985) Time course study of the effects of chronic nicotine infusion on drug response and brain receptors. J Pharmacol Exp Ther 235:619–628
Marks MJ, Stitzel JA, Romm E, Wehner JM, Collins AC (1986) Nicotinic binding sites in rat and mouse brain: comparison of acetylcholine, nicotine, and alpha-bungarotoxin. Mol Pharmacol 30:427–436
Marks MJ, Grady SR, Collins AC (1993) Downregulation of nicotinic receptor function after chronic nicotine infusion. J Pharmacol Exp Ther 266:1268–1276
Matsuyama S, Matsumoto A (2003) Epibatidine induces long-term potentiation (LTP) via activation of alpha4beta2 nicotinic acetylcholine receptors (nAChRs) in vivo in the intact mouse dentate gyrus: both alpha7 and alpha4beta2 nAChRs essential to nicotinic LTP. J Pharmacol Sci. 93:180–187
Matsuyama S, Matsumoto A, Enomoto T, Nishizaki T (2000) Activation of nicotinic acetylcholine receptors induces long-term potentiation in vivo in the intact mouse dentate gyrus. Eur J Neurosci 12:3741–3747

Matta SG, Balfour DJ, Benowitz NL et al (2007) Guidelines on nicotine dose selection for in vivo research. Psychopharmacology 190:269–319

McCallum SE, Caggiula AR, Booth S, Breese CR, Lee MJ, Donny EC, Sved AF (2000) Mecamylamine prevents tolerance but enhances whole brain [3H] epibatidine binding in response to repeated nicotine administration in rats. Psychopharmacology 150:1–8

McEchron MD, Bouwmeester H, Tseng W, Weiss C, Disterhoft JF (1998) Hippocampectomy disrupts auditory trace fear conditioning and contextual fear conditioning in the rat. Hippocampus 8:638–646

McGehee DS (1999) Molecular diversity of neuronal nicotinic acetylcholine receptors. Ann N Y Acad Sci 868:565–577

McKay BE, Placzek AN, & Dani JA (2007) Regulation of synaptic transmission and plasticity by neuronal nicotinic acetylcholine receptors. Biochem Pharmacol, 74:1120–1133

Mendrek A, Monterosso J, Simon SL, Jarvik M, Brody A, Olmstead R, Domier CP, Cohen MS, Ernst M, London ED (2006) Working memory in cigarette smokers: comparison to non-smokers and effects of abstinence. Addict Behav 31:833–844

Morris RG, Anderson E, Lynch GS, Baudry M (1986) Selective impairment of learning and blockade of long-term potentiation by an N-methyl-D-aspartate receptor antagonist, AP5. Nature 319:774–776

Myers CS, Taylor RC, Moolchan ET, Heishman SJ (2008) Dose-related enhancement of mood and cognition in smokers administered nicotine nasal spray. Neuropsychopharmacology 33:588–598

Nakazawa K, Quirk MC, Chitwood RA, Watanabe M, Yeckel MF, Sun LD et al (2002) Requirement for hippocampal CA3 NMDA receptors in associative memory recall. Science 297:211–218

Nguyen PV, Abel T, Kandel ER (1994) Requirement of a critical period of transcription for induction of a late phase of LTP. Science 265:1104–1107

Nicoll RA, Malenka RC (1999) Expression mechanisms underlying NMDA receptor-dependent long-term potentiation. Ann N Y Acad Sci 868:515–525

O'Dell LE, Bruijnzeel AW, Smith RT, Parsons LH, Merves ML, Goldberger BA, Richardson HN, Koob GF, Markou A (2006) Diminished nicotine withdrawal in adolescent rats: implications for vulnerability to addiction. Psychopharmacology 186:612–619

Olale F, Gerzanich V, Kuryatov A, Wang F, Lindstrom J (1997) Chronic nicotine exposure differentially affects the function of human alpha3, alpha4, and alpha7 neuronal nicotinic receptor subtypes. J Pharmacol Exp Ther 283:675–683

Orr-Urtreger A, Seldin MF, Baldini A, Beaudet AL (1995) Cloning and mapping of the mouse alpha 7-neuronal nicotinic acetylcholine receptor. Genomics 26:399–402

Orr-Urtreger A, Goldner FM, Saeki M et al (1997) Mice deficient in the alpha7 neuronal nicotinic acetylcholine receptor lack alpha-bungarotoxin binding sites and hippocampal fast nicotinic currents. J Neurosci 17:9165–9171

Otani S, Marshall CJ, Tate WP, Goddard GV, Abraham WC (1989) Maintenance of long-term potentiation in rat dentate gyrus requires protein synthesis but not messenger RNA synthesis immediately post-tetanization. Neuroscience 28:519–526

Park S, Knopick C, McGurk S, Meltzer HY (2000) Nicotine impairs spatial working memory while leaving spatial attention intact. Neuropsychopharmacology 22:200–209

Patterson F, Jepson C, Strasser AA, Loughead J, Perkins KA, Gur RC, Lerman C (2009) Varenicline improves mood and cognition during smoking abstinence. Biol Psychiatry 65:144–149

Patterson F, Jepson C, Loughead J, Perkins K, Strasser AA, Siegel S, Lerman C (2010) Working memory deficits predict short-term smoking resumption following brief abstinence. Drug Alcohol Depend 106:61–64

Paylor R, Nguyen M, Crawley JN, Patrick J, Beaudet A, Orr-Urtreger A (1998) Alpha 7 nicotinic receptor subunits are not necessary for hippocampal-dependent learning or sensorimotor gating: a behavioral characterization of acra7-deficient mice. Learn Mem 5:302–316

Perry DC, Xiao Y, Nguyen HN, Musachio JL, Davila-Garcia MI, Kellar KJ (2002) Measuring nicotinic receptors with characteristics of alpha4beta2, alpha3beta2 and alpha3beta4 subtypes in rat tissues by autoradiography. J Neurochem 82:468–481

Place R, Lykken C, Beer Z, Suh J, McHugh TJ, Tonegawa S et al (2012) NMDA signaling in CA1 mediates selectively the spatial component of episodic memory. Learn Mem 19:164–169

Platenik J, Kuramoto N, Yoneda Y (2000) Molecular mechanisms associated with long-term consolidation of the NMDA signals. Life Sci 67:335–364

Pocivavsek A, Icenogle L, Levin ED (2006) Ventral hippocampal alpha7 and alpha4beta2 nicotinic receptor blockade and clozapine effects on memory in female rats. Psychopharmacology 188:597–604

Polesskaya OO, Fryxell KJ, Merchant AD, Locklear LL, Ker KF, McDonald CG et al (2007) Nicotine causes age-dependent changes in gene expression in the adolescent female rat brain. Neurotoxicol Teratol 29:126–140

Porter JT, Cauli B, Tsuzuki K, Lambolez B, Rossier J, Audinat E (1999) Selective excitation of subtypes of neocortical interneurons by nicotinic receptors. J Neurosci 19:5228–5235

Portugal GS, Gould TJ (2009) Nicotine withdrawal disrupts new contextual learning. Pharmacol Biochem Behav 92:117–123

Portugal GS, Kenney JW, Gould TJ (2008) b2 Containing Acetylcholine Receptors Mediate Nicotine Withdrawal Deficits in Learning. Neurobiol Learn Mem 89:106–113

Portugal G, Wilkinson D, Kenney J, Sullivan C, Gould T (2012a) Strain-dependent effects of acute, chronic, and withdrawal from chronic nicotine on fear conditioning. Behav Genetics 42:1–18

Portugal GS, Wilkinson DS, Turner JR, Blendy JA, Gould TJ (2012b) Developmental effects of acute, chronic, and withdrawal from chronic nicotine on fear conditioning. Neurobiol Learn Mem 97:482–494

Poser S, Storm DR (2001) Role of Ca2 + -stimulated adenylyl cyclases in LTP and memory formation. Int J Dev Neurosci 19:387–394

Radcliffe KA, Fisher JL, Gray R, Dani JA (1999) Nicotinic modulation of glutamate and GABA synaptic transmission of hippocampal neurons. Ann N Y Acad Sci 868:591–610

Rapier C, Lunt GG, Wonnacott S (1990) Nicotinic modulation of [3H]dopamine release from striatal synaptosomes: pharmacological characterisation. J Neurochem 54:937–945

Raybuck JD, Gould TJ (2007) Extracellular signal-regulated kinase 1/2 involvement in the enhancement of contextual fear conditioning by nicotine. Behav Neurosci 121:1119–1124

Raybuck JD, Gould TJ (2009) Nicotine withdrawal-induced deficits in trace fear conditioning in C57BL/6 mice–a role for high-affinity b2 subunit-containing nicotinic acetylcholine receptors. Eur J Neurosci 29:377–387

Raybuck JD, Gould TJ (2010) The role of nicotinic acetylcholine receptors in the medial prefrontal cortex and hippocampus in trace fear conditioning. Neurobiol Learn Mem 94:353–363

Ribeiro EB, Bettiker RL, Bogdanov M, Wurtman RJ (1993) Effects of systemic nicotine on serotonin release in rat brain. Brain Res 621:311–318

Robinson SE, James JR, Lapp LN, Vann RE, Gross DF, Philibin SD, Rosecrans JA (2006) Evidence of cellular nicotinic receptor desensitization in rats exhibiting nicotine-induced acute tolerance. Psychopharmacology 184:306–313

Robinson SE, Vann RE, Britton AF, O'Connell MM, James JR, Rosecrans JA (2007) Cellular nicotinic receptor desensitization correlates with nicotine-induced acute behavioral tolerance in rats. Psychopharmacology 192:71–78

Rowell PP, Winkler DL (1984) Nicotinic stimulation of [3H]acetylcholine release from mouse cerebral cortical synaptosomes. J Neurochem 43:1593–1598

Runyan JD, Moore AN, Dash PK (2004) A role for prefrontal cortex in memory storage for trace fear conditioning. J Neurosci 24:1288–1295

Rush R, Kuryatov A, Nelson ME, Lindstrom J (2002) First and second transmembrane segments of α3, α4, β2, and β4 nicotinic acetylcholine receptor subunits influence the efficacy and potency of nicotine. Mol Pharmacol 61:1416–1422

Rushforth SL, Steckler T, Shoaib M (2011) Nicotine improves working memory span capacity in rats following sub-chronic ketamine exposure. Neuropsychopharmacology 36:2774–2781
Sananbenesi F, Fischer A, Schrick C, Spiess J, Radulovic J (2002) Phosphorylation of hippocampal Erk-1/2, Elk-1, and p90-Rsk-1 during contextual fear conditioning: interactions between Erk-1/2 and Elk-1. Mol Cell Neurosci 21:463–476
Scerri C, Stewart C, Breen K, Balfour D (2006) The effects of chronic nicotine on spatial learning and bromodeoxyuridine incorporation into the dentate gyrus of the rat. Psychopharmacology 184:540–546
Schwartz RD, Kellar KJ (1983) Nicotinic cholinergic receptor binding sites in the brain: regulation in vivo. Science 220:214–216
Seguela P, Wadiche J, Dineley-Miller K, Dani JA, Patrick JW (1993) Molecular cloning, functional properties, and distribution of rat brain alpha 7: a nicotinic cation channel highly permeable to calcium. J Neurosci 13:596–604
Semenova S, Contet C, Roberts AJ, Markou A (2012) Mice Llcking the b4 subunit of the nicotinic acetylcholine receptor show memory deficits, altered anxiety- and depression-like behavior, and diminished nicotine-induced analgesia. Nicotine Tob Res 14:1346–1355
Sharifzadeh M, Tavasoli M, Naghdi N, Ghanbari A, Amini M, Roghani A (2005) Post-training intrahippocampal infusion of nicotine prevents spatial memory retention deficits induced by the cyclo-oxygenase-2-specific inhibitor celecoxib in rats. J Neurochem 95:1078–1090
Sharp BM, Beyer HS (1986) Rapid desensitization of the acute stimulatory effects of nicotine on rat plasma adrenocorticotropin and prolactin. J Pharmacol Exp Ther 238:486–491
Shram MJ, Le AD (2010) Adolescent male Wistar rats are more responsive than adult rats to the conditioned rewarding effects of intravenously administered nicotine in the place conditioning procedure. Behav Brain Res 206:240–244
Silva AJ, Kogan JH, Frankland PW, Kida S (1998) CREB and memory. Annu Rev Neurosci 21:127–148
Socci DJ, Sanberg PR, Arendash GW (1995) Nicotine enhances Morris water maze performance of young and aged rats. Neurobiol Aging 16:857–860
Sorenson EM, Shiroyama T, Kitai ST (1998) Postsynaptic nicotinic receptors on dopaminergic neurons in the substantia nigra pars compacta of the rat. Neuroscience 87:659–673
Spaeth AM, Barnet RC, Hunt PS, Burk JA (2010) Adolescent nicotine exposure disrupts context conditioning in adulthood in rats. Pharmacol Biochem Behav 96:501–506
Staley JK, Krishnan-Sarin S, Cosgrove KP, Krantzler E, Frohlich E, Perry E, van Dyck CH (2006) Human tobacco smokers in early abstinence have higher levels of β2* nicotinic acetylcholine receptors than nonsmokers. J Neurosci 26:8707–8714
Strekalova T, Zorner B, Zacher C, Sadovska G, Herdegen T, Gass P (2003) Memory retrieval after contextual fear conditioning induces c-Fos and JunB expression in CA1 hippocampus. Genes Brain Behav 2:3–10
Szyndler J, Sienkiewicz-Jarosz H, Maciejak P et al (2001) The anxiolytic-like effect of nicotine undergoes rapid tolerance in a model of contextual fear conditioning in rats. Pharmacol Biochem Behav 69:511–518
Tian S, Gao J, Han L, Fu J, Li C, Li Z (2008) Prior chronic nicotine impairs cued fear extinction but enhances contextual fear conditioning in rats. Neuroscience 153:935–943
Tonegawa S, Tsien JZ, McHugh TJ, Huerta P, Blum KI, Wilson MA (1996) Hippocampal CA1-region-restricted knockout of NMDAR1 gene disrupts synaptic plasticity, place fields, and spatial learning. Cold Spring Harb Symp Quant Biol 61:225–238
Torres OV, Tejeda HA, Natividad LA, O'Dell LE (2008) Enhanced vulnerability to the rewarding effects of nicotine during the adolescent period of development. Pharmacol Biochem Behav 90:658–663
Trauth JA, Seidler FJ, McCook EC, Slotkin TA (1999) Adolescent nicotine exposure causes persistent upregulation of nicotinic cholinergic receptors in rat brain regions. Brain Res 851:9–19
Trauth JA, Seidler FJ, Slotkin TA (2000) An animal model of adolescent nicotine exposure: effects on gene expression and macromolecular constituents in rat brain regions. Brain Res 867:29–39

Tsien JZ, Huerta PT, Tonegawa S (1996) The essential role of hippocampal CA1 NMDA receptor-dependent synaptic plasticity in spatial memory. Cell 87:1327–1338

Walters CL, Cleck JN, Kuo YC, Blendy JA (2005) Mu-opioid receptor and CREB activation are required for nicotine reward. Neuron, 46:933–943

Wehner JM, Keller JJ, Keller AB et al (2004) Role of neuronal nicotinic receptors in the effects of nicotine and ethanol on contextual fear conditioning. Neuroscience 129:11–24

Wenk GL, Walker LC, Price DL, Cork LC (1991) Loss of NMDA, but not GABA-A, binding in the brains of aged rats and monkeys. Neurobiol Aging 12:93–98

Westfall TC, Grant H, Perry H (1983) Release of dopamine and 5-hydroxytryptamine from rat striatal slices following activation of nicotinic cholinergic receptors. Gen Pharmacol 14:321–325

Wilkie GI, Hutson P, Sullivan JP, Wonnacott S (1996) Pharmacological characterization of a nicotinic autoreceptor in rat hippocampal synaptosomes. Neurochem Res 21:1141–1148

Wilkinson DS, Gould TJ (2013) Withdrawal from chronic nicotine and subsequent sensitivity to nicotine challenge on contextual learning. Behav Brain Res 250:58–61

Wonnacott S (1986) A bungarotoxin binds to low-affinity nicotine binding sites in rat brain. J Neurochem 47:1706–1712

Wonnacott S (1997) Presynaptic nicotinic ACh receptors. Trends Neurosci 20:92–98

Zarei MM, Radcliffe KA, Chen D, Patrick JW, Dani JA (1999) Distributions of nicotinic acetylcholine receptor alpha7 and beta2 subunits on cultured hippocampal neurons. Neuroscience 88:755–764

Tsien JZ, Huerta PT, Tonegawa S (1996) The essential role of hippocampal CA1 NMDA receptor-dependent synaptic plasticity in spatial memory. Cell 87:1327–1338

[illegible] (2005) [illegible] Neuron 46:[illegible]

[illegible]

[illegible]

[illegible]

[illegible]

[illegible]

[illegible]

A Hierarchical Instrumental Decision Theory of Nicotine Dependence

Lee Hogarth and Joseph R. Troisi II

Abstract It is important to characterize the learning processes governing tobacco-seeking in order to understand how best to treat this behavior. Most drug learning theories have adopted a Pavlovian framework wherein the conditioned response is the main motivational process. We favor instead a hierarchical instrumental decision account, wherein expectations about the instrumental contingency between voluntary tobacco-seeking and the receipt of nicotine reward determines the probability of executing this behavior. To support this view, we review titration and nicotine discrimination research showing that internal signals for deprivation/satiation modulate expectations about the current incentive value of smoking, thereby modulating the propensity of this behavior. We also review research on cue-reactivity which has shown that external smoking cues modulate expectations about the probability of the tobacco-seeking response being effective, thereby modulating the propensity of this behavior. Economic decision theory is then considered to elucidate how expectations about the value and probability of response-nicotine contingency are integrated to form an overall utility estimate for that option for comparison with qualitatively different, nonsubstitute reinforcers, to determine response selection. As an applied test for this hierarchical instrumental decision framework, we consider how well it accounts for individual liability to smoking uptake and perseveration, pharmacotherapy, cue-extinction therapies, and plain packaging. We conclude that the hierarchical instrumental account is successful in reconciling this broad range of phenomenon precisely because it accepts that multiple diverse sources of internal and external information must be integrated to shape the decision to smoke.

Keywords Internal discriminative stimuli · External drug cue-reactivity · Smoking decisions · Hierarchical instrumental learning · Pharmacotherapy · Exposure therapy · Individual vulnerability · Nicotine dependence

---

L. Hogarth (✉)
School of Psychology, University of Exeter,
Washington Singer Building, Perry Road, Exeter Ex4 4QG, UK
e-mail: l.hogarth@exeter.ac.uk

J.R. Troisi II
Department of Psychology, Saint Anselm College, Manchester NH 03102, USA

D.J.K. Balfour and M.R. Munafò (eds.), *The Neurobiology and Genetics of Nicotine and Tobacco*, Current Topics in Behavioral Neurosciences 23,
DOI 10.1007/978-3-319-13665-3_7

Contents

1 Associative Structures Underpinning Smoking ... 166
2 Titration ... 168
3 Drug Discrimination ... 169
3.1 Human Studies ... 169
3.2 Animal Studies ... 170
4 Other Interoceptive Stimulus Effects ... 172
5 Cue-Reactivity ... 173
6 Integrating Expected Probability and Value ... 175
7 The Response-Outcome Gestalt ... 176
8 Choice Among Alternatives ... 178
9 Five Applied Tests for the Hierarchical Account ... 179
9.1 Individual Differences in Smoking Uptake ... 179
9.2 Individual Differences in Smoking Perseveration ... 179
9.3 Pharmacotherapy ... 181
9.4 Extinction Therapy ... 181
9.5 Plain Packaging ... 182
10 Conclusion ... 183
References ... 183

1 Associative Structures Underpinning Smoking

Associative learning theory seeks to characterize the psychological mechanisms that underpin acquired motivated behavior. For this reason, the associative framework has been co-opted to understand addictive behavior in both humans and animals. Such associative addiction theories generally propose a triadic structure linking three observable parts: (1) the drug (e.g., nicotine) which acts as the reinforcer or outcome; (2) voluntary instrumental motor responses such as nicotine-seeking/taking; and (3) stimulus events, both external and internal, which predict some dimension of the response or outcome. The point of divergence between the various associative addiction theories concerns the precise associative structural and functional relations embedded within the three-term relationship between the stimuli (S), response (R), and the reinforcing drug outcome (O) as well what is classified in these roles. Traditional Pavlovian views of positive and negative reinforcement theories have emphasized conditioned appetitive and aversive states, respectively (Ahmed and Koob 2005); Incentive salience theory has emphasized conditioned attentional bias to external drug stimuli (e.g., Marissen et al. 2006; c.f., Hogarth et al. 2008); Behavioral economic theories have emphasized the reinforcement of voluntary instrumental actions and the reinforcement value of the drug outcome, whereas habit/compulsion theory highlights direct associations between stimuli and responses (S-R). These theories are typically used to explain drug use in both humans and animals. However, whereas human researchers are generally happy to accept propositional knowledge of associative relationships as causal in driving behavior, animal researchers tend to eschew such anthropomorphically cognitive accounts in keeping with their radical behaviorist origins. Despite nearly half a century of research in this area, there remains little consensus as to the precise associative structure and function underpinning addictive behavior. This is perhaps not surprising given that, even within basic learning theory itself, the precise functions governing even the simplest motivated behaviors in rats remains to be fully specified (Harris et al. 2013). Our question is: what associative structure must be postulated to successfully embrace the complex range of phenomenon observed in human and animal addictive behavior?

One challenge in specifying the associative basis of addictive behavior is its apparently progressive developmental nature, where the key drivers of the behavior apparently change over an individual's drug use history (Hogarth et al. 2013a). For instance, initiation of smoking behavior might be mediated by the hedonically positive reinforcing effects of the drug, but as tolerance to nicotine's effects mounts over time, the user may ingest more nicotine on a schedule (titration), so as to prevent (or escape from) the hedonically aversive effects of nicotine withdrawal (i.e., negative reinforcement; Ahmed and Koob 2005; Baker et al. 2004; Eissenberg 2004). Of course, the brain circuitry involved in this transition may likely switch (e.g., see Everitt and Robbins 2013). With these changes from initiation to maintenance of nicotine self-administration, environmental stimuli (external stimuli) that regulate the "when-where-and-how" of nicotine self-administration continue to sustain behavior. For instance, an empty cigarette pack continues to function as a discriminative cue that prompts an extended distal chain of behavior (driving to store, request for name brand, receipt of package) that eventually culminates in a proximal chain (placing cigarette package down, placing cigarette in mouth, igniting cigarette, and the usual smoking topography). Additionally, however, subjective nicotine effects (and its withdrawal profile) are likely to change more considerably than the external stimuli over time. Moreover, other neurophysiological events (endocrine, autonomic, somatic, and so on) may also function as internal stimuli that drive behavior. The interaction among these internal and external stimuli is of particular focus in this chapter.

In this chapter, we elaborate a *hierarchical instrumental* framework (Colwill and Rescorla 1990; Rescorla 1987, 1991, 1992b) as the fundamental associative function underpinning tobacco-seeking and self-administration across the extent of a drug user's drug-taking history. The core proposition is that individuals come in to contact with the instrumental contingency between the tobacco-seeking response and the nicotine outcome, and hence propositional expectations emerge (i.e., as measured by verbal self-report) that the R will produce the O. The expected *value* of the O is generally modulated by internal states (i.e., blood nicotine level, stress, emotional states, hormonal changes, as well as others) which previously (i.e., past experience) signaled the current value of the O (incentive learning: Dickinson and Balleine 2002, 2010). For example, when rats are food restricted, food is a more valuable commodity than when they are sated. Similarly, when a person is nicotine deprived (withdrawn), nicotine is a highly valued commodity relative to other reinforcing outcomes. Stressors and negative emotional states similarly raise the expected reinforcement value of nicotine because these states anticipate that nicotine will be experienced as more reinforcing due to amelioration of these aversive states (Hogarth 2012; Hogarth and Chase 2011; Hutcheson et al. 2001).

By contrast, the expected probability that the R will lead to the O is generally modulated by external instrumental discriminative stimuli (S^Ds), due to the fact that in prior experience such S^Ds *set the occasion* (Skinner 1938) for R to produce O (Hammond 1980). The term *hierarchical* pertains to the assumption that the S modulates (i.e., retrieves) an amalgamated gestalt representation of the R-O contingency (notated S:R-O). This hierarchical position must be distinguished from binary associative accounts that attribute behavior to the learning of binary S-O, S-R, and R-O associations, and/or their summation (Bradfield and Balleine 2013; de Wit and Dickinson 2009). Indeed, there is rather compelling evidence that a hierarchical relationship is not reducible to a summation of these binary contingencies (Bradfield and Balleine 2013; Colwill and Rescorla 1990; Rescorla 1987, 1991, 1992b), but rather R-O relations are established and modulated by the presence or absence of S.

In practice, we propose that external S (discrete stimuli or contexts) evoke a learned relationship (i.e., a belief) regarding the precise R-O contingencies that are momentarily in effect (probable), and concomitantly, internal S attribute a value estimate to these R-O contingencies. These probability and value estimates are integrated to yield the *utility* calculation for each R-O contingency, whereupon the R with the greater utility is selected for execution (Vlaev et al. 2011). Stated in more behavioral terms, the response that produces the maximal reinforcing gain in one moment in time is most likely to occur during a specific interaction among internal and external states. Such integrated control of response selection not only provides a molecular account of the processes governing the maintenance of smoking behavior, but may also have utility in understanding relapse. To elaborate this account, in the following sections we discuss how internal and external nicotine cues individually and jointly modulate the frequency of nicotine-seeking, and then how the nicotine-seeking choice is selected from amongst other qualitatively different reinforcers. In the final section, we consider how well the hierarchical account can explain individual differences in vulnerability to nicotine dependence and the impact of several key therapeutic interventions.

2 Titration

Nicotine is maximally reinforcing at approximately 35 ng/ml of plasma, and less reinforcing above and below this peak (Corrigall and Coen 1989a; Donny et al. 1995; Feyerabend et al. 1985; Shoaib and Stolerman 1999). For example, Harvey and colleagues (Harvey et al. 2004) found that intravenous nicotine maintained lever-pulling in eight male smokers significantly above placebo, and rates of self-administration increased across the dose range: placebo, 0.75, 1.5, and 3.0 mg/injection. Because nicotine is optimally reinforcing within a particular dose range, internal states for nicotine load will come to reliably signal and thereby generate expectations concerning the value of further nicotine consumption (i.e., whether smoking will be punished, negatively or positively reinforcing), and these expectations will in turn modulate the propensity of the behavior (stop or start smoking, respectively; Troisi et al. 2012, 2013 for a more behavior analytic presentation of this view). Several sources of evidence that follow support this claim.

There is substantial evidence that humans and animals regulate their level of instrumental nicotine-seeking or smoking behavior in accordance with their level of nicotine deprivation/satiation, titrating toward the optimal dose (Corrigall and Coen 1989b; Epstein et al. 1991; Perkins et al. 1994, 1997a; Rusted et al. 1998; Tiffany and Drobes 1991; Willner et al. 1995). Titration may be governed by, (1) a centrally mediated nicotine stimulus (Hanson et al. 1979; Perkins et al. 1999; Rose et al. 1989, 1996), (2) by peripheral gustatory/olfactory sensations (Behm and Rose 1994; Gerhardstein et al. 1993; Rose et al. 1999), or (3) by pneumatic aspects of cigarette smoking (Chait and Griffiths 1982; Nemeth-Coslett and Griffiths 1984a, b; Wiley and Wickham 1974), which are typically correlated with level of nicotine absorption (Benowitz et al. 1990; Russell et al. 1975) for reviews see: Rose and Corrigall (1997), Scherer (1999). There are likely to be other sensory cues as well. These central and peripheral nicotine associated stimuli may regulate behavior independently (Hasenfratz et al. 1993; Nil and Battig 1989; Rose et al. 1993), but they are also apparently additive in modulating smoking behavior, thereby demonstrating an integration of multiple diverse sensory signals in the control of nicotine self-administration. For example, Westman et al. (1995) found that nicotine replacement therapy plus citric acid inhaler (to mimic sensory cues) was more effective than either alone in reducing smoking behavior, consistent with an additive integration of central and peripheral cues in modulating the propensity to engage in nicotine self-administration (but see Litvin and Brandon 2010).

3 Drug Discrimination

3.1 Human Studies

More direct evidence that internal stimuli retrieve an expectancy of R-O value which plays a role in response selection comes from the finding that the internal nicotine stimulus produces subjective effects in humans and enables discrimination of instrumental action. For instance, using a human drug discrimination methodology, Perkins et al. (1997b) trained subjects to discriminate nicotine from the placebo nasal spray using monetary reinforcement for correctly identifying each. The results showed that self-reported "head rush" correlated with the accuracy of discriminative control, and head rush could be attenuated by blocking central nicotine receptors (Perkins et al. 1999); this suggests that subjectively detected internal nicotine states can provide the basis for instrumental response selection.

On this analysis, the observation that nicotine deprivation and satiation modulate subjective craving (Tiffany and Drobes 1991) may be thought of as the internal nicotine stimulus modulating subjective report of instrumental intentions (Willner et al. 1995).

3.2 Animal Studies

Consistent with this analysis, internal nicotine discriminative stimuli (S^D) that are scheduled to signal when a food rewarded instrumental response will be effective, acquire control over the performance of that instrumental response. For instance, several studies using two-lever choice and one-lever go/no-go procedures (Stolerman et al. 1984; Troisi 2003; Troisi et al. 2010) have demonstrated that the S^D functions of nicotine govern response choice and that these effects are mediated by central receptor mechanisms. Moreover, with a one-lever procedure (Troisi 2003; Troisi et al. 2010) found that the internal nicotine S^D could be trained to augment a response if this state set the occasion that the response was effective in producing food reward, but could also be trained to inhibit the response if the state set the occasion when the response would not be reinforced. These findings suggest the nicotine stimulus can be established to retrieve both S:R-O and S:R-noO contingencies, which presumably promotes selection between alternatives in two-lever discrimination procedures.

One of the most important findings for the hierarchical analysis presented here, is that discriminative control exerted by the internal nicotine stimulus over food-reinforced responding is immediately affected by postconditioning devaluation of the food reinforcer (Pittenger and Bevins 2013; Troisi et al. 2012). In the study by Troisi et al. (2012), food deprived rats were trained with nicotine as a signal (S^D) that a response would produce food or not produce food (S^-) in separate groups. Rats were then shifted to a sated state in which nicotine functioned as an S^- for all animals (i.e., the new satiety-nicotine combination signaled an extinction contingency for the S^D group). The important finding was that when rats were returned to the deprived state, responding immediately recovered under nicotine for the S^D group but not the S^- group. These results suggest that the nicotine S^D was integrated with the food deprivation state to create a unique combination which signaled when response-food contingency was in force, such that the reinstatement of this combined internal state determined the propensity to perform the response. Thus, internal cues can be combined together to qualify the status of R-O contingencies. Somewhat related, Pittenger and Bevins (2013) showed a similar effect with a nominally Pavlovian arrangement. In this study, nicotine was scheduled to signal when dipper entry would be reinforced with liquid sucrose. Crucially, control exerted by the internal nicotine stimulus over the dipper entry response was immediately reduced by postconditioning devaluation of the sucrose by lithium chloride induced sickness. The implication of this finding is that the internal nicotine stimulus retrieved a representation of the current value of the sucrose reinforcer yielded by dipper entry, thereby reducing selection of that response.

The second important evidence for the hierarchical analysis comes from studies showing that the discriminative control exerted by the internal nicotine stimulus over one food-reinforced response (R1) can transfer to modulate a topographically different response (R2) maintained by the same food reinforcer, which has hitherto been acquired in the absence of the nicotine stimulus (Troisi et al. 2010). The hierarchical interpretation of this transfer effect is that the internal nicotine stimulus raised the expected probability of all R-O contingencies involving food, which summed with external contextual cues in which R2 had previously been reinforced, thereby enhancing selective performance of R2. Finally, the observation that the internal nicotine discriminative stimulus can acquire control over heterogeneous instrumental chains of topographically different responses (nose poke–lever press versus lever press—nose poke) indicates that the internal nicotine stimulus is capable of modulating complex instrumental performance, rather than simply exerting effects on behavior through Pavlovian processes (Grindley 1932; Troisi 2013a).

The third most important finding for the hierarchical analysis is that nicotine can function as an occasion setter in Pavlovian preparations (Palmatier and Bevins 2008). That is, if the internal nicotine stimulus is scheduled to signal when a CS will be followed by reinforcement (i.e., signal the CS–US association), the nicotine stimulus will come to modulate the conditioned response (food dipper entries) evoked by the CS, and transfer control over another CS that has been paired with the same reinforcer in a different internal state. Thus, the internal nicotine stimulus may hierarchically govern utilization of knowledge about CS–US associations. Importantly, it has been argued that that the Pavlovian occasion setting function of stimuli (where the S signals the CS–US relationship) is equivalent to an instrumental occasion setting function (where the S signals the R-O relationship). For example, Davidson et al. (1988) found that a stimulus established as a Pavlovian occasion setter transferred to control over a separately acquired instrumental response for the same outcome, and vice versa, suggesting there is a common process operating in both the Pavlovian and instrumental preparations that stimuli modulate response selection via the retrieval of R-O expectancies (Rescorla 1987). However, Pavlovian stimuli show much weaker transfer of stimulus control over instrumental performance than do S^Ds when these cues are external stimuli for natural rewards (Rescorla 1994) or drug rewards (Di Ciano and Everitt 2003) or internal nicotine stimuli for natural rewards (Troisi 2006; Troisi et al. 2010; Troisi 2013b). Thus, Pavlovian stimuli are not perfect substitutes for discriminative stimuli in the control of instrumental performance, although this may be due to differences in the extent to which these stimuli are inferred as being equivalent to the target stimulus rather than a difference in underlying learning mechanisms (Meeter et al. 2009). Although further work is needed in this area of transfer among Pavlovian and instrumental occasion setters, at present this area broadly supports the hierarchical instrumental claim that internal nicotine stimuli modulate performance by retrieving expectancies about the nature of the currently available R-O contingencies.

4 Other Interoceptive Stimulus Effects

Other internal stimuli (non-nicotine stimuli) can similarly acquire control over nicotine self-administration. For instance, caffeine has been shown to reinstate extinguished nicotine self-administration in rats (Liu and Jernigan 2012). Ethanol has also been shown to modulate nicotine self-administration in humans and rats (Troisi et al. 2013). Moreover, nicotine can be combined with other drugs to form a gestalt stimulus governing instrumental action (Stolerman et al. 1987; Troisi et al. 2013). Similarly, stress (Jarvik et al. 1989), anxiety (Harris et al. 1986), negative mood (Willner and Jones 1996), distress (Perkins et al. 2012), and fatigue (Delfino et al. 2001) have all been implicated in modulating the experienced reinforcement value of smoking/nicotine and the modulation of nicotine-seeking behavior. The implication of these data is that internal states for nicotine deprivation/satiation, as well as a plethora of mood states, may function as complex stimulus gestalts which signal the reinforcement value of nicotine, and thus acquires modulatory control over instrumental smoking behavior.

Multiple internal cues can be additive (and perhaps subtractive) and the combination may form a unique cue directing behavior. In a recent investigation, Troisi et al. (2013) trained rats to discriminate a mixture of nicotine plus ethanol from saline. For some animals, the mixture functioned as an S^D. Each element that comprised the mixture individually controlled behavior. In a subsequent phase, responding was extinguished with nicotine and ethanol separately. Recombining the elements as a mixture promoted robust recovery of discriminative responding, but only for a group in which the mixture served as S^D and not for a group in which it served as S−. These data suggested that the internal elements configured to form a unique discriminative cue. By contrast, if the elements were established separately as individual S^Ds, extinction of responding with the mixture appeared to generalize back to the elements. Thus, with respect to the compound internal gestalt cue's retrieval the R-O expectation, the whole was greater than the sum of the component parts as a function of what the whole and its part predicted. These data mirrored findings reported with exteroceptive compound CSs in a Pavlovian procedure (Bouton et al. 2012). It thus appears that multiple internal stimuli are "perceived" in unique ways in their function as signals for the status of R-O contingencies.

Another important observation is that the nicotine discrimination devaluation study summarized earlier (Troisi et al. 2012) suggested that the nicotine internal cue interacted with the hunger and no-hunger states to form a gestalt stimulus modulating behavior. We proposed that other neuroendocrine functions (internal cues) including leptin, ghrelin, and orexin likely interacted with the nicotine cue in directing behavior. Consistent with this proposition, subjective reports of nicotine vary as a function of the menstrual cycle in women smokers (Devito et al. 2014). Thus, any number of combinations and permutations of various overlapping internal S^Ds that are associated with nicotine R-O relations presumably play additive roles in modulating R-O expectancies. It may also be true that internal cues play subtractive roles in regulating the R-O relation. For instance, internal nicotine cues function effectively as S^D stimuli, and if added to a differing internal S^- cue that signals the absence of the R-O relationship, can alter the probability of the response in a summative way. Unpublished pilot data (Troisi) suggest that combining a nicotine stimulus that predicted a Pavlovian food-US with an ethanol stimulus that predicted no US neutralized responding in a counterbalanced (by drug role) Pavlovian discriminated goal-tracking procedure. Now, under the assumption that stress and other emotional states individually signaled the value of smoking in one's past, it seems plausible that if such states were evoked concurrently, they would additively increase the propensity of nicotine-seeking compared to if only one internal stimulus was present (e.g., Panlilio et al. 1996). For example, negative mood and stress might individually raise the expected value of nicotine reward; but, when the two are combined, the effect might be additive. If nicotine withdrawal is further added to this complex internal state, it is possible that smoking behavior might be inevitable. To be sure, more systematic variations that empirically test these sorts of predictions are sorely needed if we are to understand the interface between physiology, emotion, decision-making, and response selection.

To summarize, the titration studies showed that the internal nicotine stimulus can modulate craving and the performance of established nicotine self-administration behavior toward the maximally reinforcing dose. This suggests that this stimulus controls performance by modulating expectations about the current reinforcement value of the drug and is entirely consistent with incentive learning theory (Hutcheson et al. 2001). The nicotine discrimination studies showed something slightly different—that if the nicotine stimulus signals when an instrumental response will be reinforced, it will come modulate when that response is performed. Thus, normal smoking apparently establishes the nicotine stimulus as a signal for the current value of nicotine outcome, whereas discrimination studies utilize an experimentally tractable but perhaps ecologically invalid schedule in which the internal nicotine stimulus signals when a response will be reinforced (signaling R-O probability rather than R-O value; see Davidson 1993). Nevertheless, in both research domains the nicotine stimulus signaled dimensions of the R-O contingency (value and probability, respectively) and thereby modulated performance of the R. As we will argue in the section below, external discriminative stimuli in their ecologically valid contingencies tend to signal when a response will be reinforced, and thus modulate performance of the response by engaging expectations about the probability that responses will be effective (Troisi 2013c). The question addressed then, is how expected probability carried by external nicotine cues, and expected value carried by internal nicotine cues, are integrated to guide response selection.

5 Cue-Reactivity

Whereas internal stimuli generally signal the current biological value of an outcome, external stimuli signal a range of dimensions concerning the nature of the response—outcome pair, including: *what* response to perform and what outcome will be produced, *whether* the response is likely to produce the outcome, *where* the response should be performed and where the outcome will occur, *when* the response should be performed and when the outcome will occur, and *why* the response produces the outcome (i.e., a causal mechanistic mental model). In humans, embedded within these beliefs (as measured by verbal self-report) are such dimensions as the *effort*, *risks,* or *magnitude* associated with the response or outcome, as well as the different *short* and *long*-term aspects of the outcome (e.g., brief euphoria and lung cancer, respectively).

In most human laboratory nicotine studies, however, external stimuli tend to be scheduled as signals for whether a response will produce nicotine. External stimuli scheduled in this way apparently contribute to the *utility estimate* of the R-O contingency by retrieving a belief about the probability that the response will produce nicotine. The evidence for this claim comes from several sources. For instance, smoking stimuli enhance subjective craving/intentions to smoke (Tiffany and Drobes 1991), are correlated with the probability of relapse (O'Connell and Martin 1987; Shiffman 1986, 2009) and enhance smoking topography in experimental tests (Droungas et al. 1995; Elash et al. 1994; Glad and Adesso 1976; Herman 1974; Hogarth et al. 2010; Niaura et al. 1992; Payne et al. 1991; Surawy et al. 1985; but see Shiffman et al. 2013). Such cue-reactivity effects have also been found with arbitrary stimuli scheduled in the laboratory that signal when smoking is allowed (Mucha et al. 1998; Payne et al. 1990), and can be produced by simply informing participants about these discriminative contingencies (Dols et al. 2000, 2002). In one informative study (Hogarth et al. 2010), participants learned that a specific icon stimulus signaled when an instrumental tobacco-seeking response would be effective, thereby endowing the icon with control over that instrumental response. This stimulus subsequently enhanced puff probability during ad libitum smoking, which attests to the importance of the stimulus' initial discriminative function in signaling instrumental R-O contingency as a basis for transfer of control over actual consummatory behavior. A related finding by Perkins et al. (1994) is also informative. They found that a cigarette stimulus only enhanced instrumental response rate for smoking reward when the R-O probability was low, i.e., when a lean schedule was in force, but not when a rich schedule was in force. The implication is that when the estimated probability of the R-O contingency is asymptotically high, as signaled by schedule associated contextual cues, the addition of discrete smoking cues is unable to raise this estimate further—and so, these cues are motivationally ineffective. Conversely, when the contextually signaled R-O contingency is low, the addition of discrete smoking cues can raise the estimated R-O probability and thus prime the response. Similarly, in animals, the presentation of contextual stimuli that signal a lean response-drug contingency in conjunction with discrete drug paired stimulus produces additive effects in priming performance of the drug-seeking response (Remedios et al. 2014). These findings strongly encourage us to adopt a theory of external stimulus control which considers how holistic external stimulus configurations are incorporation into ongoing economic beliefs about the status of the R-O contingency (MacKillop et al. 2010).

More support for the expectancy element of stimulus control theory again comes from the transfer procedure. It has been shown that an arbitrary stimulus which signaled whether one tobacco-seeking response (R1) would be reinforced transfered control over a separately acquired tobacco-seeking response (R2), but did not transfer control over another response that produced a different reward (Hogarth et al. 2007). This outcome-specific transfer effect has also been found with smoking pictures (Hogarth 2012; Hogarth and Chase 2011, 2012). These findings suggest that external smoking stimuli enhance the expected probability of response-tobacco contingencies generally, which sums with contextual cues signaling that R2 is currently reinforced, thereby increasing the performance of R2 over alternative Rs. In direct support of this expectancy account, we recently showed that this outcome-specific transfer effect was both correlated with the extent to which stimuli evoked a self-reported expectation that the transfer response had a higher probability of being effective, and could be abolished by instructions which contradicted this expectation (Hogarth et al. 2014). Finally, compelling evidence for the causal role of cued expected probability comes from Carter and Tiffany (2001), who showed that a cue-provoked tobacco-seeking response readily came under the control of trial-by-trial verbal instructions stating what the current probability was of that response being effective. Together, this cue-reactivity literature suggests that external smoking cues prime tobacco-seeking by evoking an expectation that the response is likely to be effective, raising the utility of that response option. These external cue-elicited expectations concerning currently effective R-O contingencies must then be integrated with estimates of outcome value usually carried by internal cues in order to guide selection between response options.

6 Integrating Expected Probability and Value

We have argued that by virtue of their associative history, internal and external nicotine associated stimuli generally signal the current value and probability of the response-nicotine contingency, respectively (although as noted, these stimuli could be reversed in this role, or could qualify any other dimension of the response-outcome expectancy such as what, whether, when, why, effort, risks, magnitude, etc., if scheduled appropriately). In this section, we consider how expected value and probability estimates are integrated to initiate a discussion on how associative learning gives rise to the multidimensional beliefs that guide action selection. We will show that external and internal stimuli exert independent but additive effects on the propensity to perform the response; that is, nicotine-seeking is most likely when the response is expected to be both effective and nicotine currently has a high value. These probability and value estimates may be integrated to form an overall utility estimate for that R-O contingency, providing a common currency for comparison and choice between alternative reinforcers (Vlaev et al. 2011).

In relation to the integration of external and internal S^Ds, one well-replicated finding must be considered by any economic decision account of nicotine-seeking

(Ostlund and Balleine 2008); this is the finding that external cue-priming of tobacco-seeking is not itself sensitive to changes in the value of tobacco. In perhaps the seminal observation of this effect, Herman (1974) found that latency to smoke was decreased by the presentation of a smoking cue, and by deprivation; but, this latency-reduction effect over a noncue baseline was not itself enhanced by deprivation. Similar results have been obtained by Perkins et al. (1994) who found that although smoking cues enhanced responding for tobacco in lean schedules, and smoking deprivation enhanced responding overall, the cueing effect over noncue baseline was not enhanced by deprivation. Finally, we have found that the extent to which an external S^D for one tobacco-seeking response (R1) enhances performance of a separately trained tobacco-seeking response (R2) over baseline is not modulated by deprivation/satiety state (Hogarth and Chase 2011), although deprivation/satiety does modulate the baseline performance of R2. This insensitivity of the transfer effect to outcome devaluation has been consistently observed in basic animal learning studies (see Hogarth et al. 2013a for a review). Further evidence comes from a related transfer design in which we found that deprivation did not modulate cue-enhancement of puff probability over baseline during ad libitum smoking, but did modulate the baseline puff probability (Hogarth et al. 2010). Numerous cue-elicited craving studies confirm this autonomy of cue effects from deprivation. As noted by Tiffany et al. (2009, p. 180) when reviewing this literature, "abstinence-induced phasic craving and cue-specific cigarette craving appear to contribute additively to the total level of craving observed in a smoker at any given time (Drobes and Tiffany 1997; Maude-Griffin and Tiffany 1996; Tiffany et al. 2000)". Thus, in total, these studies have found a main effect of both cue presentation and deprivation on the propensity of tobacco-seeking (or craving)—but no interaction; this suggests that the probability estimate carried by external cues and the value estimate carried by internal states are additive in determining the propensity to engage in tobacco-seeking—not multiplicative. Stated differently, they each make an independent contribution to the likelihood of engaging in tobacco-seeking. The task therefore, is to construct an adequate human decision model which can explain how diverse signals for different dimensions of the R-O expectancy are able to converge on action selection independently (Davidson 1993; Dickinson and Balleine 2002; Rangel et al. 2008).

7 The Response-Outcome Gestalt

One of the primary claims made in this chapter is that the basic unit of decision-making is a gestalt belief concerning the nature of the currently viable R-O pairs. Previous associative models of action selection have tended to favor binary accounts where, for example, a representation of the outcome might retrieve the associated response, or the stimulus might retrieve possible responses, which retrieves their associated outcomes, which then feed back to "weigh" the responses, and so on (Balleine and Ostlund 2007; de Wit and Dickinson 2009). Our claim of a gestalt R-O expectancy (i.e., the hierarchical account) therefore warrants some empirical elaboration. Formal support for this claim comes from a rich variety of observations made outside the nicotine field, but will be briefly considered here for completeness: First, an established literature within perception research has shown that external stimuli have remarkable 'affordance' in that they rapidly activate responses that are relevant to the intended outcome, for example, doors afford the action of opening to achieve the outcome of passing through (Şahin et al. 2007). Second, more decisively is the finding that choice between competing response alternatives is determined by the net payoff of the expected reward minus the response costs, indicating that the response and the outcome must be integrated to form a net value of the R-O expectancy. Neurobiological confirmation of this account comes from the discovery of cells that encode these two dimensions of a choice (outcome magnitude and response costs), which are colocalized in the prefrontal cortex—providing an organic nexus for the R-O gestalt (Kennerley and Walton 2011). Third, so-called response-compatibility effects have shown that stimuli evoke responses more quickly if the stimulus is compatible with the outcome expected from the response with respect to spatial position (Hommel 1993; Kunde 2001; Lu and Proctor 1995), affective code (Eder et al. 2012), perceptual identity or semantic meaning (Koch and Kunde 2002). These data indicate that compatibility between the stimulus and the outcome representation facilitates response production, suggesting that such response production is entangled with conjoined R-O expectancy. Similarly, instrumental discrimination learning is faster if the stimulus signaling the response is compatible with the expected outcome with respect to spatial position (Overmier et al. 1971; Trapold 1970; Urcuioli 2005; see also Rescorla and Cunningham 1979) or perceptual identity (de Wit et al. 2012; Dwyer et al. 2010), suggesting that compatibility between the stimulus and the outcome facilitates encoding of the S:R-O relations. The final and perhaps most decisive evidence for the hierarchical position is the instrumental biconditional discrimination task developed by Rescorla in which two R-O contingencies are reversed in different contextual stimuli (specifically, S1:R1-O1, R2-O2; S2:R1-O2, R2-O1), such that the binary S-R, R-O, and S-O associations are equated. Thus, arguably, animals can only shift their choice of one response over the other (following devaluation or novel cue presentation, for example) if they retrieve knowledge of which R-O pairs are in force in each S (Colwill and Rescorla 1990; Rescorla 1987, 1991, 1992b). This hierarchical account of action control has been confirmed by more recent analysis (Bradfield and Balleine 2013). Although separating binary and hierarchical accounts of action control is extremely challenging, the studies listed here provide evidence in favor of the hierarchical position, which must at least be taken into consideration when deciding between these closely aligned positions. Moreover, based on our experience of human associative procedures, where mentalistic or teleological explanations of behavior hold more sway (but see Rachlin 1997), we would come down on the conclusion that external discriminative stimuli set the occasion for response production by evoking a gestalt

R-O relationship/association, i.e., a momentary representation or belief that a particular response will produce a particular outcome which guides the selection of that response.

8 Choice Among Alternatives

It remains unclear how the various dimensions of an R-O gestalt (what, whether, where when, why … etc.) are integrated to enable choice in comparison with other options. What is clear, is that the decision to engage in nicotine-seeking not only depends on the utility (probability + value) of the response-nicotine expectancy, but also depends on the utility of competing alternatives (Ahmed et al. 2013). The evidence for this claim comes from numerous evidence for this claim comes from numerous studies in which animals are given concurrent free choice between an addictive drug versus a qualitatively different natural reinforcer (often sucrose). These studies have shown that when the magnitude, value, or probability of the natural alternative reinforcer is increased, responding for the drug decreases and vice versa (Ahmed 2010; Banks and Negus 2012; Jimenez-Gomez and Shahan 2008; Woolverton and Anderson 2006). Importantly, if the drug response is extinguished and the naturally rewarded response continues to be reinforced choice predictably shifts toward the naturally rewarded responses; but, if the naturally rewarded response is then extinguished, *resurgence* of the original drug response is found (Quick et al. 2011; Winterbauer and Bouton 2010). This resurgence effect suggests that changing the status of the alternative response modulates the utility of the competing drug response, indicating that action choice is driven by a consideration between alternatives. The limited animal concurrent choice studies that have employed nicotine generally support this claim (Manzardo et al. 2002; Stairs et al. 2010). Similarly, numerous human studies have shown that responding for tobacco or nicotine can be decreased by increasing the utility of the competing alternative (Bickel and Madden 1999; Epstein et al. 1991; Johnson and Bickel 2003; Shahan et al. 2001). Importantly, two studies found that devaluing the natural reinforcer in a human concurrent choice procedure immediately enhanced responding for the tobacco outcome during an extinction test, indicating that the decreased value of the alternative raised *expected* relative value of the tobacco outcome, which raised choice of this response (Hogarth 2012; Hogarth and Chase 2011). Finally, clinical research on contingency management has shown that smoking cessation can be promoted if monetary payment is made contingent on abstinence (Donatelle et al. 2000; Roll and Higgins 2000; Shoptaw et al. 2002; Volpp et al. 2006), and if exercise is promoted as a competing alternative (Ussher et al. 2012). A decision model wherein action selection is determined by an evaluation of competing R-O options is well placed to accommodate these concurrent choice data.

9 Five Applied Tests for the Hierarchical Account

We have claimed that in any given moment smokers compare various R-O gestalts (one of which is smoking) and chooses the one with the highest relative utility. In the sections that follow, we consider how well such a choice model can accommodate five different observations about the nature of nicotine dependence.

9.1 Individual Differences in Smoking Uptake

There are individual differences in the liability to become dependent to nicotine following early initiation (Anthony et al. 1994; Chassin et al. 2000), but the psychological basis of this liability remains contested. The behavioral economists have consistently argued that dependence is mediated by relative reinforcement value of nicotine establishing higher rates of voluntary nicotine-seeking in some individuals (MacKillop et al. 2012). The concurrent choice studies conducted by the author (Hogarth) have strongly supported this relative reinforcement account of individual liability. We have found that nicotine dependence level in smokers with an average age of 20, and 4 years of smoking history (where nicotine dependence indexed by a variety of standard questionnaire proxies) significantly correlated with preferential choice of tobacco over a natural reward alternative (Chase et al. 2013; Hogarth 2011; Hogarth and Chase 2011, 2012; Hogarth et al. 2012b). Importantly, this preference was found in extinction tests during which free choice between responses was without consequence, and choice was sensitive to manipulations of deprivation/satiety. These data suggests that the preference was mediated by the *expected* reinforcement value of tobacco and is therefore consistent with a decision account over a stimulus-response account.

9.2 Individual Differences in Smoking Perseveration

More critically for the generality of a decision (or choice) model of dependence, is whether concurrent nicotine preference not only marks nicotine dependence in young adult smokers, but also in older more experienced smokers. Consistent with such a claim, Perkins et al. (2002) found that preferential nicotine choice in a concurrent choice procedure predicted propensity to relapse in a sample of smokers with average age of 41.5 and 22.7 years of smoking. This relationship has been confirmed in older cocaine users among whom relapse was predicted by their preferential choice of cocaine pictures (Moeller et al. 2013). Furthermore, greater hedonic experience of smoking lapses following quitting predicted the transition to full relapse (Shiffman et al. 2006). Additionally, there is consistent evidence that craving level (anticipated smoking reward) predicts the likelihood of relapse (Killen and Fortmann 1997) and reductions in craving mediate some therapeutic effects on smoking cessations (Ferguson et al. 2006; McCarthy et al. 2008; Piper et al. 2008). Together, the forgoing data suggest that greater expected reinforcement value of nicotine presents a risk factor for dependence in both the early and late portion of a smoking history, attesting to the hierarchical decision account as a general model of dependence across the life span.

A parallel domain of research, however, suggests that chronic drug exposure promotes a transition from goal-directed (intentional) to habitual control over drug-seeking. The principle evidence for this claim comes from animal studies showing that chronic drug exposure impairs the ability of animals to use knowledge of the current expected value of an outcome to determine the propensity to engage in responding for that outcome. Animals instead respond for the outcome at a level unchanged from prior training consistent with control of this response by S-R habit learning rather than goal-directed R-O knowledge (Corbit et al. 2012; Dickinson et al. 2002; Nelson and Killcross 2013). The critical question therefore, is whether there is any evidence for predominance of S-R habit learning in more dependent smokers—and if so, what proportion of the variance in dependence in this older group is accounted for by the S-R habit mechanisms versus the expected value of the R-O contingency?

There is *indirect* evidence that nicotine dependence is associated with predilection for S-R habit learning. First, trait impulsivity has been linked to higher risk of smoking relapse (Doran et al. 2004; VanderVeen et al. 2008), and impulsive smokers show greater habitual control of natural reward-seeking behavior in the outcome devaluation assay, the principal assay for demonstrating habitual control (Hogarth et al. 2012b). Similarly, in ad libitum assays of smoking topography, high impulsive smokers show a decoupling (null correlation) between subjective craving and number of puffs consumed, which is consistent with habitual control (Tiffany 1990); whereas, low impulsive smokers show a correlation between craving and number of puffs consumed, which is consistent with intentional control (Hogarth 2011). Finally, we have shown that habitual control of tobacco-seeking in the outcome devaluation assay can be produced by the presence of an irrelevant distracting reinforcer (Hogarth et al. 2013b) or by acute alcohol intoxication (Hogarth et al. 2012a), consistent with the view that smokers who have cognitive impairment may be more likely to transition to habitual control over smoking behavior (Brody et al. 2004; Patterson et al. 2010).

However, although we have consistently found that impulsivity is associated with more pronounced habit learning, nicotine dependence itself has shown no direct association with either (Hogarth 2011, 2012; Hogarth and Chase 2011). We therefore favor the view that trait impulsivity is orthogonal to trait sensitivity to drug reinforcement (dependence), such that these two traits represent independent additive sources of vulnerability to the maintenance and perseveration of smoking behavior (Belin et al. 2008). In summary, although nicotine value may drive the extent of smoking behavior across the span of a smoking history, impulsivity, or those with cognitive decline may accrue and additional predilection for habitual control over smoking which further undermines successful quitting.

9.3 Pharmacotherapy

Support for the hierarchical decision account comes from the observation that the main pharmacotherapeutic agents for smoking cessation, nicotine replacement therapy (NRT), varenicline and bupropion, reduce the expected value of smoking by mimicking an internal stimulus for satiety. The basis for this claim is that acute administration of nicotine replacement therapy (NRT), varenicline, and bupropion all reduce craving (Brandon et al. 2011; Ferguson and Shiffman 2009; Franklin et al. 2011; Hitsman et al. 2013). NRT also reduces nicotine choice and smoking behavior (Benowitz and Jacob 1990; Johnson and Bickel 2003; Perkins et al. 1992; Rose et al. 1985). Most decisively, NRT also immediately modulates nicotine choice in an extinction test, indicating that NRT modulated the expected value nicotine (Hogarth 2012). Moreover, this effect varies as a function of dependence level, with low-dose NRT augmenting tobacco choice in heavier smokers, and suppressing tobacco choice in lighter smokers. These findings are consistent with NRT mimicking a nicotine load that was lower or higher than these groups' constitutional optimum dose, respectively, causing a titration of responding towards the optimum governed by expected nicotine reward. Crucially, however, NRT had no impact on the ability of an external smoking stimulus to prime performance of the tobacco-seeking response in the extinction test, suggesting NRT did not block the smoking cues' enhancement of the expected probability of the response-tobacco contingency. Thus, pharmacotherapy appears to modulate the expected value but not probability of the response-nicotine contingency, accounting for its partial therapeutic effect.

9.4 Extinction Therapy

In contrast to pharmacotherapy, cue-extinction therapy works best when the cue is scheduled to signal that the response-tobacco contingency has a low probability in accordance with the hierarchical account. To give some background, cue-exposure therapies have commonly been of two types: Pavlovian extinction where the drug stimulus is presented without the drug outcome thus degrading the Pavlovian S-O contingency; or instrumental extinction where the 'mock' drug-taking responses are performed without producing the drug outcome thus degrading the instrumental R-O contingency. Although such extinction procedures temporarily reduce cue-evoked craving in the laboratory, they produce no long term effects on abstinence in the field (Collins and Brandon 2002; Conklin and Tiffany 2002; Price et al. 2010; Thewissen et al. 2006; Xue et al. 2012). Consistent with an insight reported by Conklin and Tiffany (2002), the hierarchical position explains this clinical failure on the grounds that degradation of the binary S-O or R-O contingencies does little to modify the hierarchical function of drug stimuli in signaling the current strength of the response-drug contingency (S:R-O). In support of this claim, we recently found that the ability of drug cues to transfer control over a separately trained drug-seeking response was not abolished by Pavlovian extinction where the stimulus was presented without the drug (S-no O), but was abolished by discriminative extinction training where the stimulus signaled that the response-drug contingency would be nonreinforced (S:R-no O) (Hogarth et al. 2014). This reduced effectiveness of Pavlovian compared to discriminative operant extinction training in abolishing the transfer effect confirmed previous studies with natural rewards in humans and animals (Delamater 1996; Gámez and Rosas 2005; Rescorla 1992a; Rosas et al. 2010). In a further study, we found that participants who abandoned their hierarchical beliefs following instructions that stimuli did *not* signal which R-O contingency was more likely to be rewarded, showed no transfer effect. This dependency of the transfer effect on hierarchical beliefs supports the claim that transfer of stimulus control over instrumental performance can be propositional in nature (Heyes and Dickinson 1990; Mitchell et al. 2009); that is, S retrieves an expectancy (that can be verbalized inself-report) pertaining to the R-O probability, which augments selection of that response. The implication for cue-extinction therapy is that one should target the belief evoked by smoking cues that under these conditions tobacco-seeking is a viable response.

9.5 Plain Packaging

Like discriminative extinction training, plain packaging also appears to degrade the ability of pack stimuli to enhance the expected probability of the response-tobacco contingency. In this most recent study by Hogarth et al. (2014), we compared the capacity of plain versus branded cigarette pack stimuli to enhance selection of a tobacco-seeking response in an extinction test. Two experiments found that whereas branded pack stimuli enhanced tobacco choice consistent with previous findings (Hogarth 2012; Hogarth and Chase 2011, 2012), plain pack stimuli failed to do so, suggesting that plain packs were unable to enhance the expected probability of the response-tobacco contingency. By contrast, there was no difference in preferential selection of plain and branded packs when these were schedules as outcomes, suggesting they were of equivalent economic value. The effect of discriminative extinction training is comparable, insofar as this intervention also abolished the capacity of the smoking stimuli to prime selection of the tobacco-seeking response in extinction—an effect that was dependent on degradation of propositional hierarchical beliefs that the stimulus signaled a greater response-tobacco probability. These findings of plain packaging and discriminative extinction are doubly dissociable against the effect of pharmacotherapy, which modified the expected value of tobacco in the preferential choice test but had no impact on cue-priming of the tobacco response in the transfer test. The implication is that greater treatment efficacy would be better achieved by combined therapies targeting both the effects of cue-priming and expected value on the control over nicotine-seeking, that is, combining pharmacotherapy with either extinction training, or plain packaging or both.

10 Conclusion

We have reviewed evidence from a multitude of domains including titration, nicotine discrimination, cue-reactivity, hierarchical learning, economic choice theory, individual liability to dependence, pharmacotherapy, cue-extinction therapy, and plain packaging. All of the data are consistent with a hierarchical instrumental decision theory wherein smokers retrieve a gestalt expectancy regarding the net utility of available response-nicotine contingencies and select that response where the utility exceeds concurrently available alternatives. Individual differences in dependence liability appear to be mediated by greater expected value of response-nicotine contingencies across the lifetime, although S-R habit learning may contribute in parallel in impulsive or cognitively impaired smokers (Heyman 2013). Pharmacotherapy functions by selectively decreasing the expected value of the response-nicotine contingency, whereas cue-extinction therapy and plain packaging function by selectively decreasing the expected probability of the response-nicotine contingency. Treatments should therefore be combined to degrade both independent forms of control over smoking behavior. The main advantage of this hierarchical instrumental decision account of nicotine dependence over other models is that it is able to successfully reconcile a broad variety of findings precisely because it accepts that multiple diverse sources of internal and external information must be integrated to shape the decision to smoke.

Acknowledgments Funding: This work was supported by the MRC (G0701456: LH) and ESCR (RES-000-22-4365: LH); and by New Hampshire IDeA Network of Biological Research Excellence (NH-INBRE; NIH Grant Number 1P20RR030360-01 from the INBRE Program of the National Center for Research Resources: JRT).

References

Ahmed SH (2010) Validation crisis in animal models of drug addiction: beyond non-disordered drug use toward drug addiction. Neurosci Biobehav Rev 35:172–184

Ahmed SH, Koob GF (2005) Transition to drug addiction: a negative reinforcement model based on an allostatic decrease in reward function. Psychopharmacology 180:473–490

Ahmed SH, Lenoir M, Guillem K (2013) Neurobiology of addiction versus drug use driven by lack of choice. Curr Opin Neurobiol 23:581–587

Anthony JC, Warner LA, Kessler RC (1994) Comparative epidemiology of dependence on tobacco, alcohol, controlled substances, and inhalants: basic findings from the national comorbidity survey. Exp Clin Psychopharmacol 2:244–268

Baker TB, Piper ME, McCarthy DE, Majeskie MR, Fiore MC (2004) Addiction motivation reformulated: an affective processing model of negative reinforcement. Psychol Rev 111:33–51

Balleine BW, Ostlund SB (2007) Still at the choice-point: action selection and initiation in instrumental conditioning. In: Balleine BW, Doya K, Doherty JO, Sakagami M (eds) Reward and decision making in corticobasal ganglia networks (Annals of the New York Academy of Sciences.). Blackwell Publishing, Malden, pp 147–171

Banks ML, Negus SS (2012) Preclinical determinants of drug choice under concurrent schedules of drug self-administration. Adv Pharmacol Sci 2012:281768

Behm F, Rose J (1994) Reducing craving for cigarettes while decreasing smoke intake using capsaicin-enhanced low tar cigarette. Exp Clin Psychopharmacol 2:143–153
Belin D, Mar AC, Dalley JW, Robbins TW, Everitt BJ (2008) High impulsivity predicts the switch to compulsive cocaine-taking. Science 320:1352–1355
Benowitz NL, Jacob P (1990) Intravenous nicotine replacement suppresses nicotine intake from cigarette smoking. J Pharmacol Exp Ther 254:1000–1005
Benowitz N, Porchet H, Jacob P (1990) Pharmacokinetics, metablolism, and pharmacodynamics of nicotine. In: Wonnacott S, Russell M, Stolerman I (eds) Nicotine psychopharmacology molecular, cellular and behavioral aspects. Oxford University Press, New York, pp 112–157
Bickel WK, Madden GJ (1999) A comparison of measures of relative reinforcing efficacy and behavioral economics: cigarettes and money in smokers. Behav Pharmacol 10:627–637
Bouton ME, Doyle-Burr C, Vurbic D (2012) Asymmetrical generalization of conditioning and extinction from compound to element and element to compound. J Exp Psychol Anim Behav Process 38:381–393
Bradfield LA, Balleine BW (2013) Hierarchical and binary associations compete for behavioral control during instrumental biconditional discrimination. J Exp Psychol Anim Behav Process 39:2–13
Brandon T, Drobes D, Unrod M, Heckman B, Oliver J, Roetzheim R, Karver S, Small B (2011) Varenicline effects on craving, cue reactivity, and smoking reward. Psychopharmacology 218:1–13
Brody AL, Mandelkern MA, Jarvik ME, Lee GS, Smith EC, Huang JC, Bota RG, Bartzokis G, London ED (2004) Differences between smokers and nonsmokers in regional gray matter volumes and densities. Biol Psychiatry 55:77–84
Carter BL, Tiffany ST (2001) The cue-availability paradigm: the effects of cigarette availability on cue reactivity in smokers. Exp Clin Psychopharmacol 9:183–190
Chait L, Griffiths R (1982) Differential control of puff duration and interpuff interval in cigarette smokers. Pharmacol Biochem Behav 17:155–158
Chase HW, MacKillop J, Hogarth L (2013) Isolating behavioral economic indices of demand in relation to nicotine dependence. Psychopharmacology 226:371–380
Chassin L, Presson CC, Pitts SC, Sherman SJ (2000) The natural history of cigarette smoking from adolescence to adulthood in a midwestern community sample: multiple trajectories and their psychosocial correlates. Health Psychol 19:223–231
Collins BN, Brandon TH (2002) Effects of extinction context and retrieval cues on alcohol cue reactivity among nonalcoholic drinkers. J Consult Clin Psychol 70:390–397
Colwill RM, Rescorla RA (1990) Evidence for the hierarchical structure of instrumental learning. Anim Learn Behav 18:71–82
Conklin CA, Tiffany ST (2002) Applying extinction research and theory to cue-exposure addiction treatments. Addiction 97:155–167
Corbit LH, Nie H, Janak PH (2012) Habitual alcohol seeking: time course and the contribution of subregions of the dorsal striatum. Biol Psychiatry 72:389–395
Corrigall W, Coen K (1989a) Nicotine maintains robust self-administration in rats on a limited-access schedule. Psychopharmacology 99:473–478
Corrigall WA, Coen KM (1989b) Nicotine maintains robust self-administration in rats on a limited-access schedule. Psychopharmacology 99:473–478
Davidson TL (1993) The nature and function of interoceptive signal to feed: towards integration of physiological and learning perspectives. Psychol Rev 100:640–657
Davidson TL, Aparicio J, Rescorla R (1988) Transfer between Pavlovian facilitators and instrumental discriminative stimuli. Anim Learn Behav 16:285–291
de Wit S, Dickinson A (2009) Associative theories of goal-directed behavior: a case for animal–human translational models. Psychol Res 73:463–476
de Wit S, Ridderinkhof KR, Fletcher P, Dickinson A (2012) Resolution of outcome-induced response conflict by humans after extended training. Psychol Res 47:1–14
Delamater AR (1996) Effects of several extinction treatments upon the integrity of Pavlovian stimulus-outcome associations. Learn Behav 24:437–449

Delfino RJ, Jamner LD, Whalen CK (2001) Temporal analysis of the relationship of smoking behavior and urges to mood states in men versus women. Nicotine Tob Res 3:235–248
Devito EE, Herman AI, Waters AJ, Valentine GW, Sofuoglu M (2014) Subjective, physiological, and cognitive responses to intravenous nicotine: effects of sex and menstrual cycle phase. Neuropsychopharmacology 39:1431–1440
Di Ciano P, Everitt BJ (2003) Differential control over drug-seeking behavior by drug-associated conditioned reinforcers and discriminative stimuli predictive of drug availability. Behav Neurosci 117:952–960
Dickinson A, Balleine BW (2002) The role of learning in the operation of motivational systems. In: Gallistel CR (ed) Stevens' handbook of experimental psychology, learning, motivation and emotion, vol 3, 3rd edn. Wiley, New York, pp 497–533
Dickinson A, Balleine B (2010) The cognitive/motivational interface. In: Kringelbach ML, Berridge KC (eds) Pleasures of the brain. The neural basis of taste, smell and other rewards. Oxford University Press, Oxford, pp 74–84
Dickinson A, Wood N, Smith JW (2002) Alcohol seeking by rats: action or habit? Q J Exp Psychol B Comp Physiol Psychol 55:331–348
Dols M, Willems B, van den Hout M, Bittoun R (2000) Smokers can learn to influence their urge to smoke. Addict Behav 25:103–108
Dols M, Hout Mvd, Kindt M, Willems B (2002) The urge to smoke depends on the expectation of smoking. Addiction 97:87–93
Donatelle RJ, Prows SL, Champeau D, Hudson D (2000) Randomised controlled trial using social support and financial incentives for high risk pregnant smokers: significant other supporter (SOS) program. Tob Control 9:iii67–iii69
Donny E, Caggiula A, Knopf S, Brown C (1995) Nicotine self-administration in rats. Psychopharmacology 122:390–392
Doran N, Spring B, McChargue D, Pergadia M, Richmond M (2004) Impulsivity and smoking relapse. Nicotine Tob Res 6:641–647
Drobes DJ, Tiffany ST (1997) Induction of smoking urge through imaginal and in vivo procedures: physiological and self-report manifestations. J Abnorm Psychol 106:15–25
Droungas A, Ehrman R, Childress A, O'Brien C (1995) Effects of smoking cues and cigarette availability on craving and smoking behavior. Addict Behav 20:657–673
Dwyer DM, Dunn MJ, Rhodes SEV, Killcross AS (2010) Lesions of the prelimbic prefrontal cortex prevent response conflict produced by action–outcome associations. Q J Exp Psychol 63:417–424
Eder A, Müsseler J, Hommel B (2012) The structure of affective action representations: temporal binding of affective response codes. Psychol Res 76:111–118
Eissenberg T (2004) Measuring the emergence of tobacco dependence: the contribution of negative reinforcement models. Addiction 99:5–29
Elash CA, Burton SM, Tiffany ST (1994) The effect of imagery-induced urges on smoking behavior. Paper presented at the annual meeting of the Midwestern Psychological Association, Chicago
Epstein LH, Bulik CM, Perkins KA, Caggiulla AR, Rodefer J (1991) Behavioral economics analysis of smoking: money and food as alternatives. Pharmacol Biochem Behav 38:715–721
Everitt BJ, Robbins TW (2013) From the ventral to the dorsal striatum: devolving views of their roles in drug addiction. Neurosci Biobehav Rev 37:1946–1954
Ferguson SG, Shiffman S (2009) The relevance and treatment of cue-induced cravings in tobacco dependence. J Subst Abuse Treat 36:235–243
Ferguson SG, Shiffman S, Gwaltney CJ (2006) Does reducing withdrawal severity mediate nicotine patch efficacy? A randomized clinical trial. J Consult Clin Psychol 74:1153–1161
Feyerabend C, Ings R, Russell M (1985) Nicotine pharmacokinetics and its application to intake from smoking. Br J Pharmacol 19:239–247
Franklin T, Wang Z, Suh JJ, Hazan R, Cruz J, Li Y, Goldman M, Detre JA, O'Brien CP, Childress AR (2011) Effects of varenicline on smoking cue-triggered neural and craving responses. Arch Gen Psychiatry 68:516–526

Gámez AM, Rosas JM (2005) Transfer of stimulus control across instrumental responses is attenuated by extinction in human instrumental conditioning. Int J Psychol Psychol Ther 5:207–222

Gerhardstein L, Wang A, Burki N (1993) Nicotine is responsible for airway irritation evoked by cigarette smoking inhalation in men. J Appl Physiol 75:1955–1961

Glad W, Adesso VJ (1976) The relative importance of socially induced tension and behavioral contagion for smoking behavior. J Abnorm Psychol 85:119–121

Grindley GC (1932) The formation of a simple habit in guinea-pigs. Br J Psychol 23:127–147

Hammond LJ (1980) The effect of contingency upon the appetitive conditioning of free-operant behavior. J Exp Anal Behav 34:297–304

Hanson H, Ivester C, Morton B (1979) Nicotine self-administration in rats. In: Krasnegor N (ed) Cigarette smoking as a dependence process. NIDA Research Monograph 23, USD-HEW, Rockville

Harris CM, Emmett-Oglesby MW, Robinson NG, Lal H (1986) Withdrawal from chronic nicotine substitutes partially for the interoceptive stimulus produced by pentylenetetrazol (PTZ). Psychopharmacology 90:85–89

Harris JA, Andrew BJ, Kwok DWS (2013) Magazine approach during a signal for food depends on Pavlovian, not instrumental, conditioning. J Exp Psychol Anim Behav Process 39:107–116

Harvey DM, Yasar S, Heishman SJ, Panlilio LV, Henningfield JE, Goldberg SR (2004) Nicotine serves as an effective reinforcer of intravenous drug-taking behavior in human cigarette smokers. Psychopharmacology 175:134–142

Hasenfratz M, Baldinger B, Battig K (1993) Nicotine or tar titration in cigarette smoking behavior. Psychopharmacology 112:253–258

Herman CP (1974) External and internal cues as determinants of the smoking behavior of light and heavy smokers. J Pers Soc Psychol 30:664–672

Heyes C, Dickinson A (1990) The intentionality of animal action. Mind Lang 5:87–104

Heyman GM (2013) Addiction and choice: theory and new data. Front Psychiatry 4:31

Hitsman B, Hogarth L, Tseng L-J, Teige JC, Shadel WG, DiBenedetti DB, Danto S, Lee TC, Price LH, Niaura R (2013) Dissociable effect of acute varenicline on tonic versus cue-provoked craving in non-treatment-motivated heavy smokers. Drug Alcohol Depend 130:135–141

Hogarth L (2011) The role of impulsivity in the aetiology of drug dependence: reward sensitivity versus automaticity. Psychopharmacology 215:567–580

Hogarth L (2012) Goal-directed and transfer-cue-elicited drug-seeking are dissociated by pharmacotherapy: evidence for independent additive controllers. J Exp Psychol Anim Behav Process 38:266–278

Hogarth L, Chase HW (2011) Parallel goal-directed and habitual control of human drug-seeking: implications for dependence vulnerability. J Exp Psychol Anim Behav Process 37:261–276

Hogarth L, Chase HW (2012) Evaluating psychological markers for human nicotine dependence: tobacco choice, extinction, and Pavlovian-to-instrumental transfer. Exp Clin Psychopharmacol 20:213–224

Hogarth L, Dickinson A, Wright A, Kouvaraki M, Duka T (2007) The role of drug expectancy in the control of human drug seeking. J Exp Psychol Anim Behav Process 33:484–496

Hogarth L, Dickinson A, Janowski M, Nikitina A, Duka T (2008) The role of attentional bias in mediating human drug seeking behavior. Psychopharmacology 201:29–41

Hogarth L, Dickinson A, Duka T (2010) The associative basis of cue elicited drug taking in humans. Psychopharmacology 208:337–351

Hogarth L, Attwood AS, Bate HA, Munafò MR (2012a) Acute alcohol impairs human goal-directed action. Biol Psychol 90:154–160

Hogarth L, Chase HW, Baess K (2012b) Impaired goal-directed behavioral control in human impulsivity. Q J Exp Psychol 65:305–316

Hogarth L, Balleine BW, Corbit LH, Killcross S (2013a) Associative learning mechanisms underpinning the transition from recreational drug use to addiction. Ann N Y Acad Sci 1282:12–24

Hogarth L, Field M, Rose AK (2013b) Phasic transition from goal-directed to habitual control over drug-seeking produced by conflicting reinforcer expectancy. Addict Biol 18:88–97
Hogarth L, Retzler C, Munafò MR, Tran DMD, Troisi JRI, Rose A, Jones A, Field M (2014) Extinction of cue-evoked drug-seeking relies on targeting hierarchical instrumental knowledge. Behav Res Ther 59:61–70
Hogarth L, Maynard OM, Munafò MR (2014) Plain cigarette packs do not exert pavlovian to instrumental transfer of control over tobacco-seeking. Addiction
Hommel B (1993) Inverting the Simon effect by intention. Psychol Res 55:270–279
Hutcheson DM, Everitt BJ, Robbins TW, Dickinson A (2001) The role of withdrawal in heroin addiction: enhances reward or promotes avoidance? Nat Neurosci 4:943–947
Jarvik ME, Caskey NH, Rose JE, Herskovic JE, Sadeghpour M (1989) Anxiolytic effects of smoking associated with four stressors. Addict Behav 14:379–386
Jimenez-Gomez C, Shahan TA (2008) Matching law analysis of rats' alcohol self-administration in a free-operant choice procedure. Behav Pharmacol 19:353–356
Johnson MW, Bickel WK (2003) The behavioral economics of cigarette smoking: the concurrent presence of a substitute and an independent reinforcer. Behav Pharmacol 14:137–144
Kennerley SW, Walton ME (2011) Decision making and reward in frontal cortex: complementary evidence from neurophysiological and neuropsychological studies. Behav Neurosci 125:297–317
Killen JD, Fortmann SP (1997) Craving is associated with smoking relapse: findings from three prospective studies. Exp Clin Psychopharmacol 5:137–142
Koch I, Kunde W (2002) Verbal response-effect compatibility. Mem Cogn 30:1297–1303
Kunde W (2001) Response-effect compatibility in manual choice reaction tasks. J Exp Psychol Hum Percept Perform 27:387–394
Litvin EB, Brandon TH (2010) Testing the influence of external and internal cues on smoking motivation using a community sample. Exp Clin Psyhopharmacol 18:61–70
Liu X, Jernigan C (2012) Effects of caffeine on persistence and reinstatement of nicotine-seeking behavior in rats: interaction with nicotine-associated cues. Psychopharmacology 220:541–550
Lu CH, Proctor RW (1995) The influence of irrelevant location information on performance: a review of the Simon and spatial Stroop effects. Psychon Bull Rev 2:174–207
MacKillop J, O'Hagen S, Lisman SA, Murphy JG, Ray LA, Tidey JW, McGeary JE, Monti PM (2010) Behavioral economic analysis of cue-elicited craving for alcohol. Addiction 105:1599–1607
MacKillop J, Few LR, Murphy JG, Wier LM, Acker J, Murphy C, Stojek M, Carrigan M, Chaloupka F (2012) High-resolution behavioral economic analysis of cigarette demand to inform tax policy. Addiction 107:2191–2200
Manzardo AM, Stein L, Belluzzi JD (2002) Rats prefer cocaine over nicotine in a two-lever self-administration choice test. Brain Res 924:10–19
Marissen MAE, Franken IHA, Waters AJ, Blanken P, van den Brink W, Hendriks VM (2006) Attentional bias predicts heroin relapse following treatment. Addiction 101:1306–1312
Maude-Griffin PM, Tiffany ST (1996) Production of smoking urges through imagery: the impact of affect and smoking abstinence. Exp Clin Psychopharmacol 4:198–208
McCarthy DE, Piasecki TM, Lawrence DL, Jorenby DE, Shiffman S, Baker TB (2008) Psychological mediators of bupropion sustained-release treatment for smoking cessation. Addiction 103:1521–1533
Meeter M, Shohamy D, Myers CE (2009) Acquired equivalence changes stimulus representations. J Exp Anal Behav 91:127–141
Mitchell CJ, De Houwer J, Lovibond PF (2009) The propositional nature of human associative learning. Behav Brain Sci 32:183–198
Moeller SJ, Beebe-Wang N, Woicik PA, Konova AB, Maloney T, Goldstein RZ (2013) Choice to view cocaine images predicts concurrent and prospective drug use in cocaine addiction. Drug Alcohol Depend 130:178–185
Mucha RF, Pauli P, Angrilli A (1998) Conditioned responses elicited by experimentally produced cues for smoking. Can J Physiol Pharmacol 76:259–268

Nelson AJD, Killcross S (2013) Accelerated habit formation following amphetamine exposure is reversed by D1, but enhanced by D2, receptor antagonists. Front Neurosci 7:76
Nemeth-Coslett R, Griffiths R (1984a) Determinants of puff duration in cigarette smokers: I. Pharmacol Biochem Behav 20:965–971
Nemeth-Coslett R, Griffiths R (1984b) Determinants of puff duration in cigarette smokers: II. Pharmacol Biochem Behav 21:903–912
Niaura RS, Abrams DB, Pedraza M, Monti P, Rosenhow DJ (1992) Smokers reactions to interpersonal interactions and presentation of smoking cues. Addict Behav 17:557–566
Nil R, Battig K (1989) Separate effects of cigarette smoke yield and smoke taste on smoking behavior. Psychopharmacology 99:54–59
O'Connell K, Martin E (1987) Highly tempting situations associated with abstinance, temporary lapse, and relapse among participants in smoking cessation program. J Consult Clin Psychol 55:367–371
Ostlund SB, Balleine BW (2008) The disunity of Pavlovian and instrumental values. Behav Brain Sci 31:456
Overmier JB, Bull JA, Trapold MA (1971) Discriminative cue properties of different fears and their role in response selection in dogs. J Comp Physiol Psychol 76:478–482
Palmatier MI, Bevins RA (2008) Occasion setting by drug states: functional equivalence following similar training history. Behav Brain Res 195:260–270
Panlilio LV, Schindler CW, Weiss SJ (1996) Cocaine self-administration increased by compounding discriminative stimuli. Psychopharmacology 125:202–208.
Patterson F, Jepson C, Loughead J, Perkins K, Strasser AA, Siegel S, Frey J, Gur R, Lerman C (2010) Working memory deficits predict short-term smoking resumption following brief abstinence. Drug Alcohol Depend 106:61–64
Payne T, Etscheidt M, Corrigan S (1990) Conditioning arbitrary stimuli to cigarette smoke intake: a preliminary study. J Subst Abuse 2:113–119
Payne TJ, Schare ML, Levis DJ, Colletti G (1991) Exposure to smoking-relevant cues: effects on desire to smoke and topographical components of smoking behavior. Addict Behav 16:467–479
Perkins KA, Grobe JE, Stiller RL, Fonte C, Goettler JE (1992) Nasal spray nicotine replacement suppresses cigarette smoking desire and behavior. Clin Pharmacol Ther 52:627–634
Perkins KA, Epstein LH, Grobe J, Fonte C (1994) Tobacco abstinence, smoking cues, and the reinforcing value of smoking. Pharmacol Biochem Behav 47:107–112
Perkins KA, Grobe J, Fonte C (1997a) Influence of acute smoking exposure on the subsequent reinforcing value of smoking. Exp Clin Psychopharmacol 5:277–285
Perkins KA, Sanders M, D'Amico D, Wilson A (1997b) Nicotine discrimination and self-administration in humans as a function of smoking status. Psychopharmacology 131:361–370
Perkins KA, Sanders M, Fonte C, Wilson AS, White W, Stiller R, McNamara D (1999) Effects of central and peripheral nicotinic blockade on human nicotine discrimination. Psychopharmacology 142:158–164
Perkins KA, Broge M, Gerlach D, Sanders M, Grobe JE, Cherry C, Wilson AS (2002) Acute nicotine reinforcement, but not chronic tolerance, predicts withdrawal and relapse after quitting smoking. Health Psychol 21:332–339
Perkins KA, Giedgowd GE, Karelitz JL, Conklin CA, Lerman C (2012) Smoking in response to negative mood in men versus women as a function of distress tolerance. Nicotine Tob Res 14:1418–1425
Piper ME, Federmen EB, McCarthy DE, Bolt DM, Smith SS, Fiore MC, Baker TB (2008) Using mediational models to explore the nature of tobacco motivation and tobacco treatment effects. J Abnorm Psychol 117:94–105
Pittenger ST, Bevins RA (2013) Interoceptive conditioning with a nicotine stimulus is susceptible to reinforcer devaluation. Behav Neurosci 127:465–473
Price KL, Saladin ME, Baker NL, Tolliver BK, DeSantis SM, McRae-Clark AL, Brady KT (2010) Extinction of drug cue reactivity in methamphetamine-dependent individuals. Behav Res Ther 48:860–865

Quick SL, Pyszczynski AD, Colston KA, Shahan TA (2011) Loss of alternative non-drug reinforcement induces relapse of cocaine-seeking in rats: role of dopamine D1 receptors. Neuropsychopharmacology 36:1015–1020
Rachlin H (1997) Four teleological theories of addiction. Psychon Bull Rev 4:462–473
Rangel A, Camerer C, Montague PR (2008) A framework for studying the neurobiology of value-based decision making. Nat Rev Neurosci 9:545–556
Remedios J, Woods C, Tardif C, Janak PH, Chaudhri N (2014) Pavlovian-conditioned alcohol-seeking behavior in rats is invigorated by the interaction between discrete and contextual alcohol cues: implications for relapse. Brain Behav 4:278–289
Rescorla RA (1987) A Pavlovian analysis of goal-directed behavior. Am Psychol 42:119–129
Rescorla RA (1991) Associative relations in instrumental learning—the 18 Bartlett memorial lecture. Q J Exp Psychol B Comp Physiol Psychol 43:1–23
Rescorla RA (1992a) Associations between an instrumental discriminative stimulus and multiple outcomes. J Exp Psychol Anim Behav Process 18:95–104
Rescorla RA (1992b) Response-outcome versus outcome-response associations in instrumental learning. Anim Learn Behav 20:223–232
Rescorla RA (1994) Control of instrumental performance by Pavlovian and instrumental stimuli. J Exp Psychol Anim Behav Process 20:44–50
Rescorla RA, Cunningham CL (1979) Spatial contiguity facilitates Pavlovian second-order conditioning. J Exp Psychol Anim Behav Process 5:152–161
Roll JM, Higgins ST (2000) A within-subject comparison of three different schedules of reinforcement of drug abstinence using cigarette smoking as an exemplar. Drug Alcohol Depend 58:103–109
Rosas JM, Paredes-Olay MC, García-Gutiérrez A, Espinosa JJ, Abad MJF (2010) Outcome-specific transfer between predictive and instrumental learning is unaffected by extinction but reversed by counterconditioning in human participants. Learn Motiv 41:48–66
Rose JE, Corrigall WA (1997) Nicotine self-administration in animals and humans: similarities and differences. Psychopharmacology 130:28–40
Rose J, Heskovic J, Trilling Y, Jarvik M (1985) Transdermal nicotine reduces cigarette craving and nicotine preference. Clin Pharmacol Ther 38:450–456
Rose J, Sampson A, Levin E, Henningfield J (1989) Mecamylamine increase nicotine preference and attenuates nicotine discrimination. Pharmacol Biochem Behav 32:399–938
Rose JE, Behm FM, Levin ED (1993) Role of nicotine dose and sensory cues in the regulation of smoke intake. Pharmacol Biochem Behav 44:891–900
Rose JE, Westman EC, Behm FM (1996) Nicotine/mecamylamine combination treatment for smoking cessation. Drug Dev Res 38:243–256
Rose JE, Westman EC, Behm FM, Johnson MP, Goldberg JS (1999) Blockade of smoking satisfaction using the peripheral nicotinic antagonist trimethaphan. Pharmacol Biochem Behav 62:165–172
Russell M, Wilson C, Patel U, Fayerband C, Cole P (1975) Plasma nicotine levels after smoking cigarettes with high median and low nicotine yields. Br Med J 2:414–416
Rusted JM, Mackee A, Williams R, Willner P (1998) Deprivation state but not nicotine content of the cigarette affects responding by smokers on a progressive ratio task. Psychopharmacology 140:411–417
Şahin E, Çakmak M, Doğar MR, Uğur E, Üçoluk G (2007) To afford or not to afford: a new formalization of affordances toward affordance-based robot control. Adapt Behav 15:447–472
Scherer G (1999) Smoking behavior and compensation: a review of the literature. Psychopharmacology 145:1–20
Shahan TA, Bickel WK, Badger GJ, Giordano LA (2001) Sensitivity of nicotine-containing and de-nicotinized cigarette consumption to alternative non-drug reinforcement: a behavioral economic analysis. Behav Pharmacol 12:277–284
Shiffman S (1986) A cluster-analytic classification of smoking relapse episodes. Addict Behav 11:295–307

Shiffman S (2009) Responses to smoking cues are relevant to smoking and relapse. Addiction 104:1617–1618
Shiffman S, Ferguson SG, Gwaltney CJ (2006) Immediate hedonic response to smoking lapses: relationship to smoking relapse, and effects of nicotine replacement therapy. Psychopharmacology 184:608–618
Shiffman S, Dunbar M, Kirchner T, Li X, Tindle H, Anderson S, Scholl S (2013) Smoker reactivity to cues: effects on craving and on smoking behavior. J Abnorm Psychol 122:264–280
Shoaib M, Stolerman I (1999) Plasma nicotine and cotinine levels following intravenous nicotine self-administration in rats. Psychopharmacology 143:318–321
Shoptaw S, Rotheram-Fuller E, Yang X, Frosch D, Nahom D, Jarvik ME, Rawson RA, Ling W (2002) Smoking cessation in methadone maintenance. Addiction 97:1317–1328
Skinner BF (1938) The behavior of organisms: an experimental analysis. Appleton-Century, Oxford
Stairs DJ, Neugebauer NM, Bardo MT (2010) Nicotine and cocaine self-administration using a multiple schedule of intravenous drug and sucrose reinforcement in rats. Behav Pharmacol 21:182–193
Stolerman I, Garcha H, Pratt J, Kumar R (1984) Role of training dose in discrimination of nicotine and related compounds in rats. Psychopharmacology 84:413–419
Stolerman IP, Rauch RJ, Norris EA (1987) Discriminative stimulus effects of a nicotine-midazolam mixture in rats. Psychopharmacology 93:250–256
Surawy B, Stepney R, Cox T (1985) Does watching others smoke increase smoking? Br J Addict 80:207–210
Thewissen R, Snijders SJBD, Havermans RC, van den Hout M, Jansen A (2006) Renewal of cue-elicited urge to smoke: implications for cue exposure treatment. Behav Res Ther 44:1441–1449
Tiffany ST (1990) A cognitive model of drug urges and drug-use behavior: role of automatic and nonautomatic processes. Psychol Rev 97:147–168
Tiffany ST, Drobes DJ (1991) The development and initial validation of a questionnaire on smoking urges. Br J Addict 86:1467–1476
Tiffany ST, Cox LS, Elash CA (2000) Effects of transdermal nicotine patches on abstinence-induced and cue-elicited craving in cigarette smokers. J Consult Clin Psychol 68:233–240
Tiffany ST, Warthen MW, Goedeker KC (2009) The functional significance of craving in nicotine dependence. In: Caggiula AR, Bevins RA (eds) The motivational impact of nicotine and its role in tobacco use (Nebraska symposium on motivation). Springer, New York, pp 1–27
Trapold MA (1970) Are expectancies based upon different positive reinforcing events discriminably different? Learn Motiv 1:129–140
Troisi JRII (2003) Nicotine vs. ethanol discrimination: extinction and spontaneous recovery of responding. Integr Physiol Behav Sci 38:104–123
Troisi JRII (2006) Pavlovian-instrumental transfer of the discriminative stimulus effects of nicotine and ethanol in rats. Psychol Rec 56:499–512
Troisi JRII (2013a) Acquisition, extinction, recovery, and reversal of different response sequences under conditional control by nicotine in rats. J Gen Psychol 140:187–203
Troisi JRII (2013b) The Pavlovian vs. operant interoceptive stimulus effects of EtOH: commentary on Besheer, Fisher, & Durant (2012). Alcohol 47:433–436
Troisi JRII (2013c) Perhaps more consideration of Pavlovian operant interactions may improve the clinical efficacy of behaviorally based drug treatment programs. Psychol Rec 63:863–894
Troisi JRII, LeMay BJ, Järbe TUC (2010) Transfer of the discriminative stimulus effects of Δ9-THC and nicotine from one operant response to another in rats. Psychopharmacology 212:171–179
Troisi JRII, Bryant E, Kane J (2012) Extinction of the discriminative stimulus effects of nicotine with a devalued reinforcer: recovery following revaluation. Psychol Rec 62:707–718
Troisi JRII, Dooley TF, Craig EM (2013) The discriminative stimulus effects of a nicotine-ethanol compound in rats: extinction with the parts differs from the whole. Behav Neurosci 127:899–912
Urcuioli PJ (2005) Behavioral and associative effects of differential outcomes in discrimination learning. Anim Learn Behav 33:1–21

Ussher MH, Taylor A, Faulkner G (2012) Exercise interventions for smoking cessation. The Cochrane Database of Systematic Reviews 1:CD002295
VanderVeen JW, Cohen LM, Cukrowicz KC, Trotter DRM (2008) The role of impulsivity on smoking maintenance. Nicotine Tob Res 10:1397–1404
Vlaev I, Chater N, Stewart N, Brown GDA (2011) Does the brain calculate value? Trends Cogn Sci 15:546–554
Volpp KG, Gurmankin Levy A, Asch DA, Berlin JA, Murphy JJ, Gomez A, Sox H, Zhu J, Lerman C (2006) A randomized controlled trial of financial incentives for smoking cessation. Cancer Epidemiol Biomark Prev 15:12–18
Westman EC, Behm FM, Rose JE (1995) Airway sensory replacement combined with nicotine replacement for smoking cessation—a randomized, placebo controlled trial using a citric acid inhaler. Chest 107:1358–1364
Wiley R, Wickham J (1974) The fabrication and application of a puff-by-puff smoking machine. Tob Sci 18:67–69
Willner P, Jones C (1996) Effects of mood manipulation on subjective and behavioral measures of cigarette craving. Behav Pharmacol 7:355–363
Willner P, Hardman S, Eaton G (1995) Subjective and behavioral evaluation of cigarette cravings. Psychopharmacology 118:171–177
Winterbauer NE, Bouton ME (2010) Mechanisms of resurgence of an extinguished instrumental behavior. J Exp Psychol Anim Behav Process 36:343–353
Woolverton WL, Anderson KG (2006) Effects of delay to reinforcement on the choice between cocaine and food in rhesus monkeys. Psychopharmacology 186:99–106
Xue YX, Luo YX, Wu P, Shi HS, Xue LF, Chen C, Zhu WL, Ding ZB, Bao YP, Shi J, Epstein DH, Shaham Y, Lu L (2012) A memory retrieval-extinction procedure to prevent drug craving and relapse. Science 336:241–245

Ussher MH, Taylor A, Faulkner G (2012) Exercise interventions for smoking cessation. The Cochrane Database of Systematic Reviews [illegible]
[illegible]
[illegible]
[illegible]
[illegible]
[illegible]
[illegible]
[illegible]
[illegible]
[illegible]
[illegible]

Smoking Abstinence and Neurocognition: Implications for Cessation and Relapse

F. Joseph McClernon, Merideth A. Addicott and Maggie M. Sweitzer

Abstract In this chapter, we review the last decade of research on the effects of smoking abstinence on various forms of neurocognition, including executive function (working memory, sustained attention, response inhibition), reward processing, and cue-reactivity. In our review we identify smoking abstinence-induced deficits in executive function mediated by effects on frontal circuitry, which in turn is known to be affected by modulation of cholinergic, dopaminergic, and other neurotransmitter systems. We also review evidence that smoking abstinence blunts reactivity to non-drug reinforcers—a finding that is consistent with results in the animal literature. Finally, our review of cue-reactivity indicates that smoking abstinence does not appear to amplify cue-provoked craving, although it may increase attentional bias to smoking-related cues. Inconsistencies across findings and potential contributing factors are discussed. In addition, we review the literature on the effects of nicotine and non-nicotine factors in neurocognition. Finally, we provide a multi-factor model and an agenda for future research on the effects of smoking abstinence on neurocognition. The model includes four distinct yet interacting factors, including: Negative Reinforcement, Drug-Reward Bias, Goal and Skill Interference, and Non-Cognitive Factors. Additional research is needed to further evaluate the scope and time-course of abstinence-induced changes in neurocognition, the mechanisms that underlie these changes and the specific role of these processes in drug reinforcement, lapse, and relapse.

Keywords Nicotine · Tobacco · Withdrawal · Smoking · Executive function · Reward processing · Cue-reactivity · Cognition · Neurocognition · fMRI

F.J. McClernon (✉) · M.A. Addicott · M.M. Sweitzer
Department of Psychiatry and Behavioral Sciences, Duke University
School of Medicine, 2608 Erwin Rd Box 3527
Lakeview Pavilion E Ste 300, Durham, NC 27705-4596, USA
e-mail: francis.mcclernon@dm.duke.edu

F.J. McClernon · M.A. Addicott · M.M. Sweitzer
Duke-UNC Brain Imaging and Analysis Center, Duke University School of Medicine,
Durham, NC, USA

D.J.K. Balfour and M.R. Munafò (eds.), *The Neurobiology and Genetics of Nicotine and Tobacco*, Current Topics in Behavioral Neurosciences 23,
DOI 10.1007/978-3-319-13665-3_8

Contents

1 Introduction 194
2 The Effects of Smoking Abstinence on Withdrawal Symptoms and Brain Physiology 195
3 The Effects of Smoking Abstinence on Neurocognition 197
3.1 Sustained Attention 198
3.2 Response Inhibition 200
3.3 Working Memory 201
3.4 Reward Processing 204
3.5 Cue-Reactivity 207
3.6 Intermediate Conclusions 209
4 Nicotine Versus Non-nicotine Factors in Neurocognition 211
4.1 Executive Function 212
4.2 Reward Processes 213
4.3 Cue-Reactivity 213
5 A Multi-part Model of Neurocognition and Relapse 215
6 Research Questions and Future Directions 216
6.1 Research on the Causes of Abstinence-Induced Changes in Neurocognition 216
6.2 The Role of Abstinence-Induced Neurocognitive Changes in Lapse and Relapse 217
6.3 The Treatment of Abstinence-Induced Changes in Neurocognition 218
7 Summary 219
References 219

1 Introduction

Tobacco use takes a staggering toll on the people, economies, and healthcare systems of the world. Each year, tobacco use accounts for almost 6 million premature deaths globally (WHO 2013) and costs $96.8 billion in lost productivity in the U.S. alone (CDC 2008). Part of the reason for this devastating impact is that the majority of smokers have difficulty quitting smoking—whereas the majority of smokers wish to quit, only a small proportion of attempts (<10 %) are successful (Fiore et al. 2008).

In this chapter, we provide an updated review of studies investigating the effects of quitting or abstaining from smoking on various aspects of neurocognition, with an emphasis on how these effects might increase the probability of lapse (i.e., a single episode of smoking after quitting) and relapse (i.e., resumed use of tobacco after quitting). We have largely restricted this review to human laboratory and neuroimaging studies published in the last 10 years that specifically evaluate the effects of 8 or more hours of smoking abstinence compared to smoking as usual. In this work, we refer to smoking abstinence and nicotine withdrawal interchangeably. Although nicotine has been shown to improve neurocognitive performance in nonsmokers (Heishman et al. 2010), this literature is less relevant to understanding the role of withdrawal-related neurocognitive effects on relapse among dependent smokers.

In addition to focusing exclusively on abstinence-induced changes in neurocognition, our review is restricted to neurocognitive processes that are frequently studied or are putatively involved in behaviors leading to lapse/relapse. These include working memory, sustained attention, inhibitory control, reward processes, and cue-reactivity. Other forms of neurocognition, including emotional information processing and emotional regulation, were not included due to the small number of studies in the literature. However, we make recommendations regarding their additional study.

As well as reviewing the research on the effects of smoking abstinence on neurocognition, we also review the small, but growing literature on the effects of nicotine versus non-nicotine components of cigarette smoke. With the passage of the Family Smoking Prevention and Tobacco Control Act in 2009, the Food and Drug Administration (FDA) now has the regulatory authority to reduce nicotine levels in cigarettes to nonzero levels. Research on the different effects of nicotine and non-nicotine components of cigarettes on neurocognition can provide valuable insights into what effect a reduction in cigarette nicotine levels might have on smokers' behavior. Moreover, understanding these different components can help elucidate which effects of abstinence on neurocognition can and cannot be alleviated by nicotine replacement therapy.

Finally, we end this chapter by presenting a multi-factor model of abstinence-induced changes in neurocognition on smoking lapse/relapse that synthesizes the current literature. We also discuss research directions that could provide a clearer picture of the role of neurocognition in smoking lapse/relapse.

2 The Effects of Smoking Abstinence on Withdrawal Symptoms and Brain Physiology

Acetylcholine is a neuromodulator that plays an important role in the integration of cognitive function in the prefrontal cortex, including attention, working memory, and response inhibition (for review, see Wallace and Bertrand 2013). Chronic tobacco use/nicotine exposure results in an upregulation of nAChRs; in contrast, other agonists typically result in a downregulation of the affected receptor system (Wonnacott 1990). The upregulation of nAChRs is thought to occur because nicotine causes receptor desensitization/inactivation, which is a reversible reduction in a receptor's response to the prolonged application of its agonist (Giniatullin et al. 2005). It has been proposed that both the activation and desensitization of nAChRs contributes to the effects of nicotine in the brain and on behavior (Picciotto et al. 2008). For instance, the activation of nAChR following the first cigarette of the day may contribute to nicotine's primary rewarding effect, but the desensitization of nAChRs may promote conditioned responses to environmental smoking cues (Picciotto et al. 2008). Although nAChRs are nicotine's direct targets, it is important to note that nicotine also indirectly affects many other neurotransmitter systems that play important roles in cognition (for review, see Picciotto et al. 2008).

By its actions on nAChRs, nicotine has putative performance-enhancing effects on cognitive function. A large meta-analysis of the acute effects of nicotine on cognitive performance found significant positive effects of nicotine on fine motor reaction time, attention, and working memory in nonsmokers or minimally-deprived smokers (Heishman et al. 2010). However, tolerance may develop to some of these acute effects following prolonged use (although studies on this subject have been inconclusive, (although studies on this subject have been inconclusive, see Heishman and Henningfield 2000; Perkins 2002; Perkins et al. 1994), and abrupt cessation of chronic tobacco use results in a well-characterized withdrawal syndrome. Withdrawal symptoms include irritability, anxiety, difficulty concentrating, increased appetite, restlessness, depressed mood, and insomnia (APA 2013). Withdrawal symptoms occur within hours of smoking cessation and can persist for weeks (Gilbert et al. 2004; Hendricks et al. 2006; Hughes 2007). Of note, the only cognitive effect of withdrawal recognized by the Diagnostic and Statistical Manual version 5 (APA 2013) is "difficulty concentrating". However, as discussed below, other aspects of withdrawal (e.g., insomnia, depressed mood) may indirectly impinge on neurocognition.

Nicotine withdrawal symptoms and changes in neurocognition are ostensibly due to changes in brain physiology following the cessation of tobacco smoking. As noted above, chronic nicotine use causes an upregulation of nAChRs and the decreased activation of these receptors following cessation likely results in significant changes in both nAChR-mediated functions and other downstream effects. Nicotine antagonist-precipitated withdrawal, for instance, has been shown to decrease dopamine output in the nucleus accumbens of the rat (Hildebrand et al. 1998), and this effect is mediated by nAChRs in the ventral tegmental area (VTA) (Hildebrand et al. 1999). Overnight abstinence following chronic nicotine exposure decreases extracellular dopamine in the striatum of rhesus monkeys (Domino and Tsukada 2009). Moreover, chronic nicotine administration produces higher extracellular dopamine dopamine clearance in the nucleus accumbens; this effect persists following nicotine abstinence, which may further decrease extracellular dopamine levels (Rahman et al. 2004). Finally, tobacco smoke contains monoamine oxidase inhibitors that reduce the breakdown of dopamine (Fowler et al. 1996a, b); during tobacco withdrawal, less monoamine oxidase inhibition may increase the metabolism of dopamine and decrease brain dopamine and other amine levels. Functionally, changes in nAChR, dopamine and other neurotransmitter systems in fronto-striatal circuits have been related to the effects of nicotine on affect, reward (Bruijnzeel and Markou 2004), and cognition (Levin et al. 2006a; Singer et al. 2004).

Changes in neurotransmitter function may be reflected in other smoking abstinence-induced changes in brain physiology. Regional cerebral blood flow (CBF) has been used as an indirect marker of brain metabolism and as such, during a resting state, brain regions with greater CBF are believed to be more active. Acute, intravenous nicotine administration increases cortical CBF in drug-naïve rats (Uchida et al. 1997) and chronic cigarette smoking reduces CBF among human subjects (Rogers et al. 1985; Yamashita et al. 1988). The acute effects of cigarette smoking or nicotine administration among dependent smokers have also been investigated. Studies report both increased regional CBF (e.g., in the prefrontal cortex, thalamus, and visual cortex) and decreased regional CBF (e.g., in the amygdala) (Domino et al. 2000, 2004; Mathew and Wilson 1991; Rose et al. 2003); for example, nicotine administered after overnight withdrawal has been shown to reduce regional CBF in the temporal cortex and amygdala, but increase CBF in the thalamus (Zubieta et al. 2001). These regions may represent areas of significant nicotine binding, but may also relate to the anticipation of smoking or other arousal-related effects. A study by Wang and colleagues correlated changes in cortical and subcortical CBF from a smoking as usual to an abstinent condition with changes in self-reported craving (Wang et al. 2007). A later study by Addicott and colleagues found that correlations between regional CBF and self-reported craving were related to the non-nicotine chemical and sensorimotor effects of smoking (i.e., smoking a very-low nicotine cigarette), rather than being related to nicotine itself (Addicott et al. 2014). These findings suggest that the changes in regional CBF following smoking abstinence likely reflect changes in the sensation of craving, and in turn, changes in craving are more strongly related to the administration of non-nicotine versus nicotine components of smoking (Addicott et al. 2014).

Changes in electroencephalography (EEG) measures have also been assessed in response to nicotine and smoking abstinence. In nonsmokers, nicotine administration results in greater frontal alpha power (Fisher et al. 2012). Among smokers, studies have reported that tobacco withdrawal decreases the dominant alpha frequency (Herning et al. 1983; Pickworth et al. 1989). Furthermore, a smoking cessation study observed that EEG activity slowed significantly (i.e., greater power in lower frequency bands) and did not return to baseline levels over the course of 31 days (Gilbert et al. 2004). Collectively, EEG, CBF, and brain neurochemistry (see chapter entitled Imaging Tobacco Smoking with PET and SPECT; volume 24) studies suggest that brain physiology changes in significant ways following smoking abstinence.

3 The Effects of Smoking Abstinence on Neurocognition

As noted in the introduction, this review is specifically focused on changes in neurocognition that occur following smoking abstinence. In the literature, abstinence-induced changes in neurocognition are usually investigated during a short-term (8–48 h) abstinence condition compared to a smoking as usual condition. Less frequently, abstinence-induced changes in neurocognition are evaluated during a smoking cessation attempt compared to smokers who continue to smoke as usual. In this chapter, we limit our review to those studies which have employed abstinence challenges of at least 8 h or have examined changes in neurocognition following a smoking cessation attempt. In addition, we limit this review to those forms of neurocognition with evidence bases that are large enough to draw conclusions, these are: executive functions including sustained attention, working memory and response inhibition; reward processing, and cue-reactivity. The literature from animal studies, when available, is also briefly reviewed in order to provide additional context for interpreting human clinical studies.

3.1 *Sustained Attention*

Sustained attention is the ability to maintain readiness to detect rare and unpredictable signals over an extended period of time; it is assessed via indices of behavioral performance such as detection accuracy and false alarms (for review, see Sarter et al. 2001). Tasks of sustained attention in humans typically require participants to maintain alertness and attention in order to respond appropriately to targets presented on a computer screen over durations of 10–20 min. Stimuli and required responses are usually simplistic and require minimal cognitive processing in order to isolate sustained attentional processes from other elements of executive function such as response inhibition (e.g., Stroop task) or working memory (e.g., *n*-back task). Sustained attention is mediated by the basal forebrain cholinergic system, which makes this cognitive function especially relevant to nicotine and nicotine withdrawal (Levin et al. 2006b). Hypothetically, nicotine withdrawal produces deficits in sustained attention by disrupting cholinergic function in the frontal lobes.

3.1.1 Behavioral Studies (Rodent)

Sustained attention in rodents is commonly measured with the 5-choice serial reaction time task (5-CSRTT), which is analogous to a continuous performance task for human subjects (see below) and requires animals to rapidly and accurately respond to a visual stimulus presented in pseudorandom order across an array of horizontal apertures. Two studies have demonstrated deficits in sustained attention during nicotine withdrawal using this task. Shoaib and Bizarro administered nicotine via osmotic minipumps to rats for 7 continuous days and then abruptly withdrew the drug. After 10–16 h of nicotine abstinence, performance on the 5-CSRTT revealed increased omission errors and decreased response accuracy. However, the rats recovered from these behavioral deficits after 34 h of abstinence (Shoaib and Bizarro 2005). In a similar study by Semenova and colleagues, rats were administered nicotine for 7 days, and performance on the 5-CSRTT was tested during nicotine withdrawal. During the first day of withdrawal, dramatic changes in performance were observed, including fewer correct responses, more omissions, and longer response latencies; however, an overall summary score of accuracy was not affected (Semenova et al. 2007).

3.1.2 Behavioral Studies (Human)

Several studies have examined the effects of smoking withdrawal on sustained attention in humans, with largely consistent evidence of impairment relative to smoking as usual. Several studies have observed increased omission errors or disruptions in response time variability on continuous performance tasks (CPT)—a widely used measure of sustained visual attention—following overnight smoking abstinence (Bohadana et al. 2000; Harrison et al. 2009; Kollins et al. 2013; McClernon et al. 2008; Sacco et al. 2005), although two studies found no effect (Ashare and Hawk 2012; Jacobsen et al. 2005). Other studies have demonstrated consistent evidence of sustained attention deficits during abstinence using a Rapid Visual Information Processing (RVIP) task, in which a series of digits appear one at a time on a computer screen and subjects are required to press a response key when target sequences of digits appear. Four studies utilizing the RVIP task have reported decreased accuracy (the number of correctly identified targets) during nicotine withdrawal. For example, McClernon and colleagues reported that subjects' accuracy was impaired following overnight abstinence, although commission errors and reaction time were not affected (McClernon et al. 2008). Interestingly, deficits in sustained attention as measured with the RVIP may arise as early as 30 min into withdrawal (Hendricks et al. 2006). Moreover, they may persist for weeks following smoking abstinence. Gilbert and colleagues examined RVIP performance of 54 female smokers before and during a 31-day abstinence period and reported that accuracy was impaired throughout the abstinence period (Gilbert et al. 2004). In contrast with rodent studies suggesting a fast recovery of attentional deficits (Shoaib and Bizarro 2005), these findings suggest the effects of withdrawal on attention can last longer than a month, and more research is needed to determine whether or not these attentional deficits ever resolve.

3.1.3 Neuroimaging Studies (Human)

To our knowledge, only one recent neuroimaging study has examined the effects of smoking withdrawal on tasks of sustained attention. Kozink et al. (2010b) investigated the neural correlates of attention in smokers either 24-h abstinent or following smoking as usual using a RVIP task. The authors reported that abstinence decreased sustained (block-related) activation in some frontal regions, but increased transient (event-related) activation in response to RVIP targets throughout multiple cortical areas involved in attention (Kozink et al. 2010b). Kozink and colleagues suggested that nicotine abstinence negatively affects the efficiency of brain regions controlling attention, and additional cortical areas are recruited to help offset deficits in target detection. The disrupted activation in frontal regions during abstinence supports the hypothesis that nicotine withdrawal interferes with frontal regulation of attentional control (Kozink et al. 2010b).

3.1.4 Summary

Disruptions in the cholinergic system, which may occur during nicotine withdrawal, are hypothesized to be responsible for abstinence-induced decreases in sustained attention. Based on our review of the literature, there is evidence of impaired sustained attention during smoking abstinence as measured in both rodents and humans. Performance decrements may arise shortly after smoking abstinence and may not resolve for weeks following cessation among some smokers. Sustained attention performance may be a reasonable behavioral assay for the nicotine withdrawal symptom "difficulty concentrating" described in DSM-5.

3.2 Response Inhibition

Response inhibition is the ability to inhibit a prepotent or overlearned response; this ability is an integral feature of executive control. Behavioral measures of inhibitory control are tested with tasks that require participants to withhold a response to some, but not other stimuli presented in a series (e.g., Go/No-Go task) or withhold a response to a Stop signal presented at varying delays following a Go signal (e.g., Stop-Signal task). The number of commission errors (i.e., failures to inhibit a response) represents impairments in response inhibition in both types of tasks. In a changing environment, the ability to inhibit or suppress actions that are no longer required is a function of the prefrontal cortex (for review, see Ridderinkhof et al. 2004). Functional neuroimaging studies have shown that the correct inhibition of responses activates regions in the prefrontal cortex and the anterior cingulate (Rubia et al. 2001). Furthermore, abnormal prefrontal function has been implicated in drug addiction in general (Goldstein and Volkow 2002), and tobacco addiction in particular (Brody 2006). Hypothetically, nicotine withdrawal would impair response inhibition.

3.2.1 Behavioral Studies (Rodents)

Few studies have examined effects of nicotine withdrawal on animal models of response inhibition. In one study examining a rodent model of nicotine dependence, 7 days of chronic nicotine exposure impaired response inhibition and early nicotine withdrawal (12–84 h) actually improved response inhibition compared to placebo controls (Kolokotroni et al. 2012). These results are surprising and appear to contradict the effects of withdrawal on response inhibition proposed in human subjects; however, given that animals are typically required to inhibit a response to a cue in order to obtain a reward, the rodent model of response inhibition may be more of a measure of reward motivation than the inhibition of a prepotent response (see Kolokotroni et al. 2012).

3.2.2 Behavioral Studies (Humans)

The effects of smoking abstinence on inhibitory control have been assessed in human smokers using Go/No-Go (Harrison et al. 2009; Kozink et al. 2010a; McClernon et al. 2008) and Stop Signal tasks (Ashare and Hawk 2012). Utilizing a

Go/No-Go task, Kozink and colleagues observed that nicotine abstinence decreased the number of correctly inhibited No-Go trials, but did not affect reaction time or accuracy on Go trials (Kozink et al. 2010a). Another study by McClernon and colleagues used a cued Go/No-Go task in which cues were presented before Go and No-Go targets. The cues accurately predicted their respective targets in 80 % of trials. In this task, subjects were instructed to withhold a response when a Go cue was followed by a No-Go target. The authors reported that nicotine abstinence did not affect accuracy of performance, but did result in slower reaction times during Go trials (McClernon et al. 2008). Two other studies did not report performance accuracy, but showed that abstinence increased subjects' reaction times during stop-signal trials (Ashare and Hawk 2012) and during Go trials (Harrison et al. 2009).

3.2.3 Neuroimaging Studies (Human)

In a study by Kozink and colleagues, smokers underwent fMRI scanning while performing a Go/No-Go task. Compared with smoking as usual, abstinence was associated with increased blood-oxygen level-dependent (BOLD) activation in the right inferior frontal cortex during No-Go trials, despite decreased performance accuracy. Activation in the pre-supplementary area was not affected. Kozink and colleagues suggested that inhibitory control is associated with inferior frontal cortical activity, and that smoking abstinence increases attentional demands, rather than disrupts inhibitory motor processes per se, to allow for successful response inhibition (Kozink et al. 2010a).

3.2.4 Summary

Given the small number of studies reviewed here, it is difficult to draw a conclusion regarding the effects of smoking abstinence on response inhibition. Only two studies reported on response inhibition accuracy and their results were inconsistent (Kozink et al. 2010a; McClernon et al. 2008). The other studies reported slower reaction times during task performance. Taken together, it appears that nicotine abstinence does interfere with overall response inhibition performance, but it remains unclear whether this is a disruption of cognitive motor planning or attentional demands (Kozink et al. 2010a). Moreover, it is not known whether abstinence-induced deficits in response inhibition are predictive of lapse or relapse.

3.3 Working Memory

Working memory is the ability to actively maintain information for a short period of time in a way that is useful for the performance of goal-directed behavior. Measures of working memory can incorporate both visuospatial or auditory information, and can require information to be retained and updated for a brief period of time across several trials (continuous working memory) or immediately recalled on a current trial (discrete working memory). The prefrontal cortex is thought to underlie working memory performance; activity in the dorsolateral prefrontal cortex in particular appears to represent the maintenance of information and representations that are no longer present in the environment (for review see Curtis and D'Esposito 2003). Pharmacological studies have shown that acetylcholine plays an important role in encoding new memories, and nAChR agonists enhance the encoding of new information (Hasselmo 2006). Dopamine also modulates working memory function: selective dopamine D1 receptor antagonists administered in the dorsolateral prefrontal cortex can induce errors in memory-guided behavior (Sawaguchi and Goldman-Rakic 1991). Hypothetically, nicotine withdrawal impairs working memory performance by interrupting acetylcholine function and its downstream effects in the prefrontal cortex.

3.3.1 Behavioral Studies (Rodents)

Studies using rodent models have confirmed that acute nicotine administration into the hippocampus and basolateral amygdala can enhance short- and long-term memory formation and memory retrieval (Barros et al. 2004, 2005). However, very few animal studies have investigated the effects of nicotine withdrawal. An early study using rats reported that 2 weeks of chronic nicotine administration improved spatial memory performance, but nicotine withdrawal did not impair performance (Levin and Rose 1990). Conversely, a more recent study using adolescent mice reported learning and memory deficits during acute nicotine exposure, and performance was improved during short-term withdrawal (Abreu-Villaca et al. 2007).

3.3.2 Behavioral Studies (Humans)

In the last decade, several studies have examined the effects of smoking abstinence on working memory with evidence of abstinence-induced impairments (Bohadana et al. 2000; Falcone et al. 2013; Jacobsen et al. 2005; Mendrek et al. 2006; Merritt et al. 2012) but also null findings (Jacobsen et al. 2007; Rubinstein et al. 2009; Wesnes et al. 2013; Xu et al. 2005) or evidence of moderating factors (Loughead et al. 2009; Sacco et al. 2005).

A variety of different working memory tasks have been utilized across studies, which could account for some of the variability in results. Most commonly, the *n*-back task has been employed as a measure of continuous working memory that requires participants to identify targets that match a stimulus presented 'n' trials previously. The majority of studies utilizing this task have observed impaired accuracy or slowed reaction time during smoking withdrawal compared with smoking as usual, with greatest impairment often at higher *n*-back levels (i.e., when working memory loads are greater) (Bohadana et al. 2000; Falcone et al. 2013;

Jacobsen et al. 2005; Mendrek et al. 2006). For example, across blocks of 0-, 1-, 2-, and 3-back conditions, smokers made more errors during abstinence than after smoking as usual; these differences appeared to be more pronounced in the 3-back condition (Mendrek et al. 2006). In another study, Jacobsen 2005 reported that adolescent smokers' performance accuracy decreased during abstinence compared to smoking as usual (Jacobsen et al. 2005).

Studies failing to find significant effects of smoking abstinence on working memory have also reported insufficient psychometric properties of the measures employed (Wesnes et al. 2013) or have used tasks assessing discreet working memory with minimal working memory load (Rubinstein et al. 2009). Furthermore, other studies demonstrating null effects were conducted among adolescents smoking five or fewer cigarettes per day (Jacobsen et al. 2007; Rubinstein et al. 2009) or with a very small sample of smokers (e.g., $n = 8$) (Xu et al. 2005). Another study found no effect of overnight abstinence on visual working memory performance among healthy smokers, but did observe declines in performance among individuals with schizophrenia (Sacco et al. 2005). Additional evidence that abstinence-induced deficits in working memory may be particularly problematic for some individuals was found by Loughead and colleagues; in their study, catechol-O-methyltransferase (*COMT*) genotype was found to interact with abstinence on *n*-back performance, such that individuals who were homozygous for the val allele (i.e., those with putative deficits in frontal dopamine neurotransmission) showed greater abstinence-induced deficits in performance at higher working memory load conditions (Loughead et al. 2009).

3.3.3 Neuroimaging Studies (Human)

At least four studies have investigated the effects of nicotine withdrawal on working memory using fMRI (Falcone et al. 2013; Jacobsen et al. 2007; Loughead et al. 2009; Xu et al. 2005), and all of these studies used the *n*-back task. Falcone and colleagues showed that, relative to smoking as usual, smoking abstinence resulted in reduced accuracy and attenuated brain activation in three regions of interest: the medial frontal/cingulate gyrus, and right and left dorsolateral prefrontal cortex (Falcone et al. 2014). Xu and colleagues reported an interaction between working memory load and abstinence in the left dorsolateral prefrontal cortex: during smoking as usual, task-related activity increased as task load increased (i.e., from the 1-back to the 3-back condition, while during abstinence there was no relationship between task-related activation and task load (Xu et al. 2005). Lastly, consistent with behavioral findings noted above, neither Loughead and colleagues nor Jacobsen and colleagues found an effect of smoking abstinence on working memory-related BOLD activation (Jacobsen et al. 2007; Loughead et al. 2009). However, Loughead and colleagues again found an interaction between *COMT* genotype and abstinence condition. Smokers who were homozygous for the val allele (which is associated with increased catechol-O-methyltransferase enzyme activity in the prefrontal cortex, leading to lower levels of prefrontal dopamine) exhibited reduced BOLD response in the dorsolateral prefrontal cortex during abstinence (Loughead et al. 2009).

3.3.4 Summary

Although there is some inconsistency in the literature on the effects of smoking abstinence on working memory performance, the majority of recent studies do find some evidence of abstinence-induced deficits, although findings may be limited to heavier smokers and to tasks targeting continuous working memory with a high memory load. Knowledge of the neurobiology of acetylcholine and dopamine and the role they play in learning and memory (for review, see Hasselmo 2006) is consistent with these findings. Taken together, these studies provide some support for working memory deficits during nicotine withdrawal.

3.4 Reward Processing

Reward processing is a fundamental aspect of animal and human behavior. Pursuit of primary rewards, such as food, water, and sex, enable survival and propagation of the species. From a behavioral perspective, rewards can be any stimulus that acts as a reinforcer—that is, anything that increases the probability of the behavior, which lead to the reward, occurring again in the future (McClure et al. 2004). This reinforcement may or may not co-occur with a subjective hedonic sensation of pleasure or a conscious motivation to obtain the reward (Berridge et al. 2009). A wide variety of tasks have been designed to assess different aspects of reward processing, including reinforcement tasks measuring behavioral responses or subjective ratings assessing affective responses. While it is acknowledged that different types of tasks may be mediated by slightly different neural underpinnings, several regions have consistently emerged as key nodes in the functional networks supporting multiple aspects of reward processing, including the ventral tegmentum with dopaminergic projections to the ventral and dorsal striatum, medial prefrontal cortex, and orbitofrontal cortex (Apicella et al. 1991; Breiter et al. 2001; Delgado et al. 2003; Elliott et al. 2003; Haber and Knutson 2010; Hikosaka and Watanabe 2000; Roesch and Olson 2003; Thut et al. 1997). Hypothetically, nicotine withdrawal disrupts dopaminergic functioning within these regions, thereby contributing to deficits in reward processing.

3.4.1 Behavioral Studies (Rodents)

Evidence from animal studies suggests that acute nicotine enhances—while withdrawal from nicotine attenuates—the incentive value of other reinforcers (Besheer and Bevins 2003; Chaudhri et al. 2006; Donny et al. 2003; Thiel et al. 2009;

Weaver et al. 2012). For example, nicotine administration lowers intracranial self-stimulation (ICSS) thresholds (Kenny and Markou 2006), suggesting that nicotine acutely renders reward systems hypersensitive to non-drug rewards. Conversely, nicotine withdrawal increases ICSS thresholds (Epping-Jordan et al. 1998; Skjei and Markou 2003), suggesting a withdrawal-induced hyposensitivity to reward in which greater levels of stimulation were necessary to provide reinforcement.

3.4.2 Behavioral Studies (Humans)

Recent human laboratory studies are generally consistent with findings from the animal literature, suggesting mildly attenuated reward functioning during withdrawal compared with smoking satiation. For example, abstinent smokers demonstrate less interference from pleasure-related words during a modified Stroop task than satiated smokers (Dawkins et al. 2006; Powell et al. 2004), suggesting a reduced attentional bias toward positive stimuli. Similarly, Leventhal and colleagues recently found that smokers high in trait anhedonia exhibited less interference from happy faces, compared to neutral faces, in a gender naming task during a withdrawn condition (Leventhal et al. 2012). This pattern of findings was specific to happy faces, with no effect for faces with anger, fear, or surprise expressions. In two other studies, abstinent smokers reported lower ratings of happiness following exposure to happy film clips compared with satiated smokers (Dawkins et al. 2007; Dawkins and Powell 2011); lower ratings of sadness among abstinent smokers were observed for one study but were not replicated, suggesting that differences in affective response were mostly limited to positive affect rather than a generalized affective blunting. Studies from the Powell and Dawkins research group (and often among the same samples of smokers as those noted above) have also demonstrated lower subjective ratings of anticipated enjoyment in pleasurable activities and reduced responsiveness to financial reward in a card sorting task during abstinence conditions compared with satiation or nicotine administration (Dawkins et al. 2006), although the validity of the task used for the latter findings has been called into question (Kalamboka et al. 2009).

Perkins and Karelitz conducted studies to determine the effect of withdrawal on reward processing using a progressive ratio task to examine reinforced responding for several different rewards. In this task, participants moved a cursor on the screen and pressed a button to "pick" apples from a field in order to earn a variety of rewards, including music, money, and negative reinforcement from termination of an aversive noise. Withdrawn smokers responded more for music after smoking a nicotinized cigarette compared with a denicotinized cigarette or nonsmoking condition, and these effects were observed for both dependent and non-dependent smokers (Perkins and Karelitz 2013b). In the second study, the researchers varied the level of smoking exposure prior to task completion to examine effects of nicotine exposure (smoking a full cigarette, plus two puffs immediately before the task) versus simple relief from withdrawal (smoking two puffs before the task only). Although reported level of withdrawal was comparable under both conditions, reinforcement enhancement for music occurred only for the group smoking a full cigarette (Perkins and Karelitz 2013a). Together, these studies provide evidence that acute nicotine exposure enhances responsiveness to reward, while the direct effects of withdrawal are less clear.

3.4.3 Neuroimaging Studies (Human)

Despite the importance of reward circuitry for understanding addiction, relatively few fMRI studies have investigated neural response to non-drug rewards among smokers. Only a handful of these have examined effects of smoking abstinence, and findings are mixed. One study found evidence of attenuated reward-related activation among smokers compared with nonsmokers, but no effect of withdrawal from nicotine (Buhler et al. 2010). In another study examining risky versus safe choices in a probabilistic decision-making task, smoking withdrawal was associated with a combination of heightened and attenuated activation across different reward-related regions and across different phases of the task (Addicott et al. 2012), but smoking abstinence did not affect behavioral outcomes.

Two recent studies by our group have demonstrated some evidence of attenuated striatal activation to monetary reward during abstinence. In an initial pilot study, abstinence-induced deficits in striatal activation were seen for only a subset of highly dependent smokers, while no overall main effect of abstinence was observed (Patterson et al. 2010). In a second, larger study examining reward-related activation to both smoking and monetary rewards, smoking abstinence was associated with attenuated anticipatory activation to monetary reward in the bilateral caudate, right insula, and brain stem (Sweitzer et al. 2014). Interestingly, these same regions were associated with heightened anticipatory activation to smoking reward during abstinence, suggesting a dissociated response depending on reward type. Furthermore, satiation was associated with greater response to monetary reward compared with smoking reward in the medial prefrontal cortex, and this pattern reversed during abstinence. These findings raise important questions about the specificity of potential reward processing deficits during abstinence. However, replication is needed given discrepancies with other studies (e.g., Buhler et al. 2010).

3.4.4 Summary

There is growing animal and human behavioral evidence of abstinence-induced deficits in processing of non-drug rewards. These findings are consistent with known changes in mesolimbic dopamine functioning as a result of chronic smoking and withdrawal. Recent neuroimaging studies present inconsistent results, but the number of studies is limited and additional research may clarify these discrepancies. Further research is also needed to determine if neural or behavioral changes in reward processing during abstinence are an important mechanism contributing to continued smoking behavior and relapse during a quit attempt.

3.5 *Cue-Reactivity*

Exposure to smoking-related stimuli increases self-reported craving to smoke (Carter and Tiffany 1999) and has long been theorized to precipitate relapse among smokers trying to quit (Shiffman et al. 1996, 2007). Neuroimaging studies employing cue-reactivity paradigms have attempted to elucidate the neurobiological basis of cue-reactivity. Typically, such paradigms involve passive presentation of smoking-related images (e.g., a pack of cigarettes or a person holding a lit cigarette) as well as neutral images matched for size of visual complexity (e.g., a pack of playing cards or a person holding a pencil). Neuroimaging studies have demonstrated that presentation of smoking cues relative to neutral cues results in robust activation of mesocorticolimbic circuitry, including precuneus, anterior cingulate cortex, superior frontal gyrus, amygdala, and ventral striatum/nucleus accumbens, which underlies the neurocognitive processes of reward, attention, and goal-directed behavior (David et al. 2007; Due et al. 2002; Franklin et al. 2007; McBride et al. 2006; McClernon et al. 2007; Wilson et al. 2005). Hypothetically, increased salience of smoking-related stimuli during withdrawal may contribute to heightened cue-reactivity.

3.5.1 Behavioral Studies (Humans)

A number of recent studies have examined the effects of smoking abstinence on reactivity to smoking-related versus neutral cues. MacKillop and colleagues observed additive (but not interactive) effects of smoking vs abstinence and cue-type on behavioral economic indices of the relative value of cigarettes (MacKillop et al. 2012). Similarly, Bailey and colleagues found that smokers exhibited heightened craving in response to smoking versus neutral cues, and while smoking abstinence resulted in a generalized increase in craving, cue-reactivity (reactivity to smoking—neutral cues) was not modulated (Bailey et al. 2010). Interestingly, despite the lack of an interactive effect on self-reported craving levels, abstinence enhanced smokers' skin conductance responses to smoking versus neutral cues, but only when a cigarette was immediately available. This finding is consistent with previous research indicating that smoking opportunity increases attentional bias to smoking-related stimuli (Wertz and Sayette 2001) and suggests that traditional cue-reactivity studies that present stimuli divorced from any predictive relationship with smoking may miss important effects.

Other studies have examined effects of smoking abstinence on the degree to which smoking-related stimuli distract smokers from some central task (i.e., attentional bias); these studies have produced mixed findings. In one study that used a modified Stroop task, both withdrawn and satiated smokers exhibited decreased attentional bias toward smoking words relative to neutral words after smoking a cigarette in-session compared with a non-smoking session—an effect that correlated with reductions in craving (Waters et al. 2009). However, these results were not replicated (Canamar and London 2012). Another study used a dot probe task to measure attentional bias toward drug cues by assessing reaction time to a dot stimulus presented in a location matching one of two previously presented cues (one smoking and one neutral); attentional bias toward smoking cues was observed among abstinent but not satiated smokers (Freeman et al. 2012). Together, these studies indicate minimal effects of withdrawal on cue-induced craving but support possible withdrawal-induced changes in attentional bias toward smoking cues.

While most studies of cue-reactivity have examined overnight or 24-h abstinence, studies evaluating prolonged abstinence suggest that cue-induced craving continues to be dissociable from generalized craving (see chapter entitled Neurobiological Bases of Cue- and Nicotine-induced Reinstatement of Nicotine Seeking: Implications for the Development of Smoking Cessation Medications; volume 24). In a study by Carter and colleagues, smokers who were abstinent for 2 weeks exhibited decreased generalized craving compared to smokers who continue to smoke; but exhibited comparable cue-provoked craving (Carter et al. 2009). Similarly, smokers provided with contingencies for abstinence across a 2-week period demonstrated greater abstinence and lower generalized craving than smokers reinforced for abstinence only on day 14; however, both groups maintained comparable levels of cue-provoked craving (Bradstreet et al. 2014). In another study, Bedi and colleagues evaluated cue-reactivity among smokers who were abstinent for varying amounts of time and observed increasing cue-induced craving with increasing durations of abstinence (i.e., incubation) (Bedi et al. 2011). Together, these findings suggest that, consistent with prior research, the effects of abstinence on generalized craving and cue-reactivity are additive and dissociable.

3.5.2 Neuroimaging Studies (Humans)

Only a few studies have examined neural response to smoking rewards among smokers tested in both abstinent and non-abstinent state, and the results have been mixed. Some studies have demonstrated heightened BOLD response to smoking cues (McClernon et al. 2009), or to cued anticipation of intravenous nicotine (Gloria et al. 2009), in reward-related brain areas during withdrawal relative to satiety. However, other studies have shown minimal effect of abstinence on response to smoking reward (McClernon et al. 2005) or opposite effects. For example, David and colleagues observed greater ventral striatal response to smoking versus neutral cues during satiation compared with abstinence (David et al. 2007). In another study, activation to smoking versus neutral cues was seen among satiated but not abstinent smokers in areas of anterior and posterior cingulate, precuneus, and ventral pallidum, although a direct comparison revealed no significant differences (McBride et al. 2006). A recent meta-analysis of neuroimaging studies of smoking cues found some evidence of increased reward-related activation during abstinence relative to satiety, but effects were relatively small and sample sizes were limited (Engelmann et al. 2012).

Interestingly, McBride and colleagues found more robust changes in cue-reactivity when manipulating smoking expectancy rather than abstinence state, such that cue-reactivity was greater across multiple regions when participants expected to smoke immediately after the scan. However, counter to findings observed for skin conductance described above, only a small region of the dorsolateral prefrontal cortex exhibited an interaction with abstinence state, with greater activation when smoking was expected and participants were non-abstinent.

Two recent studies have used alternative task designs to assess brain activation in response to cues predicting actual smoking rewards delivered after the scan, rather than traditional cue-reactivity paradigms (Buhler et al. 2010; Sweitzer et al. 2014). However, results are again inconsistent. In a study by Buhler and colleagues, smokers rapidly pressed a button to earn cigarettes and varying amounts of monetary reward. Anticipatory activation in mesocorticolimbic regions differed from that of occasional smokers and was positively correlated with behavioral response rates, but there was no effect of abstinence (Buhler et al. 2010). By contrast, in a study by Sweitzer and colleagues, smokers exhibited a robust increase in anticipatory activation to smoking rewards during abstinence throughout reward-related brain regions, including bilateral striatum, medial prefrontal cortex, and insula—the same regions that also exhibited decreased anticipatory activation to monetary reward (Sweitzer et al. 2014). Both these studies were relatively well-powered compared to earlier cue-reactivity studies, and discrepant findings are difficult to reconcile. However, these paradigms provide an interesting new direction for evaluating neural response to smoking reward, and further research is warranted to determine the conditions under which abstinence alters brain function and influences behavior.

3.5.3 Summary

Overall, behavioral studies of cue-reactivity are largely consistent in demonstrating additive effects of abstinence and smoking cues on self-reported craving, but they have not demonstrated any interaction effect. Neuroimaging results are somewhat inconsistent, which may be the result of small sample sizes or methodological differences across studies. There is some indication that smoking availability may be an important but understudied factor in cue-reactivity research. In particular, subjective craving, physiological responses (e.g., skin conductance), and neural reactivity to smoking cues are all heightened when smoking is immediately available, and there is some suggestion that abstinence may further amplify this effect. Furthermore, attentional bias toward smoking stimuli appears to be strengthened during abstinence from smoking; further research is needed to determine the extent to which these factors may contribute to smoking behavior.

3.6 Intermediate Conclusions

We have reviewed the literature published in the last decade on the effects of smoking abstinence on executive function (sustained attention, working memory, response inhibition), reward processing, and cue-reactivity. The literature on executive function is consistent with earlier work (Heishman et al. 1994) suggesting that smoking abstinence results in diminished performance on tasks requiring executive function/cognitive control. Imaging studies provide evidence that these effects are due to modulation in frontal cortical control areas (i.e., dlPFC) and probably related to abstinence-induced changes in dopaminergic neurotransmission. Modulation of executive function by smoking abstinence has been most frequently assessed following brief periods of abstinence (i.e., overnight, 24 h), though some evidence suggests the effects arise within minutes of smoking abstinence and can persist for weeks.

With respect to reward processes, there is growing evidence that smoking/nicotine abstinence decreases the salience and/or reward value of non-drug rewards. While there is some evidence that smoking abstinence blunts brain reactivity to non-drug rewards, particularly among highly dependent smokers, much work remains to be done to evaluate the neural basis of the observed behavioral effects. Little is known regarding the time course of abstinence effects on reward processing and whether these effects resolve over time.

Finally, with respect to cue-reactivity, the literature continues to support the notion that craving is additively influenced by smoking cues and smoking abstinence, with little evidence that smoking abstinence amplifies cue-reactivity. The effects of smoking abstinence on brain reactivity to smoking-related cues are mixed and include studies providing evidence of both increased and decreased activation. Whereas evidence of amplification effects for smoking-related cues is lacking, there is emerging data that reactivity to cues that are specifically and explicitly linked to smoking availability is amplified by smoking abstinence. It is unknown whether cued-availability subsides over the course of smoking abstinence.

It is evident from the literature reviewed above that there is significant inconsistency of findings on the effects of smoking abstinence on neurocognition. This may be due to a number of methodological factors including variability in abstinence durations, amount of training prior to cognitive testing, and sample characteristics.

3.6.1 Abstinence duration

We have argued (McClernon et al. 2009) for testing of abstinence-induced deficits at 24-h post cessation since other commonly employed durations (e.g., overnight abstinence) may not require foregoing any cigarettes among some smokers. This is particularly true given that rates of heavy smoking (i.e., $\geq$ 30 cigarettes/day) are declining (CDC 2012) and researchers are using increasingly lower minimum-use cutoffs for study inclusion.

3.6.2 Pre-test training

Very few of the reviewed studies on executive function required participants to reach some minimum level or stability of performance prior to testing. As such, these studies, even when counterbalanced, may have decreased effect sizes associated with participants continuing to learn the task during testing and, thus, increased variability in performance. The magnitude and duration of effects observed in Gilbert and colleagues' study, in which participants practiced the RVIP to stable performance prior to quitting smoking, may be due in part to this principle (Gilbert et al. 2004).

3.6.3 Sample characteristics

Finally, as noted above, among individuals who smoke daily, the prevalence of heavy smoking (≥30 cigarettes/day) has gone down over the last 10 years, whereas the prevalence of light smoking (1–9 cigarettes/day) has continued to go up (CDC 2012). Therefore, we might expect an increasing proportion of study participants to be light smokers who may experience different (and less severe) profiles of disrupted neurocognition during abstinence compared to heavier smokers.

In addition to the above, there remains significant variability in the types of tasks that are used to assess various forms of neurocognition, which may produce inconsistent findings and confusion when attempting to draw conclusions from this literature. The *n*-back task, for instance, is referred to as a working memory task, but requires sustained attention in order to maintain performance. However, examining brain function on the task after controlling for the 0-back condition, as is often done, results in foregoing an examination of sustained attention effects. Similarly, the RVIP is often referred to a sustained attention task, but it has a significant working memory load because participants are required to identify 3 even or 3 odd numbers in a row. As such, when studies employing the RVIP task observe effects of abstinence on performance, unless a no-working memory load control task is employed, it is unknown whether these effects are due to effects of withdrawal on sustained attention or working memory. Studies of cue-reactivity are similarly problematic as there is significant variation in the types (e.g., photographic, video, imagined) of stimuli employed, and among fMRI studies, there is variation in the levels of analysis (e.g., event-related or block design) and scanning parameters.

4 Nicotine Versus Non-nicotine Factors in Neurocognition

The research reviewed above suggests that smoking abstinence (versus smoking as usual) impacts neurocognition. Relatively less research has sought to specifically investigate the differential influence of nicotine and non-nicotine factors on various neurocognitive processes. Such research is important for a number of reasons. First, whereas nicotine replacement therapy (NRT) remains a first-line treatment for nicotine dependence, relapse rates are still high in response to NRT (Bohadana et al. 2000). This has led some to hypothesize that non-nicotine factors play a critical role in promoting relapse and, for this reason, examination of the differential effects of nicotine and non-nicotine effects on withdrawal signs and symptoms is important (Rose 2006). Second, the FDA, following passage of the Family Smoking Prevention and Tobacco Control Act in 2009, has the regulatory authority to reduce nicotine levels in cigarettes to potentially reduce smoking reinforcement (in the interest of improving public health). Reducing nicotine in cigarettes could potentially have consequences for smoking behavior that vary across smokers and these consequences are worth anticipating before any nicotine reduction policy is enacted.

4.1 Executive Function

A number of studies evaluating the effects of nicotine versus non-nicotine factors on executive function have been conducted, with most of these studies focusing on sustained attention and/or working memory. These studies generally support the hypothesis that abstinence-induced decrements in executive function are due to abstinence from, and are reversed by, nicotine. Nicotine, as compared to placebo, has been shown to improve sustained attention/working memory performance among abstinent smokers across multiple studies (Atzori et al. 2008; Beaver et al. 2011; Bohadana et al. 2000; Foulds et al. 1996). In the largest study to date, Kleykamp and colleagues observed improvements in working memory reaction time and accuracy following administration of 21 mg nicotine versus a placebo patch in overnight-deprived smokers (Kleykamp et al. 2011). There are also complementary studies examining the effects of smoking nicotine-free or very-low nicotine content cigarettes on continuous working memory performance. Two studies examined RVIP performance in overnight deprived smokers and showed better continuous working memory performance following smoking a nicotinized versus denicotinized or very-low nicotine content cigarettes (Gilbert et al. 1999; Wesnes and Warburton 1983). Using a cross-over design, in which smokers smoked denicotinized cigarettes (or were abstinent) while wearing a nicotine (or placebo) patch, we observed that working memory deficits (measured with the *n*-back task) were due to abstinence from nicotine but not non-nicotine factors (McClernon et al. 2014). Whereas these studies point to nicotine abstinence as the cause of smoking-abstinence-induced deficits, several studies have shown null effects (Bohadana et al. 2000; Sweet et al. 2010) or improved performance under placebo conditions (Cook et al. 2003). Thus additional research is warranted.

4.2 Reward Processes

Several studies have examined the effects of nicotine on reward processes by manipulating nicotine levels in smoked cigarettes or lozenges. Two of the studies described above from the same group compared placebo versus nicotine lozenge administration among overnight abstinent smokers and found reduced attentional bias, lower enjoyment of activities, and lower reward responsiveness in the placebo condition (Dawkins et al. 2006; Powell et al. 2004). These findings support the idea that abstinence effects on reward processing are largely nicotine dependent.

As noted above, Perkins and Karelitz (2013a, b) observed greater responding for music after smoking a nicotinized versus denicotinized cigarette among dependent and non-dependent smokers. Similarly, but using a different paradigm, abstinent smokers administered a denicotinized cigarette rated unfamiliar faces as less attractive than smokers administered a nicotinized cigarette (Attwood et al. 2009). This study was later replicated by the same group, although effects were only observed at a trend level (Attwood et al. 2012). These latter two studies, which were conducted in samples of non-dependent smokers, along with the Perkins and Karelitz data, not only underscore the role of nicotine in abstinence-induced changes in reward processing, but also suggest that nicotine may have acute effects on reinforcement outside its withdrawal-alleviating effects.

4.3 Cue-Reactivity

The nicotine patch was found to decrease generalized (i.e., abstinence-induced) craving but had no effect on cue-induced craving in 6-h abstinent smokers (Tiffany et al. 2000). Similarly, a high-dose nicotine patch (35 mg) did not attenuate cue-provoked craving during the first day of a smoking cessation attempt (Waters et al. 2004). Furthermore, Morisette et al. (2012) found that across a 5-h abstinent absorption period, smokers who were administered a nicotine patch (compared to a placebo patch) reported lower craving in response to both smoking and neutral cues, again with no specific effect for cue type. The effects of nicotine on cue-reactivity, however, may depend on the form of delivery or rate of absorption because nicotine gum was found to reduce cue-induced craving among abstinent smokers only 15 min post administration (Shiffman et al. 2003). Further, this effect was only observed among smokers who evidenced cue-induced craving in the first place (i.e., "reactors"), who tended to be lighter smokers who had smoked for fewer years. Among imaging studies of cue-reactivity, one study compared the BOLD response to smoking cues among overnight abstinent smokers who were administered either a nicotine or a placebo patch, and found increased reactivity to smoking cues in bilateral ventral striatum and left amygdala in the nicotine patch condition (Xu et al. 2014). In another study, prolonged abstinence with nicotine

Table 1 Summary of Multi-Factor Model of Neurocognition and Relapse

Factor	Summary of Evidence Base	Future Research Questions
Negative Reinforcement	Smoking abstinence results in decrements in executive function that interfere with daily function and as such may be aversive. These effects are reliably reversed by smoking and smokers lapse/ relapse in order to reverse these decrements	Do abstinence-induced deficits in neurocognition increase or predict smoking reinforcement? What is the timecourse of post-quit neurocognitive deficits? And do they resolve to baseline levels over time? What neurochemical changes lead to post-quit executive function deficits? And is pharmacological intervention, beyond nicotine replacement needed and/or efficacious?
Drug Reward Bias: Drug Reward Hypersensitivity	Smoking abstinence appears to have minimal and/or inconsistent influence on behavioral and brain indices of cue-reactivity, though recent evidence suggests that cue-reactivity may increase over the course of abstinence (i.e. incubation). Recent evidence also suggests that smoking abstinence may enhance attentional bias to smoking cues	What role does cue-reactivity play in real-world smoking lapse and relapse? Does incubation of cue-reactivity play a role in long-term abstinence? Does reactivity to cues that signal smoking availability/probability better predict cessation outcomes?
Drug Reward Bias: Non-Drug Reward Hyposensitivity	Preliminary data suggest that smoking abstinence may blunt reinforcement, reward and reactivity to non-drug rewards. Loss of pleasure in non-drug rewards following abstinence may contribute to negative reinforcement	Do individual differences in post-quit non-drug reward hyposensitivity predict cessation outcomes? Are their interventions that increase non-drug reward sensitivity? Do lab based-measures of non-drug reward hyposensitivity predict changes in real-world pursuit of these reinforcers?
Goal and Skill Interference	The abstinence-induced changes in neurocognition listed above may disrupt the maintenance of short and long-term goals associated with smoking cessation; and interfere with the enactment of coping skills necessary for avoiding lapse and relapse	Do executive function deficits and changes in reward sensitivity decrease smokers' ability to maintain their short and long term smoking cessation goals and enact coping skill/strategies? Can this be modeled in the lab or assessed in real-time during smoking cessation attempts?
Non-Cognitive Factors	A range of other factors associated with smoking abstinence may impinge on neurocognition in ways that bias the smoker toward relapse. These include fatigue, somatic symptoms, emotional dysregulation and negative mood	Do indices of other withdrawal-related changes in arousal, somatic symptoms and mood correlate with changes in neurocognition? Does treating these other factors improve neurocognition in ways that promote smoking abstinence?

patch use (i.e., ~50 days abstinence) was associated with greater cue-reactivity compared with pre-quit levels (Janes et al. 2009).

In sum, the effects of smoking abstinence on executive function and non-drug reward processes are likely modulated in large part by nicotine. By contrast, nicotine does not appear to exert a significant influence on cue-reactivity; studies of non-nicotine factors on cue-reactivity are needed. There is also a need for further studies of the effects of non-nicotine constituents of tobacco smoke on neurocognition.

5 A Multi-part Model of Neurocognition and Relapse

Based on our review of the literature, we propose a multi-factor model in which the effects of smoking abstinence on neurocognition increase lapse and relapse probability in abstinent smokers (see Table 1). In the *first component*, smoking abstinence leads to deficits in executive function including sustained attention, working memory, and inhibitory control that interfere to varying degrees with daily functioning. These deficits are associated with decreased frontal activation, which evidence suggests are due to abstinence-induced changes in fronto-striatal dopamine signaling. These deficits are reliably and quickly reversed by additional smoking, thereby contributing to a powerful **negative reinforcement** of smoking behavior and relapse. In the *second component*, smoking abstinence biases cognition in ways that increase the perceived salience of smoking-related stimuli in the environment relative to non-smoking-related reinforcers, the latter of which may even be devalued. In this way, smoking withdrawal increases **drug-reward bias**. *Third*, successfully maintaining abstinence requires the activation and maintenance of long-term goals; and the implementation of skills for coping with withdrawal and craving. To the extent that the factors above (disrupted executive function/drug-reward bias) result in **goal and skill interference**, they further increase the probability of relapse. *Finally*, there are a broad range of factors associated with smoking withdrawal that can interfere with neurocognition by competing for attentional demands, depleting cognitive resources, and/or otherwise slowing or interfering with cognition. These **non-cognitive factors** include sleep deprivation, craving, somatic symptoms of withdrawal, and changes in affective or mood-related functioning. Each of these factors likely interacts with changes in neurocognition. Abstinence-induced insomnia, for instance, may modulate decision making (Venkatraman et al. 2011) and disrupt neurocognition (Shekleton et al. 2014), which may in turn decrease the ability to engage in coping strategies.

6 Research Questions and Future Directions

6.1 *Research on the Causes of Abstinence-Induced Changes in Neurocognition*

Whereas the effects of smoking abstinence on neurocognition have been studied for more than 30 years, the precise causes for these effects remain to be fully elucidated. Animal studies—which have not been the focus of the current review—have provided data on the effects of nicotine withdrawal on brain dopamine and other neurotransmitter systems that subserve neurocognition. Studies have demonstrated, for instance, that cessation of nicotine administration results in substantial decreases in striatal dopamine levels (Domino and Tsukada 2009; Rahman et al. 2004), due in part to upregulated dopamine reuptake associated with nicotine administration (Fowler et al. 1996a, b). However, even within the animal literature studies linking abstinence-induced changes in neurotransmitter systems to changes in cognitive functions including learning, studies on memory and attention are rare (see Levin 2006, p. 458 for a review of this literature). There are even fewer studies that have sought to link withdrawal-induced changes in neurocognition to variability in nicotine self-administration, reinforcement, or relapse-like behavior.

Human research on the causes of abstinence-induced changes in neurocognition is similarly lacking. Whereas fMRI studies have begun to elucidate the brain circuits involved in changes in neurocognition, we are only now beginning to converge on enough data for meta-analyses of these effects. A recent meta-analysis of cue-reactivity, for instance, was able to identify only 11 fMRI studies of cue-reactivity that shared roughly comparable cue exposure paradigms and conducted whole-brain analyses allowing for pooled effects, and only a subset of those examined the effects of smoking abstinence. No such analyses have been carried out for other forms of cognition including working memory or inhibitory control and the variability in research methods, tasks, and measures will likely hamper such efforts even when sufficiently large numbers of studies are available.

Beyond fMRI, relatively little human research has sought to identify mechanisms that underlie abstinence-induced changes in neurocognition. Although PET studies have identified changes in nicotinic and other neuroreceptor availability (see chapter entitled Imaging Tobacco Smoking with PET and SPECT; volume 24), these studies have not typically sought associations between changes in smoking state and changes in cognitive performance. A small number of studies have sought to relate genetic markers associated with variability in neurotransmitter function to abstinence-induced changes in neurocognition (Loughead et al. 2009), and more work of this type—especially if it could also include direct assessment of neurochemical changes—is warranted.

6.2 The Role of Abstinence-Induced Neurocognitive Changes in Lapse and Relapse

Despite the research presented above showing that smoking abstinence modulates neurocognition, surprising little is known regarding the specific role of these deficits in relapse. Only a small number of studies have examined the effects of abstinence on neurocognition for more than 24 h following cessation. Studies that have looked beyond this 24 h window (Bradstreet et al. 2014; Gilbert et al. 2004) have observed that deficits in cognitive performance and brain function resolve only after prolonged abstinence (i.e., after one month). In contrast, studies of prolonged abstinence have provided some evidence that cue-reactivity remains stable or increases over the course of abstinence (Bedi et al. 2011). Despite these suggestive studies, more research is needed to establish the time course of neurocognitive changes across various domains using modern neuroimaging techniques. The differential effects of nicotine replacement and other therapies on neurocognitive trajectories are similarly lacking, but might provide clues as to the differential efficacy of interventions and could suggest ways in which to combine therapies to more comprehensively address multiple forms of neurocognition.

Likewise, more studies attempting to predict smoking cessation outcomes from either baseline or abstinence-induced changes in neurocognition are needed. To date, only a small number of such studies have been published, with most of these focused on cue-reactivity. Studies seeking to predict smoking cessation outcomes from cue-reactivity measures have produced mixed and often counterintuitive effects (Conklin et al. 2012; Perkins 2012; Powell et al. 2010; Wray et al. 2013). In one study, for instance, greater self-reported cue-reactivity was associated with an increased likelihood of initiating a quit attempt (Conklin et al. 2012); in another, greater pre-quit brain reactivity to smoking cues was associated with better cessation outcomes (McClernon et al. 2007). Another small-scale study, however, observed increased brain cue-reactivity among future relapsers, including in the anterior insula (Janes et al. 2010). Other studies have examined other neurocognitive predictors of cessation outcomes. Patterson and colleagues found that deficits in working memory following 3 days of abstinence were predictive of poorer smoking cessation outcomes among untreated smokers, while no association was seen among smokers treated with varenicline (Patterson et al. 2010). Another study found that, among smokers making an unaided quit attempt, deficits on measures of response inhibition and increased cue-reactivity were predictive of abstinence rates at 1 week, 1 month, and 3 months post quit (Powell et al. 2010). Finally, our own work has investigated abstinence-induced decrements in striatal activation to monetary rewards as a predictor of cessation outcomes (Sweitzer et al. in preparation). In that study, smokers were scanned at two time points, once after smoking as usual and once after 24 h of abstinence; they then completed a 3-week quit attempt supported by contingency management. Smokers who lapsed during the quit attempt exhibited significant decreases in striatal activation to monetary reward during abstinence relative to satiation, while those who maintained continuous abstinence exhibited no change. These results provide preliminary evidence for the importance of neurocognitive deficits in predicting cessation outcomes. However, much more work is clearly needed before definitive conclusions can be drawn.

Finally, in addition to the gaps identified above, a larger and potentially more important gap exists in our understanding of the specific role of cognitive factors in the sequence of events and behaviors leading up to lapse and relapse. Reasons for this gap include several factors: (a) lapse-like behavior can be difficult to model in the laboratory; (b) assessing neurocognition in real-time leading up to smoking lapse is challenging in the context of laboratory and real-world experiments, and (c) relapse, as a phenomenon, is not conducive to repeatability (i.e., smokers may not be keen on quitting over and over again so we can study relapse). For these reasons, whereas it is well established that abstinence results in deficits in executive function, is not known if or whether these deficits play a role in lapse behavior, either directly (negative reinforcement) or indirectly (goal and skill interference). In order to answer these questions, better models of lapse and relapse are needed for conducting controlled studies (McKee 2009; Sweitzer et al. 2013); and new methods for assessing real-time cognition and neurophysiology leading up to real-world lapses are needed.

6.3 The Treatment of Abstinence-Induced Changes in Neurocognition

If abstinence-induced changes in neurocognition contribute to smoking lapse and relapse, then interventions to counter or diminish these changes can potentially improve smoking cessation outcomes. As presented above, abstinence from nicotine appears to account for deficits in sustained attention, since administration of nicotine to smokers reverses these deficits (Parrott and Roberts 1991). As such, if deficits in these forms of neurocognition are linked—either directly or indirectly—to lapse and relapse, it will become vitally important that nicotine replacement therapy continues to be incorporated in smoking cessation interventions. Moreover, this may also mean that greater emphasis needs to be placed on achieving nicotine replacement at levels that replicate nicotine administration during smoking in order to achieve minimal disruption of neurocognition. This also means that other forms of procognitive therapies—whether behavioral (i.e., cognitive training; Lancaster and Stead 2005) or pharmacological (Cahill et al. 2014)—will need to demonstrate effectiveness above and beyond the effects of nicotine replacement, which remains relatively inexpensive, widely available, and with a favorable side-effect profile. In addition and related to this, additional work is needed to evaluate exactly which forms of neurocognition are remediated by nicotine and which are not. Efforts to modulate those that cannot be addressed with current first-line pharmacotherapies (nicotine, varenicline, bupropion) ought to receive the greatest attention in terms of treatment development and evaluation.

7 Summary

In this chapter, we have reviewed the last decade of research on the effects of smoking abstinence on various forms of neurocognition including executive function (working memory, sustained attention, response inhibition), reward processes, and cue-reactivity. Our review has identified that smoking abstinence results in deficits in executive function that are mediated in part by effects on frontal circuitry known to be affected by modulation of cholinergic, dopaminergic, and other neurotransmitter systems. We also identified evidence that smoking abstinence blunts reactivity to non-drug reinforcers—a finding that is consistent with findings in the animal literature. Finally, our review of cue-reactivity indicates that, consistent with the earlier literature, smoking abstinence has an additive effect on cue-provoked craving. However, the more recent literature suggests that smoking abstinence may amplify reactivity to cues that signal drug availability and that reactivity to smoking (relative to neutral cues) may increase with longer durations of abstinence. Across all domains examined, we identified inconsistencies in findings that are likely due to a number of methodological factors including variability in abstinence durations, the amount of practice prior to testing, and the variety of tasks used to measure the same or similar functions.

In addition to the above, we reviewed the literature on the effects of nicotine and non-nicotine factors in neurocognition. Available evidence suggests that abstinence from nicotine plays a significant role in the effects of smoking abstinence on executive function and responsiveness to non-drug reward. The effects of smoking abstinence on cue-reactivity appear to be due more to non-nicotine sensory and behavioral factors. Additional work is needed to further delineate the effects of nicotine and non-nicotine factors on neurocognition in order to inform treatment development and public policy.

Finally, we provided a multi-factor model and an agenda for future research on the effects of smoking abstinence on neurocognition. The model includes four distinct yet interacting factors, including: Negative Reinforcement, Drug-Reward Bias, Goal and Skill Interference, and Non-Cognitive Factors. Additional research is needed to further evaluate the scope and time course of abstinence-induced changes in neurocognition, the mechanisms that underlie these changes and the specific role of these processes in drug reinforcement, lapse, and relapse.

Acknowledgments Funding: NIDA grants R01 DA025876 (FJM), R01 DA024838 (FJM) and K01 DA033347 (MAA).

References

Abreu-Villaca Y, Medeiros AH, Lima CS, Faria FP, Filgueiras CC, Manhaes AC (2007) Combined exposure to nicotine and ethanol in adolescent mice differentially affects memory and learning during exposure and withdrawal. Behav Brain Res 181:136–146

Addicott MA, Baranger DA, Kozink RV, Smoski MJ, Dichter GS, McClernon FJ (2012) Smoking withdrawal is associated with increases in brain activation during decision making and reward anticipation: a preliminary study. Psychopharmacology 219:563–573

Addicott MA, Froeliger B, Kozink RV, Van Wert DM, Westman EC, Rose JE, McClernon FJ (2014) Nicotine and non-nicotine smoking factors differentially modulate craving, withdrawal and cerebral blood flow as measured with arterial spin labeling. Neuropsychopharmacology

APA (2013) Diagnostic and statistical manual of mental disorders, 5th edn, DSM-5. American Psychiatric Association, Arlington

Apicella P, Ljungberg T, Scarnati E, Schultz W (1991) Responses to reward in monkey dorsal and ventral striatum. Exp Brain Res 85:491–500

Ashare RL, Hawk LW Jr (2012) Effects of smoking abstinence on impulsive behavior among smokers high and low in ADHD-like symptoms. Psychopharmacology 219:537–547

Atzori G, Lemmonds CA, Kotler ML, Durcan MJ, Boyle J (2008) Efficacy of a nicotine (4 mg)-containing lozenge on the cognitive impairment of nicotine withdrawal. J Clin Psychopharmacol 28:667–674

Attwood AS, Penton-Voak IS, Munafo MR (2009) Effects of acute nicotine administration on ratings of attractiveness of facial cues. Nicotine Tob. Res. 11:44–48

Attwood AS, Penton-Voak IS, Goodwin C, et al. (2012) Effects of acute nicotine and alcohol on the rating of attractiveness in social smokers and alcohol drinkers. Drug Alcohol Depend 125:43–48

Bailey SR, Goedeker KC, Tiffany ST (2010) The impact of cigarette deprivation and cigarette availability on cue-reactivity in smokers. Addiction 105:364–372

Barros DM, Ramirez MR, Dos Reis EA, Izquierdo I (2004) Participation of hippocampal nicotinic receptors in acquisition, consolidation and retrieval of memory for one trial inhibitory avoidance in rats. Neuroscience 126:651–656

Barros DM, Ramirez MR, Izquierdo I (2005) Modulation of working, short- and long-term memory by nicotinic receptors in the basolateral amygdala in rats. Neurobiol Learn Mem 83:113–118

Beaver JD, Long CJ, Cole DM, Durcan MJ, Bannon LC, Mishra RG, Matthews PM (2011) The effects of nicotine replacement on cognitive brain activity during smoking withdrawal studied with simultaneous fMRI/EEG. Neuropsychopharmacology 36:1792–1800

Bedi G, Preston KL, Epstein DH, Heishman SJ, Marrone GF, Shaham Y, de Wit H (2011) Incubation of cue-induced cigarette craving during abstinence in human smokers. Biol Psychiatry 69:708–711

Berridge KC, Robinson TE, Aldridge JW (2009) Dissecting components of reward: 'liking', 'wanting', and learning. Curr Opin Pharmacol 9:65–73

Besheer J, Bevins RA (2003) Impact of nicotine withdrawal on novelty reward and related behaviors. Behav Neurosci 117:327–340

Bohadana A, Nilsson F, Rasmussen T, Martinet Y (2000) Nicotine inhaler and nicotine patch as a combination therapy for smoking cessation—a randomized, double-blind, placebo-controlled trial. Arch Intern Med 160:3128–3134

Bradstreet MP, Higgins ST, McClernon FJ, Kozink RV, Skelly JM, Washio Y, Lopez AA, Parry MA (2014) Examining the effects of initial smoking abstinence on response to smoking-related stimuli and response inhibition in a human laboratory model. Psychopharmacology 231:2145–2158

Breiter HC, Aharon I, Kahneman D, Dale A, Shizgal P (2001) Functional imaging of neural responses to expectancy and experience of monetary gains and losses. Neuron 30:619–639

Brody AL (2006) Functional brain imaging of tobacco use and dependence. J Psychiatr Res 40:404–418

Bruijnzeel AW, Markou A (2004) Adaptations in cholinergic transmission in the ventral tegmental area associated with the affective signs of nicotine withdrawal in rats. Neuropharmacology 47:572–579

Buhler M, Vollstadt-Klein S, Kobiella A, Budde H, Reed LJ, Braus DF, Buchel C, Smolka MN (2010) Nicotine dependence is characterized by disordered reward processing in a network driving motivation. Biol Psychiatry 67:745–752

Cahill K, Stevens S, Lancaster T (2014) Pharmacological treatments for smoking cessation. J Am Med Assoc 311:193–194

Canamar CP, London E (2012) Acute cigarette smoking reduces latencies on a Smoking Stroop test. Addict Behav 37:627–631
Carter BL, Lam CY, Robinson JD, Paris MM, Waters AJ, Wetter DW, Cinciripini PM (2009) Generalized craving, self-report of arousal, and cue reactivity after brief abstinence. Nicotine Tob Res 11:823–826
Carter BL, Tiffany ST (1999) Meta-analysis of cue-reactivity in addiction research. Addiction 94:327–340
Chaudhri N, Caggiula AR, Donny EC, Palmatier MI, Liu X, Sved AF (2006) Complex interactions between nicotine and nonpharmacological stimuli reveal multiple roles for nicotine in reinforcement. Psychopharmacology 184:353–366
CDC (2008) Smoking-attributable mortality, years of potential life lost, and productivity losses—United States, 2000-2004. Morbidity and Mortality Weekly Report 57:1226–1228. Available at: http://www.cdc.gov/mmwr/preview/mmwrhtml/mm5745a3.htm, accessed Nov 21, 2014
CDC (2012) Current cigarette smoking among adults—United States, 2011. Morbidity and Mortality Weekly Report 61:889–894. Available at http://www.cdc.gov/mmwr/preview/mmwrhtml/mm6144a2.htm, accessed Nov 21, 2014
Conklin CA, Parzynski CS, Salkeld RP, Perkins KA, Fonte CA (2012) Cue reactivity as a predictor of successful abstinence initiation among adult smokers. Exp Clin Psychopharmacol 20:473–478
Cook MR, Gerkovich MM, Graham C, Hoffman SJ, Peterson RC (2003) Effects of the nicotine patch on performance during the first week of smoking cessation. Nicotine Tob Res 5:169–180
Curtis CE, D'Esposito M (2003) Persistent activity in the prefrontal cortex during working memory. Trends Cogn Sci 7:415–423
David SP, Munafo MR, Johansen-Berg H, Mackillop J, Sweet LH, Cohen RA, Niaura R, Rogers RD, Matthews PM, Walton RT (2007) Effects of acute nicotine abstinence on cue-elicited ventral striatum/nucleus accumbens activation in female cigarette smokers: a functional magnetic resonance imaging study. Brain Imaging Behav 1:43–57
Dawkins L, Acaster S, Powell JH (2007) The effects of smoking and abstinence on experience of happiness and sadness in response to positively valenced, negatively valenced, and neutral film clips. Addict Behav 32:425–431
Dawkins L, Powell J (2011) Effects of nicotine and alcohol on affective responses to emotionally toned film clips. Psychopharmacology 216:197–205
Dawkins L, Powell JH, West R, Powell J, Pickering A (2006) A double-blind placebo controlled experimental study of nicotine: I—effects on incentive motivation. Psychopharmacology 189:355–367
Delgado MR, Locke HM, Stenger VA, Fiez JA (2003) Dorsal striatum responses to reward and punishment: effects of valence and magnitude manipulations. Cogn Affect Behav Neurosci 3:27–38
Domino EF, Minoshima S, Guthrie S, Ohl L, Ni LS, Koeppe RA, Zubieta JK (2000) Nicotine effects on regional cerebral blood flow in awake, resting tobacco smokers. Synapse 38:313–321
Domino EF, Ni LS, Xu YJ, Koeppe RA, Guthrie S, Zubieta JK (2004) Regional cerebral blood flow and plasma nicotine after smoking tobacco cigarettes. Prog Neuro-Psychopharmacol 28:319–327
Domino EF, Tsukada H (2009) Nicotine sensitization of monkey striatal dopamine release. Eur J Pharmacol 607:91–95
Donny EC, Chaudhri N, Caggiula AR, Evans-Martin FF, Booth S, Gharib MA, Clements LA, Sved AF (2003) Operant responding for a visual reinforcer in rats is enhanced by noncontingent nicotine: implications for nicotine self-administration and reinforcement. Psychopharmacology 169:68–76
Due DL, Huettel SA, Hall WG, Rubin DC (2002) Activation in mesolimbic and visuospatial neural circuits elicited by smoking cues: evidence from functional magnetic resonance imaging. Am J Psychiatry 159:954–960

Elliott R, Newman JL, Longe OA, Deakin JF (2003) Differential response patterns in the striatum and orbitofrontal cortex to financial reward in humans: a parametric functional magnetic resonance imaging study. J Neurosci 23:303–307

Engelmann JM, Versace F, Robinson JD, Minnix JA, Lam CY, Cui Y, Brown VL, Cinciripini PM (2012) Neural substrates of smoking cue reactivity: a meta-analysis of fMRI studies. Neuroimage 60:252–262

Epping-Jordan MP, Watkins SS, Koob GF, Markou A (1998) Dramatic decreases in brain reward function during nicotine withdrawal. Nature 393:76–79

Falcone M, Wileyto EP, Ruparel K, Gerraty RT, Laprate L, Detre JA, Gur R, Loughead J, Lerman C (2014) Age-related differences in working memory deficits during nicotine withdrawal. Addiction Biol 19:907−917

Fiore MC, Jaen CR, Baker TB, et al. (2008) Treating tobacco use and dependence: 2008 update. Clinical practice guideline. Rockville, MD: US Department of Health and Human Services, Public Health Service

Fisher DJ, Daniels R, Jaworska N, Knobelsdorf A, Knott VJ (2012) Effects of acute nicotine administration on resting EEG in nonsmokers. Exp Clin Psychopharmacol 20:71–75

Foulds J, Stapleton J, Swettenham J, Bell N, McSorley K, Russell MA (1996) Cognitive performance effects of subcutaneous nicotine in smokers and never-smokers. Psychopharmacology 127:31–38

Fowler JS, Volkow ND, Wang GJ, Pappas N, Logan J, MacGregor R, Alexoff D, Shea C, Schlyer D, Wolf AP, Warner D, Zezulkova I, Cilento R (1996a) Inhibition of monoamine oxidase B in the brains of smokers. Nature 379:733–736

Fowler JS, Volkow ND, Wang GJ, Pappas N, Logan J, Shea C, Alexoff D, MacGregor RR, Schlyer DJ, Zezulkova I, Wolf AP (1996b) Brain monoamine oxidase A inhibition in cigarette smokers. Proc Natl Acad Sci USA 93:14065–14069

Franklin TR, Wang Z, Wang J, Sciortino N, Harper D, Li Y, Ehrman R, Kampman K, O'Brien CP, Detre JA, Childress AR (2007) Limbic activation to cigarette smoking cues independent of nicotine withdrawal: a perfusion fMRI study. Neuropsychopharmacology 32:2301–2309

Freeman TP, Morgan CJ, Beesley T, Curran HV (2012) Drug cue induced overshadowing: selective disruption of natural reward processing by cigarette cues amongst abstinent but not satiated smokers. Psychol Med 42:161–171

Gilbert DG, McClernon FJ, Rabinovich NE, Dibb WD, Plath LC, Hiyane S, Jensen RA, Meliska CJ, Estes SL, Gehlbach BA (1999) EEG, physiology, and task-related mood fail to resolve across 31 days of smoking abstinence: relations to depressive traits, nicotine exposure, and dependence. Exp Clin Psychopharmacol 7:427–443

Gilbert D, McClernon J, Rabinovich N, Sugai C, Plath L, Asgaard G, Zuo Y, Huggenvik J, Botros N (2004) Effects of quitting smoking on EEG activation and attention last for more than 31 days and are more severe with stress, dependence, DRD2 A1 allele, and depressive traits. Nicotine Tob Res 6:249–267

Giniatullin R, Nistri A, Yakel JL (2005) Desensitization of nicotinic ACh receptors: shaping cholinergic signaling. Trends Neurosci 28:371–378

Gloria R, Angelos L, Schaefer HS, Davis JM, Majeskie M, Richmond BS, Curtin JJ, Davidson RJ, Baker TB (2009) An fMRI investigation of the impact of withdrawal on regional brain activity during nicotine anticipation. Psychophysiology 46:681–693

Goldstein RZ, Volkow ND (2002) Drug addiction and its underlying neurobiological basis: neuroimaging evidence for the involvement of the frontal cortex. Am J Psychiatry 159:1642–1652

Haber SN, Knutson B (2010) The reward circuit: linking primate anatomy and human imaging. Neuropsychopharmacology 35:4–26

Harrison EL, Coppola S, McKee SA (2009) Nicotine deprivation and trait impulsivity affect smokers' performance on cognitive tasks of inhibition and attention. Exp Clin Psychopharmacol 17:91–98

Hasselmo ME (2006) The role of acetylcholine in learning and memory. Curr Opin Neurobiol 16:710–715

Heishman SJ, Taylor RC, Henningfield JE (1994) Nicotine and smoking: a review of effects on human performance. Exp Clin Psychopharmacol 2:345–395

Heishman SJ, Henningfield JE (2000) Tolerance to repeated nicotine administration on performance, subjective, and physiological responses in nonsmokers. Psychopharmacology 152:321–333

Heishman SJ, Kleykamp BA, Singleton EG (2010) Meta-analysis of the acute effects of nicotine and smoking on human performance. Psychopharmacology 210:453–469

Hendricks PS, Ditre JW, Drobes DJ, Brandon TH (2006) The early time course of smoking withdrawal effects. Psychopharmacology 187:385–396

Herning RI, Jones RT, Bachman J (1983) EEG changes during tobacco withdrawal. Psychophysiology 20:507–512

Hikosaka K, Watanabe M (2000) Delay activity of orbital and lateral prefrontal neurons of the monkey varying with different rewards. Cereb Cortex 10:263–271

Hildebrand BE, Nomikos GG, Hertel P, Schilstrom B, Svensson TH (1998) Reduced dopamine output in the nucleus accumbens but not in the medial prefrontal cortex in rats displaying a mecamylamine-precipitated nicotine withdrawal syndrome. Brain Res 779:214–225

Hildebrand BE, Panagis G, Svensson TH, Nomikos GG (1999) Behavioral and biochemical manifestations of mecamylamine-precipitated nicotine withdrawal in the rat: Role of nicotinic receptors in the ventral tegmental area. Neuropsychopharmacology 21:560–574

Hughes JR (2007) Effects of abstinence from tobacco: valid symptoms and time course. Nicotine Tob Res 9:315–327

Jacobsen LK, Krystal JH, Mencl WE, Westerveld M, Frost SJ, Pugh KR (2005) Effects of smoking and smoking abstinence on cognition in adolescent tobacco smokers. Biol Psychiatry 57:56–66

Jacobsen LK, Pugh KR, Constable RT, Westerveld M, Mencl WE (2007) Functional correlates of verbal memory deficits emerging during nicotine withdrawal in abstinent adolescent cannabis users. Biol Psychiatry 61:31–40

Janes AC, Frederick B, Richardt S, Burbridge C, Merlo-Pich E, Renshaw PF, Evins AE, Fava M, Kaufman MJ (2009) Brain fMRI reactivity to smoking-related images before and during extended smoking abstinence. Exp Clin Psychopharmacol 17:365–373

Janes AC, Pizzagalli DA, Richardt S, de BFB, Chuzi S, Pachas G, Culhane MA, Holmes AJ, Fava M, Evins AE, Kaufman MJ (2010) Brain reactivity to smoking cues prior to smoking cessation predicts ability to maintain tobacco abstinence. Biol Psychiatry 67:722–729

Kalamboka N, Remington B, Glautier S (2009) Nicotine withdrawal and reward responsivity in a card-sorting task. Psychopharmacology 204:155–163

Kenny PJ, Markou A (2006) Nicotine self-administration acutely activates brain reward systems and induces a long-lasting increase in reward sensitivity. Neuropsychopharmacology 31:1203–1211

Kleykamp BA, Jennings JM, Eissenberg T (2011) Effects of transdermal nicotine and concurrent smoking on cognitive performance in tobacco-abstinent smokers. Exp Clin Psychopharmacol 19:75–84

Kolokotroni KZ, Rodgers RJ, Harrison AA (2012) Effects of chronic nicotine, nicotine withdrawal and subsequent nicotine challenges on behavioural inhibition in rats. Psychopharmacology 219:453–468

Kollins SH, English JS, Roley ME, O'Brien B, Blair J, Lane SD, McClernon FJ (2013) Effects of smoking abstinence on smoking-reinforced responding, withdrawal, and cognition in adults with and without attention deficit hyperactivity disorder. Psychopharmacology 227:19–30

Kozink RV, Kollins SH, McClernon FJ (2010a) Smoking withdrawal modulates right inferior frontal cortex but not presupplementary motor area activation during inhibitory control. Neuropsychopharmacology 35:2600–2606

Kozink RV, Lutz AM, Rose JE, Froeliger B, McClernon FJ (2010b) Smoking withdrawal shifts the spatiotemporal dynamics of neurocognition. Addict Biol 15:480–490

Lancaster T, Stead LF (2005) Individual behavioural counselling for smoking cessation. Cochrane Db Syst Rev 2. Art. No.: CD001292. doi:10.1002/14651858.CD001292.pub2

Leventhal AM, Munafo M, Tidey JW, Sussman S, Monterosso JR, Sun P, Kahler CW (2012) Anhedonia predicts altered processing of happy faces in abstinent cigarette smokers. Psychopharmacology 222:343–351

Levin ED, Rose JE (1990) Anticholinergic sensitivity following chronic nicotine administration as measured by radial-arm maze performance in rats. Behav Pharmacol 1:511–520

Levin, Edward D. (2006) Neurotransmitter interactions and cognitive function. Vol. 98. Birkhauser Verlag, Switzerland

Levin ED, Limpuangthip J, Rachakonda T, Peterson M (2006a) Timing of nicotine effects on learning in zebrafish. Psychopharmacology 184:547–552

Levin ED, McClernon FJ, Rezvani AH (2006b) Nicotinic effects on cognitive function: behavioral characterization, pharmacological specification, and anatomic localization. Psychopharmacology 184:523–539

Loughead J, Wileyto EP, Valdez JN, Sanborn P, Tang K, Strasser AA, Ruparel K, Ray R, Gur RC, Lerman C (2009) Effect of abstinence challenge on brain function and cognition in smokers differs by COMT genotype. Mol Psychiatry 14:820–826

MacKillop J, Brown CL, Stojek MK, Murphy CM, Sweet L, Niaura RS (2012) Behavioral economic analysis of withdrawal- and cue-elicited craving for tobacco: an initial investigation. Nicotine Tob Res 14:1426–1434

Mathew RJ, Wilson WH (1991) Substance abuse and cerebral blood flow. Am J Psychiatry 148:292–305

McBride D, Barrett SP, Kelly JT, Aw A, Dagher A (2006) Effects of expectancy and abstinence on the neural response to smoking cues in cigarette smokers: an fMRI study. Neuropsychopharmacology 31:2728–2738

McClernon FJ, Hiott FB, Huettel SA, Rose JE (2005) Abstinence-induced changes in self-report craving correlate with event-related FMRI responses to smoking cues. Neuropsychopharmacology 30:1940–1947

McClernon FJ, Hiott FB, Liu J, Salley AN, Behm FM, Rose JE (2007) Selectively reduced responses to smoking cues in amygdala following extinction-based smoking cessation: results of a preliminary functional magnetic resonance imaging study. Addict Biol 12:503–512

McClernon FJ, Kollins SH, Lutz AM, Fitzgerald DP, Murray DW, Redman C, Rose JE (2008) Effects of smoking abstinence on adult smokers with and without attention deficit hyperactivity disorder: results of a preliminary study. Psychopharmacology 197:95–105

McClernon FJ, Kozink RV, Lutz AM, Rose JE (2009) 24-h smoking abstinence potentiates fMRI-BOLD activation to smoking cues in cerebral cortex and dorsal striatum. Psychopharmacology 204:25–35

McClernon FJ, Froeliger B, Rose JE, Kozink RV, Addicott MA, Sweitzer M, Westman EC, Van Wert DM (2014) Nicotine and non-nicotine smoking factors differentially modulate working memory and associated brain function (under review)

McClure SM, York MK, Montague PR (2004) The neural substrates of reward processing in humans: the modern role of FMRI. Neuroscientist 10:260–268

McKee SA (2009) Developing human laboratory models of smoking lapse behavior for medication screening. Addict Biol 14:99–107

Mendrek A, Monterosso J, Simon SL, Jarvik M, Brody A, Olmstead R, Domier CP, Cohen MS, Ernst M, London ED (2006) Working memory in cigarette smokers: comparison to non-smokers and effects of abstinence. Addict Behav 31:833–844

Merritt PS, Cobb AR, Cook GI (2012) Sex differences in the cognitive effects of tobacco abstinence: a pilot study. Exp Clin Psychopharmacology 20:258–263

Morisette SB, Gulliver SB, Kamholz BW, Spiegel DA, Tiffany ST, Barlow DH (2012) Transdermal nicotine during cue reactivity in adult smokers with and without anxiety disorders. Psychol Addictive Behav 26:507–518

Parrott AC, Roberts G (1991) Smoking deprivation and cigarette reinstatement: effects upon visual attention. J Psychopharmacol 5:404–409

Patterson F, Jepson C, Loughead J, Perkins K, Strasser AA, Siegel S, Frey J, Gur R, Lerman C (2010) Working memory deficits predict short-term smoking resumption following brief abstinence. Drug Alcohol Depend 106:61–64

Perkins KA (2002) Chronic tolerance to nicotine in humans and its relationship to tobacco dependence. Nicotine Tob Res 4:405–422

Perkins KA (2012) Subjective reactivity to smoking cues as a predictor of quitting success. Nicotine Tob Res 14:383–387

Perkins KA, Grobe JE, Fonte C, Goettler J, Caggiula AR, Reynolds WA, Stiller RL, Scierka A, Jacob RG (1994) Chronic and acute tolerance to subjective, behavioral and cardiovascular effects of nicotine in humans. J Pharmacol Exp Ther 270:628–638

Perkins KA, Karelitz JL (2013a) Influence of reinforcer magnitude and nicotine amount on smoking's acute reinforcement enhancing effects. Drug Alcohol Depend 133:167–171

Perkins KA, Karelitz JL (2013b) Reinforcement enhancing effects of nicotine via smoking. Psychopharmacology 228:479–486

Picciotto MR, Addy NA, Mineur YS, Brunzell DH (2008) It is not "either/or": activation and desensitization of nicotinic acetylcholine receptors both contribute to behaviors related to nicotine addiction and mood. Prog Neurobiol 84:329–342

Pickworth WB, Herning RI, Henningfield JE (1989) Spontaneous EEG changes during tobacco abstinence and nicotine substitution in human volunteers. J Pharmacol Exp Ther 251:976–982

Powell JH, Pickering AD, Dawkins L, West R, Powell JF (2004) Cognitive and psychological correlates of smoking abstinence, and predictors of successful cessation. Addict Behav 29:1407–1426

Powell J, Dawkins L, West R, Powell J, Pickering A (2010) Relapse to smoking during unaided cessation: clinical, cognitive and motivational predictors. Psychopharmacology 212:537–549

Rahman S, Zhang J, Engleman EA, Corrigall WA (2004) Neuroadaptive changes in the mesoaccumbens dopamine system after chronic nicotine self-administration: a microdialysis study. Neuroscience 129:415–424

Ridderinkhof KR, van den Wildenberg WP, Segalowitz SJ, Carter CS (2004) Neurocognitive mechanisms of cognitive control: the role of prefrontal cortex in action selection, response inhibition, performance monitoring, and reward-based learning. Brain Cogn 56:129–140

Roesch MR, Olson CR (2003) Impact of expected reward on neuronal activity in prefrontal cortex, frontal and supplementary eye fields and premotor cortex. J Neurophysiol 90:1766–1789

Rogers RL, Meyer JS, Judd BW, Mortel KF (1985) Abstention from cigarette-smoking improves cerebral perfusion among elderly chronic smokers. J Am Med Assoc 253:2970–2974

Rose JE (2006) Nicotine and nonnicotine factors in cigarette addiction. Psychopharmacology 184:274–285

Rose JE, Behm FM, Westman EC, Mathew RJ, London ED, Hawk TC, Turkington TG, Coleman RE (2003) PET studies of the influences of nicotine on neural systems in cigarette smokers. Am J Psychiatry 160:323–333

Rubia K, Russell T, Overmeyer S, Brammer MJ, Bullmore ET, Sharma T, Simmons A, Williams SCR, Giampietro V, Andrew CM, Taylor E (2001) Mapping motor inhibition: conjunctive brain activations across different versions of go/no-go and stop tasks. Neuroimage 13:250–261

Rubinstein ML, Benowitz NL, Auerback GM, Moscicki AB (2009) Withdrawal in adolescent light smokers following 24-hour abstinence. Nicotine Tob Res 11:185–189

Sacco KA, Termine A, Seyal A, Dudas MM, Vessicchio JC, Krishnan-Sarin S, Jatlow PI, Wexler BE, George TP (2005) Effects of cigarette smoking on spatial working memory and attentional deficits in schizophrenia: involvement of nicotinic receptor mechanisms. Arch Gen Psychiatry 62:649–659

Sarter M, Givens B, Bruno JP (2001) The cognitive neuroscience of sustained attention: where top-down meets bottom-up. Brain Res Brain Res Rev 35:146–160

Sawaguchi T, Goldman-Rakic PS (1991) D1 dopamine receptors in prefrontal cortex: involvement in working memory. Science 251:947–950

Semenova S, Stolerman IP, Markou A (2007) Chronic nicotine administration improves attention while nicotine withdrawal induces performance deficits in the 5-choice serial reaction time task in rats. Pharmacol Biochem Behav 87:360–368

Shekleton JA, Flynn-Evans EE, Miller B, Epstein LJ, Kirsch D, Brogna LA, Burke LM, Bremer E, Murray JM, Gehrman P, Lockley SW, Rajaratnam SM (2014) Neurobehavioral performance impairment in insomnia: relationships with self-reported sleep and daytime functioning. Sleep 37:107–116

Shiffman S, Paty JA, Gnys M, Kassel JA, Hickcox M (1996) First lapses to smoking: within-subjects analysis of real-time reports. J Consult Clin Psychol 64:366–379

Shiffman S, Shadel WG, Niaura R, Khayrallah MA, Jorenby DE, Ryan CF, Ferguson CL (2003) Efficacy of acute administration of nicotine gum in relief of cue-provoked cigarette craving. Psychopharmacology 166:343–350

Shiffman S, Balabanis MH, Gwaltney CJ, Paty JA, Gnys M, Kassel JD, Hickcox M, Paton SM (2007) Prediction of lapse from associations between smoking and situational antecedents assessed by ecological momentary assessment. Drug Alcohol Depend 91:159–168

Shoaib M, Bizarro L (2005) Deficits in a sustained attention task following nicotine withdrawal in rats. Psychopharmacology 178:211–222

Singer S, Rossi S, Verzosa S, Hashim A, Lonow R, Cooper T, Sershen H, Lajtha A (2004) Nicotine-induced changes in neurotransmitter levels in brain areas associated with cognitive function. Neurochem Res 29:1779–1792

Skjei KL, Markou A (2003) Effects of repeated withdrawal episodes, nicotine dose, and duration of nicotine exposure on the severity and duration of nicotine withdrawal in rats. Psychopharmacology 168:280–292

Sweet LH, Mulligan RC, Finnerty CE, Jerskey BA, David SP, Cohen RA, Niaura RS (2010) Effects of nicotine withdrawal on verbal working memory and associated brain response. Psychiatry Res 183:69–74

Sweitzer MM, Denlinger RL, Donny EC (2013) Dependence and withdrawal-induced craving predict abstinence in an incentive-based model of smoking relapse. Nicotine Tob Res 15:36–43

Sweitzer MM, Geier CF, Joel DL, McGurrin P, Denlinger RL, Forbes EE, Donny EC (2014) Dissociated effects of anticipating smoking versus monetary reward in the caudate as a function of smoking abstinence. Biol Psychiatry 76:681–688

Thiel KJ, Sanabria F, Neisewander JL (2009) Synergistic interaction between nicotine and social rewards in adolescent male rats. Psychopharmacology 204:391–402

Thut G, Schultz W, Roelcke U, Nienhusmeier M, Missimer J, Maguire RP, Leenders KL (1997) Activation of the human brain by monetary reward. NeuroReport 8:1225–1228

Tiffany ST, Cox LS, Elash CA (2000) Effects of transdermal nicotine patches on abstinence-induced and cue-elicited craving in cigarette smokers. J Consult Clin Psychol 68:233–240. http://dx.doi.org/10.1037/0022-006X.68.2.233

Uchida S, Kagitani F, Nakayama H, Sato A (1997) Effect of stimulation of nicotinic cholinergic receptors on cortical cerebral blood flow and changes in the effect during aging in anesthetized rats. Neurosci Lett 228:203–206

Venkatraman V, Huettel SA, Chuah LY, Payne JW, Chee MW (2011) Sleep deprivation biases the neural mechanisms underlying economic preferences. J Neurosci 31:3712–3718

Wallace TL, Bertrand D (2013) Importance of the nicotinic acetylcholine receptor system in the prefrontal cortex. Biochem Pharmacol 85:1713–1720

Wang Z, Faith M, Patterson F, Tang K, Kerrin K, Wileyto EP, Detre JA, Lerman C (2007) Neural substrates of abstinence-induced cigarette cravings in chronic smokers. J Neurosci 27:14035–14040

Waters AJ, Shiffman S, Sayette MA, Paty JA, Gwaltney CJ, Balabanis MH (2004) Cue-provoked craving and nicotine replacement therapy in smoking cessation. J Consult Clin Psychol 72:1136–1143

Waters AJ, Carter BL, Robinson JD, Wetter DW, Lam CY, Kerst W, Cinciripini PM (2009) Attentional bias is associated with incentive-related physiological and subjective measures. Exp Clin Psychopharmacol 17:247–257

Weaver MT, Sweitzer M, Coddington S, Sheppard J, Verdecchia N, Caggiula AR, Sved AF, Donny EC (2012) Precipitated withdrawal from nicotine reduces reinforcing effects of a visual stimulus for rats. Nicotine Tob Res 14:824–832

Wertz JM, Sayette MA (2001) Effects of smoking opportunity on attentional bias in smokers. Psychol Addict Behav 15:268–271

Wesnes K, Warburton DM (1983) Effects of smoking on rapid information processing performance. Neuropsychobiology 9:223–229

Wesnes KA, Edgar CJ, Kezic I, Salih HM, de Boer P (2013) Effects of nicotine withdrawal on cognition in a clinical trial setting. Psychopharmacology 229:133–140

World Health Organization (2013) Noncommunicable diseases fact sheet. Available at: http://www.who.int/mediacentre/factsheets/fs355/en/, accessed Nov 21, 2014

Wilson SJ, Sayette MA, Delgado MR, Fiez JA (2005) Instructed smoking expectancy modulates cue-elicited neural activity: a preliminary study. Nicotine Tob Res 7:637–645

Wonnacott S (1990) The paradox of nicotinic acetylcholine receptor upregulation by nicotine. Trends Pharmacol Sci 11:216–219

Wray JM, Gass JC, Tiffany ST (2013) A systematic review of the relationships between craving and smoking cessation. Nicotine Tob Res 15:1167–1182

Xu J, Mendrek A, Cohen MS, Monterosso J, Rodriguez P, Simon SL, Brody A, Jarvik M, Domier CP, Olmstead R, Ernst M, London ED (2005) Brain activity in cigarette smokers performing a working memory task: effect of smoking abstinence. Biol Psychiatry 58:143–150

Xu X, Clark US, David SP, Mulligan RC, Knopik VS, McGeary J, Mackillop J, McCaffery J, Niaura RS, Sweet LH (2014) The effects of nicotine deprivation and replacement on BOLD-fMRI response to smoking cues as a function of DRD4 VNTR genotype. Nicotine Tob Res 16:939–947

Yamashita K, Kobayashi S, Yamaguchi S, Kitani M, Tsunematsu T (1988) Effect of smoking on regional cerebral blood-flow in the normal aged volunteers. Gerontology 34:199–204

Zubieta JK, Lombardi U, Minoshima S, Guthrie S, Ni LS, Ohl LE, Koeppe RA, Domino EF (2001) Regional cerebral blood flow effects of nicotine in overnight abstinent smokers. Biol Psychiatry 49:906–913

Index

A
A-82695, 110
ABT-089, 114
ABT-418, 110, 111, 116
Abuse, 116
ACh binding site, 3–6
ADHD, 104, 117, 118
Adolescence, 151
Alertness, 108
α7, 113–115, 121, 123
α4β2, 114
α4β2*, 113–116, 121
αBungarotoxin, 11, 13
Alzheimer's, 104
Alzheimer's disease, 118
Amygdala, 91
Anxiety, 172
AR-R17779, 115
Arousal, 120
Attention, 103, 106, 117
Aversive effects, 106
AZD-3480, 118, 121

B
BOLD, 121
Bupropion, 181

C
Candidate gene studies, 21
catechol-O-methyltransferase (COMT), 203
Cell-signaling cascades, 150
Cerebral blood flow, 123
5-choice serial reaction time task (5-CSRTT), 110, 115, 117, 119, 198
Chronic exposure, 112, 113
Concurrent choice, 178
Cue-reactivity, 173
Cue-reactivity paradigms, 207
CYP2A6, 37, 39, 40, 42–44, 47

D
Default network, 123
Dependence, 119
Desensitization, 112, 117
Dihydro-β-erythroidine, 114
Distress, 172
DMXB-A, 118, 123
DMXBA, 124
Dopamine, 37–39, 46, 47, 49, 53–56, 58–67, 116, 196
Dopaminergic, 93, 97, 98
Drug discrimination, 169

E
Electroencephalography (EEG), 120, 197
Event-related potentials, 120
Exome sequencing, 31
Extinction of fear, 142
Extinction therapy, 181

F
Fatigue, 172
Fear conditioning, 140
Frontal cortex, 91, 92, 96
Functional Magnetic Resonance Imaging, 121

G
GABA, 119
Genetic variation, 39, 40, 48–50, 52–60, 64, 65, 67, 68, 70, 71
Genome-wide association studies, 21
Genome sequencing, 31
Glutamate, 119, 120
Glutaminergic, 89, 95
Glycine, 119
Go/No-Go, 200
GTS-21, 115, 118

D.J.K. Balfour and M.R. Munafò (eds.), *The Neurobiology and Genetics of Nicotine and Tobacco*, Current Topics in Behavioral Neurosciences 23,
DOI 10.1007/978-3-319-13665-3

H
Habit/compulsion, 166
Habit learning, 180
Heritability, 20
Heterogeneous instrumental chains, 171
Hierarchical instrumental, 167
Hippocampus, 88, 90, 92, 95, 96
Hippocampus-dependent learning, 140
Histamine, 119
5-HT, 119

I
Imputation, 22
Incentive learning, 167
Incentive salience, 166
Individual differences, 179
intracranial self-stimulation (ICSS), 205
Isoarecolone, 116

L
Learning, 106
Lobeline, 110
Locomotor, 113
Locomotor stimulant, 114

M
Mecamylamine, 117
MEM3454, 115
Memory, 90–94, 97, 104, 106, 117
Methyllycaconitine, 114
Mismatch negativity, 121
MK801, 115
Mood, 172

N
nAChR agonists, 105
nAChRs, 37, 38, 47, 52, 53, 55, 56, 58, 138, 195
nAChR subtype, 1–3, 9, 10, 12, 13, 113, 115, 116, 119
n-back task, 203
Negative reinforcement, 167, 215
neurocognition, 197
Nicotine, 2, 6, 8, 9, 37–39, 42–60, 62, 63, 65–71, 139
Nicotine discriminative stimuli, 170
Nicotine withdrawal symptoms, 196
Nicotinic, 88–98
Nicotinic receptor subtypes, 14
NMDA receptors, 143
non-nicotine factors, 211

O
Oscillations, 120

P
P50, 121
Pavlovian, 166
Pharmacotherapy, 181
Phenotype refinement, 26
Plain packaging, 182
PNU-120596, 115
PNU-282987, 115
Polymorphisms, 118
Posner, 109
Prazosin, 120
Prefrontal cortex, 145
Processing speed, 106
progressive ratio task, 205
Propositional knowledge, 166
Propranolol, 120

R
Raclopride, 119
Rapid Visual Information Processing (RVIP) task, 199
Recall-by-genotype, 30
Regional cerebral blood flow, 196
Reinforcement value, 166
Replacement therapy, 181
Response inhibition, 200
Reward processing, 204
RG3487, 115, 118

S
SCH23390, 119
Schizophrenia, 104, 118, 121, 123
Selective attention, 108
Sensory gating, 121
Serotonergic, 93, 95, 97, 98
Serotonin, 37, 38, 53, 67–71
Set shifting, 109, 115
SIB-1553A, 115
SIB 1765F, 115
Side effects, 106
Smokers, 105
Smoking, 37–40, 42–49, 51–71, 105
Smoking perseveration, 179
Spatial learning, 144
SSR-180711, 115
Stimulus detection, 108
Stoichiometry, 9
Stress, 172

Stroop effect, 108
Subunit composition, 1, 9, 10, 12, 14
Subunit compositions, 12
Sustained attention, 107, 198

T
TC-5619, 118
Thalamus, 92, 93
Three-term relationship, 166
Titration, 168
Tolerance, 106, 112, 113, 167
Trace fear conditioning, 145

U
Utility, 168

V
Validity effect, 111
Varenicline, 122, 181
Vigilance, 107

W
Withdrawal, 105, 123
Working memory, 107, 144, 201

MIX
Papier aus verantwortungsvollen Quellen
Paper from responsible sources
FSC® C105338

If you have any concerns about our products,
you can contact us on
ProductSafety@springernature.com

In case Publisher is established outside the EU,
the EU authorized representative is:
Springer Nature Customer Service Center GmbH
Europaplatz 3, 69115 Heidelberg, Germany

Printed by Libri Plureos GmbH
in Hamburg, Germany